CYANOBACTERIA
Antibacterial Activity

CYANOBACTERIA
Antibacterial Activity

Kiran Kapoor

Oxford Book Company
Jaipur ▪ New Delhi

Published By

Oxford Book Company

Regd. Off. "Bony Residency", Gate No. 2
Opp. Tilak Public School, Vishweswariya Nagar,
Gopalpura Road, Jaipur - 302018, India
Ph: 0141 - 2761280, Telefax: 2761381
email - oxfordbookjpr@gmail.com
website - oxfordbookcompany.com

Delhi Office: 102, 1st Floor, Satyam House,
4327/3, Ansari Road, Darya Ganj,
New Delhi - 110002, India
Ph: 011-45652440

Distributed by
KEYA INTERNATIONAL PRESS LLP
E-mail: keyainternationalpressllp@gmail.com

Published in India, 2025

ISBN : 9789380179308

Typeset by Shivangi Computers, Jaipur

Printed and bound in India by Replika Press Pvt. Ltd.

Preface

Cyanobacteria, also known as blue-green algae, blue-green bacteria or Cyanophyta, is a phylum of bacteria that obtain their energy through photosynthesis. The name "cyanobacteria" comes from the colour of the bacteria. They are a significant component of the marine nitrogen cycle and an important primary producer in many areas of the ocean, but are also found in habitats other than the marine environment; in particular cyanobacteria are known to occur in both freshwater, hypersaline inland lakes and in arid areas where they are a major component of biological soil crusts. Stromatolites of fossilised oxygen-producing cyanobacteria have been found from 2.8 billion years ago.

Cyanobacteria are found in almost every conceivable environment, from oceans to fresh water to bare rock to soil. They can occur as planktonic cells or form phototrophic biofilms in fresh water and marine environments, they occur in damp soil, or even temporarily moistened rocks in deserts. A few are endosymbionts in lichens, plants, various protists, or sponges and provide energy for the host. Some live in the fur of sloths, providing a form of camouflage.

Cyanobacteria are very important organisms for the health and growth of many plants. They are one of very few groups of organisms that can convert inert atmospheric nitrogen into an organic form, such as nitrate or ammonia. It is these "fixed" forms of nitrogen which plants need for their growth, and must obtain from the soil. Fertilisers work the way they do in part because they contain additional fixed nitrogen which plants can then absorb through their roots.

The book begins an introduction to biological aspects of cyanobacteria and gives comprehensive treatment of antibacterial activity of cyanobacteria based on extensive research done in this field. It also highlights the other aspects of biological activity like anti-fungal, anti-viral, anti-inflammatory and cytotoxic characteristics of cyanobacteria. The book is written for research students and teachers of the universities and colleges who want to study microbiology, medicine and pharmaceutical sciences with respect to antibacterial effects.

Kiran Kapoor

Contents

1

Introduction to Cyanobacteria

Cyanobacteria are aquatic and photosynthetic, that is, they live in the water, and can manufacture their own food. Because they are bacteria, they are quite small and usually unicellular, though they often grow in colonies large enough to see. They have the distinction of being the oldest known fossils, more than 3.5 billion years old, in fact! It may surprise you then to know that the cyanobacteria are still around; they are one of the largest and most important groups of bacteria on earth.

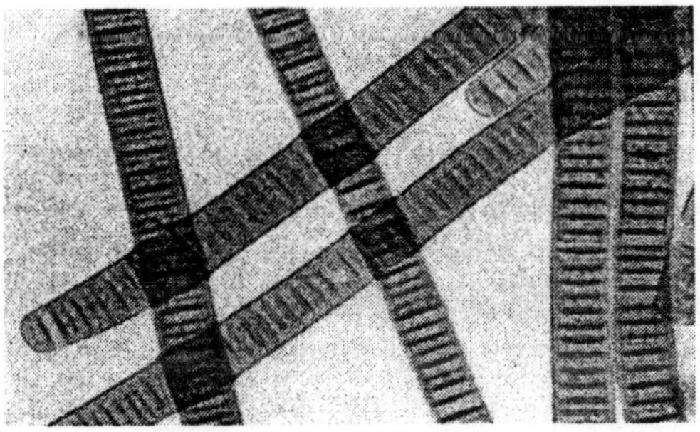

Figure 1. Cyanobacteria

Many Proterosoic oil deposits are attributed to the activity of cyanobacteria. They are also important providers of nitrogen fertiliser in the cultivation of rice and beans. The cyanobacteria have also been tremendously important in shaping the course of evolution and ecological change throughout earth's history. The oxygen atmosphere that we depend on was generated by numerous cyanobacteria during the Archaean and Proterozoic Eras. Before that time, the atmosphere had a very different chemistry, unsuitable for life as we know it today. The other great contribution of the cyanobacteria is the origin of plants. The chloroplast with which plants make food for themselves is actually a cyanobacterium living within the plant's cells. Sometime in the late Proterozoic, or in the early Cambrian, cyanobacteria began to take up residence within certain eukaryote cells, making food for the eukaryote host in return for a home. This event is known as endosymbiosis, and is also the origin of the eukaryotic mitochondrion. Because they are photosynthetic and aquatic, cyanobacteria are often called "blue-green algae". This name is convenient for talking about organisms in the water that make their own food, but does not reflect any relationship between the cyanobacteria and other organisms called algae. Cyanobacteria are relatives of the bacteria, not eukaryotes, and it is only the *chloroplast* in eukaryotic algae to which the cyanobacteria are related.

FOSSIL RECORD

The cyanobacteria have an extensive fossil record. The oldest known fossils, in fact, are cyanobacteria from Archaean rocks of western Australia, dated 3.5 billion years old. This may be somewhat surprising, since the oldest *rocks* are only a little older: 3.8 billion years old!

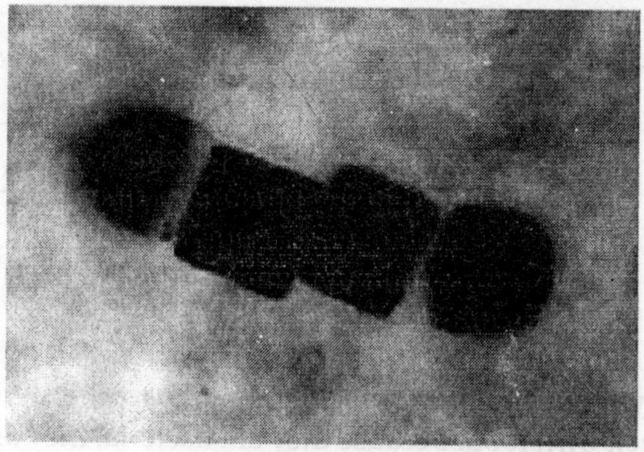

Figure 2. Fossil Cyanobacteria

Cyanobacteria are among the easiest microfossils to recognise. Morphologies in the group have remained much the same for billions of years, and they may leave chemical fossils behind as well, in the form of breakdown products from pigments. Small fossilised cyanobacteria have been extracted from Precambrian rock, and studied through the use of SEM and TEM (scanning and transmission electron microscopy).

Many Proterozoic oil deposits are attributed to the activity of cyanobacteria, such as *Gloeocapsomorpha*. Small concentrically layered structures called pisolites are also the result of fossilised bacteria. Cyanobacteria are otherwise rarely preserved in rocks other than chert, though some possible blue-green bacteria have been recovered from shale. At right is a layered stromatolite, produced by the activity of ancient cyanobacteria. The layers were produced as calcium carbonate precipitated over the growing mat of bacterial filaments; photosynthesis in the bacteria depleted carbon dioxide in the surrounding water, initiating the precipitation. The minerals, along with grains of sediment precipitating from the water, were then trapped within the sticky layer of mucilage that surrounds the bacterial colonies, which then continued to grow upwards through the sediment to form a new layer. As this process occured over and over again, the layers of sediment were created. This process still occurs today; Shark Bay in western Australia is well known for the stromatolite "turfs" rising along its beaches.

In some cases, the stromatolites were infiltrated with a mineral-rich solution which fossilised the bacteria along with the layers, but more often only the layers are preserved. The oldest stromatolites date to the Early Archaean, and they became abundant by the end of the Archaean. In the Proterozoic, stromatolites were widespread on earth, and were ecologically important as the first reefs. By the close of the Proterozoic, the abundance of stromatolites decreased markedly, though cyanobacteria continued to leave a fossil record, such as *Langiella* and *Kidstoniella* known from the Lower Devonian Rhynie chert.

The cyanobacteria have also been tremendously important in shaping the course of evolution and ecological change throughout earth's history. The oxygen atmosphere that we depend on was generated by numerous cyanobacteria photosynthesizing during the Archaean and Proterozoic Era. Before that time, the atmosphere had a very different chemistry, unsuitable for life as we know it today.

LIFE HISTORY AND ECOLOGY

Though cyanobacteria do not have a great diversity of form, and though they are microscopic, they are rich in chemical diversity. Cyanobacteria get their name from the bluish pigment phycocyanin, which they use to capture light for photosynthesis. They also contain chlorophyll a, the same photosynthetic pigment that plants use. In fact the chloroplast in plants is a symbiotic cyanobacterium, taken up by a green algal ancestor of the plants sometime in the Precambrian. However, not all "blue-green" bacteria are blue; some common forms are red or pink from the pigment phycoerythrin. These bacteria are often found growing on greenhouse glass, or around sinks and drains. The Red Sea gets its name from occasional blooms of a reddish species of *Oscillatoria*, and African flamingos get their pink color from eating *Spirulina*. Whatever their color, cyanobacteria are photosynthetic, and so can manufacture their own food. This has caused them to be dubbed "blue-green algae", though they have no relationship to any of the various eukayotic algae. The term "algae" merely refers to any aquatic organisms capable of photosynthesis, and so applies to several groups.

MORPHOLOGY

Cyanobacteria are among the easiest microfossils to recognise. They are larger than other bacteria, and morphologies in the group have remained much the same for billions of years. Like other bacteria, cyanobacteria have no nucleus or internal membrane systems. In many species, however, the external membrane has been folded to increase total surface area. The "packets" created in the cell membrane by these foldings are the surfaces across which the chemical reactions of photosynthesis take place. All cyanobacteria are unicellular, though many grow in colonies or filaments, often surrounded by a gelatinous or mucilaginous sheath. *Nostoc*, pictured below at left, may produce spherical colonies as much as three or four centimeters in diameter. These are large enough to be mistaken for amphibian eggs. Fossilised cyanobacterial mounds are called stromatolites.

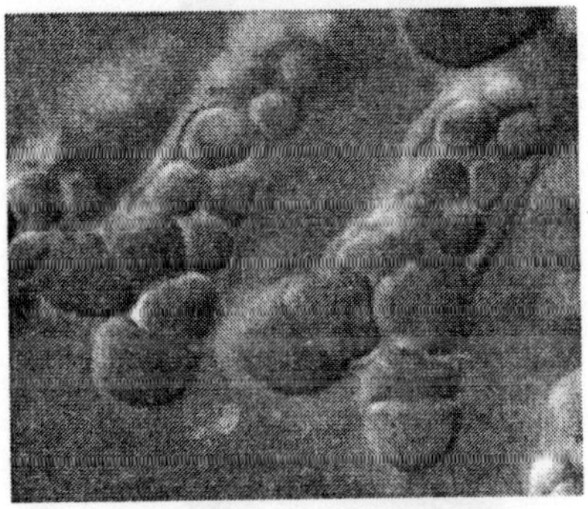

Figure 3. Hyella stella: a Cyanobacterium that lives in Marine Limestone

Despite the fact that they lack flagella, many cyanobacteria are capable of moving about. *Oscillatoria*, pictured above at right, gets its name from the "swaying" motion of its filaments when observed under the microscope. No one has yet determined how these filaments are able to do this.

Forms

Cyanobacteria are found in almost every conceivable environment, from oceans to fresh water to bare rock to soil. They can occur as planktonic cells or form phototrophic biofilms in fresh water and marine environments, they occur in damp soil, or even temporarily moistened rocks in deserts. A few are endosymbionts in lichens, plants, various protists, or sponges and provide energy for the host. Some live in the fur of sloths, providing a form of camouflage.

Cyanobacteria include unicellular and colonial species. Colonies may form filaments, sheets or even hollow balls. Some filamentous colonies show the ability to differentiate into several different cell types: vegetative cells, the normal, photosynthetic cells that are formed under favorable growing conditions; akinetes, the climate-resistant spores that may form when environmental conditions become harsh; and thick-walled heterocysts, which contain the enzyme nitrogenase, vital for nitrogen fixation. Heterocysts may also form under the appropriate environmental conditions (anoxic) wherever nitrogen is necessary.

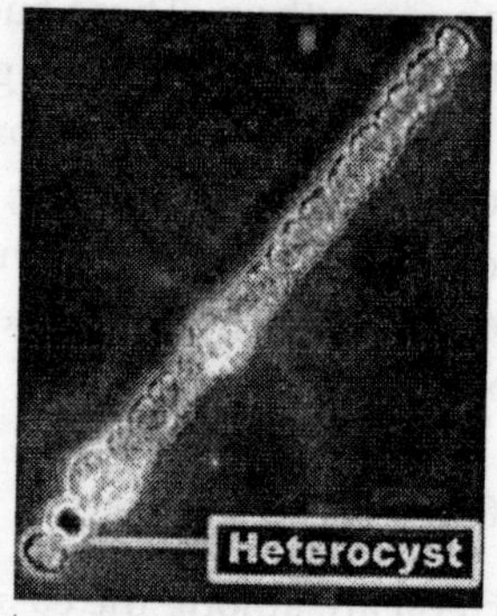

Figure 4. Heterocyst formation

Heterocyst-forming species are specialised for nitrogen fixation and are able to fix nitrogen gas into ammonia (NH_3), nitrites (NO_2) or nitrates (NO_3) which can be absorbed by plants and converted to protein and nucleic acids. The rice paddies of Asia, which produce about 75% of the world's rice, could not do so were it not for healthy populations of nitrogen-fixing cyanobacteria in the rice paddy fertiliser.

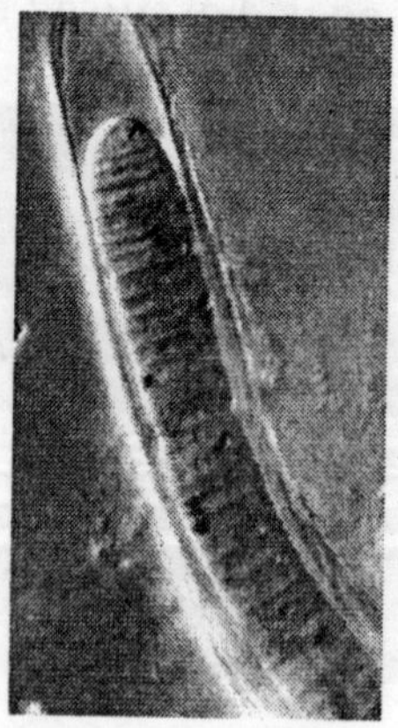

Figure 5. Lyngbya majuscula: A Marine Species

Many cyanobacteria also form motile filaments, called hormogonia, that travel away from the main biomass to bud and form new colonies elsewhere. The cells in a hormogonium are often thinner than in the vegetative state, and the cells on either end of the motile chain may be tapered.

In order to break away from the parent colony, a hormogonium often must tear apart a weaker cell in a filament, called a necridium.

Each individual cell of a cyanobacterium typically has a thick, gelatinous cell wall. They differ from other gram-negative bacteria in that the quorum sensing molecules autoinducer-2 and acyl-homoserine lactones are absent. They lack flagella, but hormogonia and some unicellular species may move about by gliding along surfaces. In water columns some cyanobacteria float by forming gas vesicles, like in archaea.

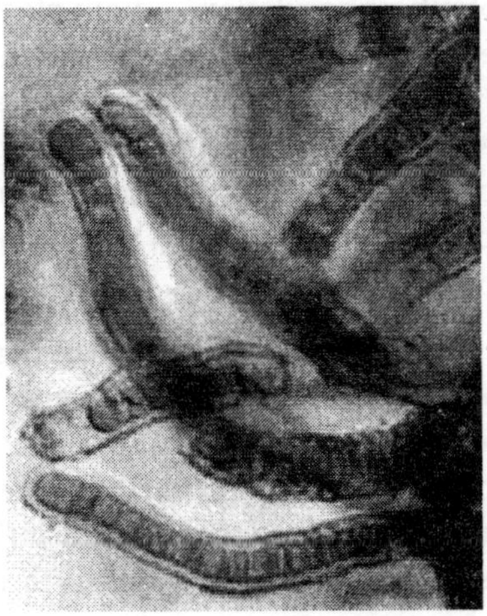

Figure 6. Scytonema endolithicum

Some of these organisms contribute significantly to global ecology and the oxygen cycle. The tiny marine cyanobacterium Prochlorococcus was discovered in 1986 and accounts for more than half of the photosynthesis of the open ocean. Many cyanobacteria even display the circadian rhythms that were once thought to exist only in eukaryotic cells.

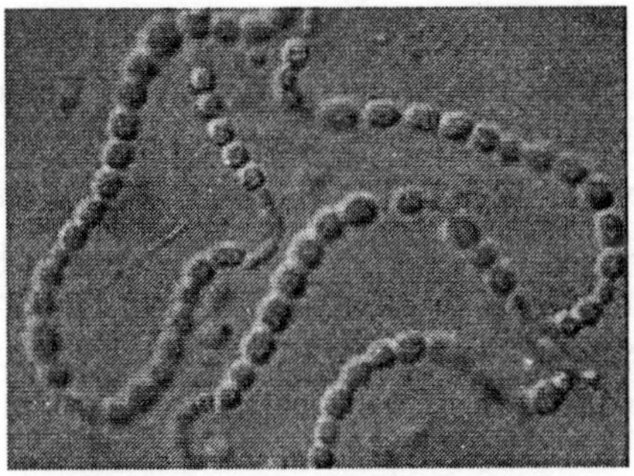

Figure 7. Microscopic View of a Nostoc Colony

PHOTOSYNTHESIS

Cyanobacteria have an elaborate and highly organised system of internal membranes which function in photosynthesis. Photosynthesis in cyanobacteria generally uses water as an electron donor and produces oxygen as a by-product, though some may also use hydrogen sulfide as occurs among other photosynthetic bacteria. Carbon dioxide is reduced to form carbohydrates via the Calvin cycle. In most forms the photosynthetic machinery is embedded into folds of the cell membrane, called thylakoids. The large amounts of oxygen in the atmosphere are considered to have been first created by the activities of ancient cyanobacteria. Due to their ability to fix nitrogen in aerobic conditions they are often found as symbionts with a number of other groups of organisms such as fungi (lichens), corals, pteridophytes (Azolla), angiosperms (Gunnera) etc.

Cyanobacteria are the only group of organisms that are able to reduce nitrogen and carbon in aerobic conditions, a fact that may be responsible for their evolutionary and ecological success. The water-oxidising photosynthesis is accomplished by coupling the activity of photosystem (PS) II and I (Z-scheme). In anaerobic conditions, they are also able to use only PS I — cyclic photophosphorylation — with electron donors other than water (hydrogen sulfide, thiosulphate, or even molecular hydrogen) just like purple photosynthetic bacteria. Furthermore, they share an archaeal property, the ability to reduce elemental sulfur by anaerobic respiration in the dark. Their photosynthetic electron transport shares the same compartment as the components of respiratory electron transport. Actually, their plasma membrane contains only components of the respiratory chain, while the thylakoid membrane hosts both respiratory and photosynthetic electron transport.

Attached to thylakoid membrane, phycobilisomes act as light harvesting antennae for the photosystems. The phycobilisome components (phycobiliproteins) are responsible for the blue-green pigmentation of most cyanobacteria. The variations to this theme is mainly due to carotenoids and phycoerythrins which give the cells the red-brownish coloration. In some cyanobacteria, the color of light influences the composition of phycobilisomes. In green light, the cells accumulate more phycoerythrin, whereas in red light they produce more phycocyanin. Thus the bacteria appear green in red light and red in green light. This process is known as complementary chromatic adaptation and is a way for the cells to maximise the use of available light for photosynthesis.

A few genera, however, lack phycobilisomes and have chlorophyll *b* instead (Prochloron, Prochlorococcus, Prochlorothrix). These were originally grouped together as the prochlorophytes or chloroxybacteria, but appear to have developed in several different lines of cyanobacteria. For this reason they are now considered as part of the cyanobacterial group.

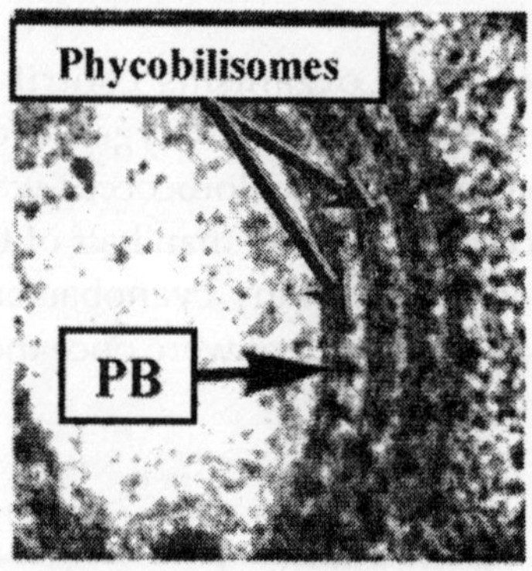

Figure 8. Phycobilisomes

Relationship to Chloroplasts

Chloroplasts found in eukaryotes (algae and plants) likely evolved. from an endosymbiotic relation with cyanobacteria. This endosymbiotic theory is supported by various structural and genetic similarities. Primary chloroplasts are found among the green plants, where they contain chlorophyll *b*, and among the red algae and glaucophytes, where they contain phycobilins. It now appears that these chloroplasts probably had a single origin, in an ancestor of the clade called Primoplantae. Other algae likely took their chloroplasts from these forms by secondary endosymbiosis or ingestion. It was once thought that the mitochondria in eukaryotes also developed from an endosymbiotic relationship with cyanobacteria; however, it is now suspected that this evolutionary event occurred when aerobic bacteria were engulfed by anaerobic host cells. Mitochondria are believed to have originated not from cyanobacteria but from an ancestor of Rickettsia.

Relationship to Earth history

The biochemical capacity to use water as the source for electrons in photosynthesis evolved once, in a common ancestor of extant cyanobacteria. The geologic record indicates that this transforming event took place early in our planet's history, at least 2450-2320 million years ago (mya), and probably much earlier. Geobiological interpretation of Archean (>2500 mya) sedimentary rocks remains a challenge; available evidence indicates that life existed 3500 mya, but the question of when oxygenic photosynthesis evolved continues to engender debate and research. A clear paleontological window on cyanobacterial evolution opened about 2000 mya, revealing an already diverse biota of blue-greens. Cyanobacteria remained principal primary producers throughout the Proterozoic Eon (2500-543 mya), in part because the redox structure of the oceans favored photautotrophs capable of nitrogen fixation. Green algae joined blue-greens as major primary producers on continental shelves near the end of the Proterozoic, but only with the Mesozoic (251-65 mya) radiations of dinoflagellates, coccolithophorids, and diatoms did primary production in marine shelf waters take modern form. The most common cyanobacterial structures in the fossil record include stromatolites and oncolites. Cyanobacteria remain critical to marine ecosystems as primary producers in oceanic gyres, as agents of biological nitrogen fixation, and, in modified form, as the plastids of marine algae.

Classification

The cyanobacteria were traditionally classified by morphology into five sections, referred to by the numerals I-V. The first three - Chroococcales, Pleurocapsales, and Oscillatoriales - are not supported by phylogenetic studies. However, the latter two - Nostocales and Stigonematales - are monophyletic, and make up the heterocystous cyanobacteria. The members of Chroococales are unicellular and usually aggregate in colonies. The classic taxonomic criterion has been the cell morphology and the plane of cell division. In Pleurocapsales, the cells have the ability to form internal spores (baeocytes). The rest of the sections include filamentous species. In Oscillatoriales, the cells are uniseriately arranged and do not form specialised cells (akinetes and heterocysts). In Nostocales and Stigonematales the cells have the ability to develop heterocysts in certain conditions. Stigonematales, unlike Nostocales, includes species with truly branched trichomes. Most taxa included in the phylum or division Cyanobacteria have not yet been validly published under the Bacteriological Code. Except:

– The classes Chroobacteria, Hormogoneae and Gloeobacteria

– The orders Chroococcales, Gloeobacterales, Nostocales, Oscillatoriales, Pleurocapsales and Stigonematales

– The families Prochloraceae and Prochlorotrichaceae

– The genera Halospirulina, Planktothricoides, Prochlorococcus, Prochloron, Prochlorothrix.

IMPORTANCE OF CYANOBACTERIA

Cyanobacteria are very important organisms for the health and growth of many plants. They are one of very few groups of organisms that can convert inert atmospheric nitrogen into an organic form, such as nitrate or ammonia. It is these "fixed" forms of nitrogen which plants need for their growth, and must obtain from the soil. Fertilisers work the way they do in part because they contain additional fixed nitrogen which plants can then absorb through their roots.

Nitrification cannot occur in the presence of oxygen, so nitrogen is fixed in specialised cells called heterocysts. These cells have an especially thickened wall that contains an anaerobic environment. You can see these larger cells among the filaments of *Nostoc*, shown at right. Many plants, especially legumes, have formed symbiotic relations with nitrifying bacteria, providing specialised tissues in their roots or stems to house the bacteria, in return for organic nitrogen. This has been used to great advantage in the cultivation of rice, where the floating fern *Azolla* is actively distributed among the rice paddies. The fern houses colonies of the cyanobacterium *Anabaena* in its leaves, where it fixes nitrogen. The ferns then provide an inexpensive natural fertiliser and nitrogen source for the rice plants when they die at the end of the season. Cyanobacteria also form symbiotic relationships with many fungi, forming complex symbiotic "organisms" known as lichens.

The cyanobacterium *Spirulina*, shown at right, has long been valued as a food source; it is high in protein, and can be cultivated in ponds quite easily. In tropical countries, it may be a very important part of the diet, and was eaten regularly by the Aztecs; it is also served in several Oriental dishes. In the US, the popularity of *Spirulina* is primarily as a "health food", being sold in stores as a dried powder or in tablet form. Many other species of cyanobacteria produce populations that are toxic to humans and animals. Blue-green pond scums have been linked to the poisoning of cattle and dogs, and occasionally people. It is therefore not recommended that wild populations be gathered and eaten without some knowledge of the organisms involved. Cyanobacteria may cause other problems as well; a species of *Lyngbya* is responsible for one of the skin irritations commonly known as "swimmer's itch."

The unicellular cyanobacterium Synechocystis sp. PCC6803 was the third prokaryote and first photosynthetic organism whose genome was completely sequenced. It continues to be an important model organism. The smallest genomes have been found in Prochlorococcus spp. and the largest in Nostoc *punctiforme*. Those of Calothrix spp. are estimated at 12-15 Mb, as large as yeast.Some cyanobacteria are sold as food, notably Aphanizomenon flos-aquae and *Arthrospira platensis* (Spirulina).

Certain cyanobacteria produce cyanotoxins like anatoxin-a, anatoxin-as, aplysiatoxin, cylindrospermopsin, domoic acid, microcystin LR, nodularin R (from Nodularia), or saxitoxin. Sometimes a mass-reproduction of cyanobacteria results in algal blooms. These toxins can be neurotoxins, hepatotoxins, cytotoxins, and endotoxins, and can be dangerous to animals and humans. Several cases of human poisoning have been documented but a lack of knowledge prevents an accurate assessment of the risks.

REFERENCES

Fogg, G.E., et.al.. 1973. *The Blue-green Algae.* Academic Press, London and New York.

Whitton, B.A. Phylum Cyanophyta (Cyanobacteria). in *The Freshwater Algal Flora of the British Isles.* Cambridge University Press, Cambridge.

2

Ecology of Cyanobacteria

The cyanobacteria provide an extraordinarily wide-ranging contribution to human affairs in everyday life and are of economic importance. Both the beneficial and detrimental features of the cyanobacteria are of considerable significance. They are important primary producers and their general nutritive value is high. The nitrogen-fixing species contribute globally to soil and water fertility. The use of cyanobacteria in food production and in solar energy conversion holds promising potential for the future. However, cyanobacteria may also be a source of considerable nuisance in many situations. Abundant growth of cyanobacteria in water reservoirs creates severe practical problems for water supplies. The development of strains containing toxins is a common experience in polluted inland water systems all over the world, as well as in some coastal waters. Thus cyanobacterial toxins, or "cyanotoxins", have become a concern for human health.

Prior to the first acute cyanotoxin poisoning of domestic animals documented in the scientific literature, reports of cyanobacteria poisonings were largely anecdotal. Perhaps one of the earliest is from the Han dynasty of China. About 1,000 years ago, General Zhu Ge-Ling, while on a military campaign in southern China, reported losing troops from poisonings whilst crossing a river. He reported that the river was green in colour at the time and that his troops drank from the green water. Codd reported that human awareness of toxic blooms existed in the twelfth century at the former Monasterium Virdis Stagni, located near the eutrophic, freshwater Soulseat Loch near Stranraer in south west Scotland. In more recent times, several investigators have noted that local people in China, Africa, North and South America and Australia, who use water from water bodies where green scums are present, will dig holes (soaks) near the water's edge in order to filter the water through the ground and thus prevent the green material from contaminating drinking-water supplies. This practice is similar to that of developing wells next to surface waters in order to use the filtering capacity of the soil to remove organisms and some chemicals from the surface waters - a technique known as bankside filtration.

CYANOBACTERIA IN WATER RESOURCES

The hydrological cycle represents a complex interconnection of diverse water types with different characteristics, each subject to different uses. Recent developments have shown the importance of water resource management in an integrated manner and of recognising interconnections, especially between human activities and water quality.

Most of the world's available freshwater exists as groundwater. This ready supply of relatively clean and accessible water has encouraged use of this resource, and in many regions groundwater provides drinking water of excellent quality. However, in some areas, geological conditions do not allow the use of groundwater or the supplies are insufficient. Thus, where groundwater supplies are insufficient or of unsuitable quality, surface water must be used for purposes such as drinking-water supply. Compared with surface waters, groundwaters have a high volume and low throughput. Over-abstraction is therefore common.

Cyanobacteria are a frequent component of many freshwater and marine ecosystems. Those species that live dispersed in the water are part of the phytoplankton whilst those that grow on sediments form part of the phytobenthos. Under certain conditions, especially where waters are rich in nutrients and exposed to sunlight, cyanobacteria may multiply to high densities - a condition referred to as a water bloom.

The composition of freshwaters is dependent on a number of environmental factors, including geology, topography, climate and biology. Many of these factors vary over different time scales such as daily, seasonally, or even over longer timespans. Large natural variations in water quality may therefore be observed in any given water system. Eutrophication is the enhancement of the natural process of biological production in rivers, lakes and reservoirs, caused by increases in levels of nutrients, usually phosphorus and nitrogen compounds. Eutrophication can result in visible cyanobacterial or algal blooms, surface scums, floating plant mats and benthic macrophyte aggregations. The decay of this organic matter may lead to the depletion of dissolved oxygen in the water, which in turn can cause secondary problems such as fish mortality from lack of oxygen and liberation of toxic substances or phosphates that were previously bound to oxidised sediments. Phosphates released from sediments accelerate eutrophication, thus closing a positive feedback cycle. Some lakes are naturally eutrophic but in many others the excess nutrient input is of anthropogenic origin, resulting from municipal wastewater discharges or run-off from fertilisers and manure spread on agricultural areas. Losses of nutrients due to erosion and run-off from soils may be low in relation to agricultural input and yet high in relation to the eutrophication they cause, because concentrations of phosphorus of less than 0.1 mg l^{-1} are sufficient to induce a cyanobacterial bloom.

Hydrological differences between rivers, impoundments and lakes have important consequences for nutrient concentrations and thus for cyanobacterial growth. Rivers generally have a significant flushing rate. The term "self-purification" was adopted to describe the rapid degradation of organic compounds in rivers where turbulent mixing effectively replenishes consumed oxygen. This term has been applied, mistakenly, to any process of removing undesirable substances from water but does not actually eliminate the contaminants, including processes such as adsorption to sediments or dilution. Substances bound to sediments may accumulate, be released back into the water, and may be carried downstream. This process is important for phosphorus. Lakes generally have long water retention times compared with rivers, and by their nature lakes tend to accumulate sediments and the chemicals associated with them. Sediments therefore act as sinks for important nutrients such as phosphorus, but if conditions change the sediments may also serve as sources, liberating the nutrient back into the water where it can stimulate the growth of cyanobacteria and algae.

Surface water systems world-wide are now often highly regulated in efforts to control water availability, whether for direct use in irrigation, hydropower generation or drinking water supplies or to guard against the consequences of floods and droughts. Many major rivers may be viewed as a cascade of impoundments. This trend in regulation of flow has an impact upon the quality and the quantity of water. It alters sediment transport and, as a result, the transport of substances attached to sediments, such as plant nutrients which may enhance cyanobacterial growth. By increasing retention times and surface areas exposed to sunlight, impoundments change the growth conditions for organisms and promote opportunities for cyanobacterial growth and water-bloom formation through modifications to river discharges. For many estuarine and coastal systems, human impact on hydrological conditions and nutrient concentrations is also now extensive.

Changes in the nature and scale of human activities have consequences both for the qualitative and quantitative properties of water resources. Historically, the development of society has involved a change from rural and agricultural to urban and industrial water uses, which is reflected in both water demands and water pollution. The general trend has been an increase in concentrations of pollutants in surface waters together with increases in urbanisation. Construction of sewerage first enhanced this trend by concentrating pollutants from latrines. After some decades, construction of sewage treatment systems began extensively in the 1950s. Originally these systems comprised only a biological step which degraded the organic material which otherwise had led to dramatic oxygen depletion in the receiving water bodies. Pathogens were also reduced to some extent, but phosphate remained unaffected. Upgrading treatment systems to remove phosphorus only began in the 1960s and also had the side-effect of further reducing pathogens. A resultant decline in eutrophication, and thus of cyanobacterial blooms, is lagging behind the decline of phosphorus inputs to freshwaters because phytoplankton growth becomes nutrient-controlled only below threshold concentrations.

It is unclear whether the historical shift in water demand from rural to urban will continue in the future, although a number of influences are apparent. The anticipated food crises of the early twenty first century will place increasing demands upon irrigated agriculture - a process that already accounts for about 70 per cent of water demand world-wide. By contrast, many industries have successfully developed processes with substantial water economy measures, and their demand upon water resources per unit of activity is now decreasing in some countries. Domestic water consumption tends to increase with population and affluence, but development of lower consumption appliances and control of losses from water mains may stabilise, or even reduce, demand in the future. Nevertheless, overall trends point to an increasing total demand for water, driven principally by global population growth.

CYANOBACTERIAL BLOOMS

Eutrophication was recognised as a pollution problem in many western European and North American lakes and reservoirs in the middle of the twentieth century. Since then, it has become more widespread, especially in some regions; it has caused deterioration in the aquatic environment and serious problems for water use, particularly in drinking-water treatment. A recent survey showed that in the Asia Pacific Region, 54 per cent of lakes are eutrophic; the proportions for Europe, Africa, North America and South America are 53 per cent, 28 per cent, 48 per cent and 41 per cent respectively. Eutrophication also affects slow flowing rivers, particularly if they have extended low-flow periods during a dry season. Practical measures for prevention of nutrient loading from wastewater and from agriculture have been developed. In some regions preventative measures are being implemented more and more. During the 1990s, increasing introduction of nutrient removal during sewage treatment in North America and in north western Europe has begun to show success in reducing phosphorus concentrations; in a few water bodies, algal and cyanobacterial blooms have actually declined. Technical measures for reduction of nutrients already present in lakes are also available but have not been widely applied.

Wherever conditions of temperature, light and nutrient status are conducive, surface waters may host increased growth of algae or cyanobacteria. Where such proliferation is dominated by a single species, the phenomenon is referred to as an algal or cyanobacterial bloom. Problems associated with cyanobacteria are likely to increase in areas experiencing population growth with a lack of concomitant sewage treatment and in regions with agricultural practices causing nutrient losses to water bodies through over-fertilisation and erosion.

There are important differences in algal and cyanobacterial growth between tropical and temperate areas. A characteristic pattern of seasonal succession of algal and cyanobacterial communities is, for example, diatoms in association with rapidly growing small flagellates in winter and spring, followed by green algae in late spring and early summer, and then by species which cannot easily be eaten by zooplankton, such as dinoflagellates, desmids and large yellow-green algae (in moderately turbulent waters also diatoms) in late summer and autumn. In eutrophic and hypertrophic waters, cyanobacteria often dominate the summer phytoplankton. As winter approaches, in most water bodies, increasing turbulence and the lack of light during the winter leads to their replacement by diatoms. In the tropics, seasonal differences in environmental factors are often not great enough to induce the replacement of cyanobacteria by other phytoplankton species. If cyanobacteria are present or even dominant for most of the year, the practical problems associated with high cyanobacterial biomass and the potential health threats from their toxins increase. High cyanobacterial biomass may also contribute to aesthetic problems, impair recreational use, and affect the taste of treated drinking water.

Phosphorus is the major nutrient controlling the occurrence of water blooms of cyanobacteria in many regions of the world, although nitrogen compounds are sometimes relevant in determining the amount of cyanobacteria present. However, in contrast to planktonic algae, some cyanobacteria are able to escape nitrogen limitation by fixing atmospheric nitrogen. The lack of nitrate or ammonia, therefore, favours the dominance of these species. Thus, the availability of nitrate or ammonia is an important factor in determining which cyanobacterial species become dominant.

Cyanobacterial blooms are monitored using biomass measurements coupled with the examination of the species present. A widely-used measure of algal and cyanobacterial biomass is the chlorophyll *a* concentration. Peak values of chlorophyll *a* for an oligotrophic lake are about 1-10 $\mu g\ l^{-1}$, while in a eutrophic lake they can reach 300 $\mu g\ l^{-1}$. In cases of hypereutrophy, such as Hartbeespoort Dam in South Africa, maxima of chlorophyll *a* can be as high as 3,000 $\mu g\ l^{-1}$.

Trophic state classifications, such as that adopted by the Organisation for Economic Cooperation and Development (OECD), combine information concerning nutrient status and algal biomass. They provide a basis for the evaluation of status and trends for management and they facilitate international information exchange and comparison.

TOXIC CYANOBACTERIA

The contamination of water resources and drinking water supplies by human excreta remains a major human health concern, just as it has been for centuries. By contrast, the importance of toxic substances, such as metals and synthetic organic compounds, has only emerged in the latter half of the twentieth century. Although eutrophication has been recognised as a growing concern since the 1950s, only recently have cyanobacterial toxins become widely recognised as a human health problem arising as a consequence of eutrophication. The importance of such toxins, relative to other water-health issues, can currently only be estimated. A significant proportion of cyanobacteria produce one or more of a range of potent toxins. If water containing high concentrations of toxic cyanobacteria or their toxins is ingested, they present a risk to human health. Some cyanobacterial substances may cause skin irritation on contact.

The relationship between water resources and health is complex. The most well recognised relationship is the transmission of infectious and toxic agents through consumption of water. Drinking water has therefore played a prominent role in concerns for water and human health. Diseases arising from the consumption of contaminated water are generally referred to as "waterborne". Globally, the waterborne diseases of greatest importance are those caused by bacteria, viruses and parasites, such as cholera, typhoid, hepatitis A, cryptosporidiosis and giardiasis. Most of the pathogens involved are derived from human faeces and the resulting diseases are generally referred to as "faecal-oral" diseases; however they can also be spread by means other than contaminated water, such as by contaminated food. Waterborne diseases also include some caused by toxic chemicals, although many of these may only cause health effects some time after exposure has occurred and may therefore be difficult to associate directly with the cause.

The second major area of interaction between water and human health concerns its role in personal and domestic hygiene, through which it contributes to the control of disease. Because hygiene is a key measure in the control of faecal-oral disease, such diseases are also "water hygiene" diseases. Other water hygiene diseases include skin and eye infections and infestations, such as tinea, scabies, pediculosis and trachoma. All of these diseases occur less frequently when adequate quantities of water are available for personal and domestic hygiene. It is important to note that the role of water in control of water hygiene diseases depends on availability and use, and water quality is therefore a secondary consideration in this context.

"Water contact diseases" are the third group of water-related diseases and occur through skin contact. The most important example world-wide is schistosomiasis (bilharzia). In infected persons, eggs of *Schistosoma* spp. are excreted in faeces or urine. The schistosomes require a snail intermediate host and go on to infect persons in contact with water by penetrating intact skin. The disease is of primary importance in areas where collection of water requires wading or direct contact with contaminated surface waters such as lakes or rivers. The water contact diseases also include those diseases arising from non-infectious agents in the water, that may give rise, for example, to allergies and to skin irritation or to dermatitis.

The fourth principal connection between water and human health concerns "water habitat vector" diseases. These are diseases transmitted by insect vectors that spend all or part of their lives in or near water. The best-known examples are malaria and filariasis.

The classification of water-related disease into four groups (waterborne disease, water hygiene disease, water contact disease and water habitat vector disease) was originally developed in order to associate groups of disease more clearly with the measures for their transmission and control and has contributed greatly to furthering this understanding. Because of its importance to the global burden of disease, the classification is based upon infectious disease. Nevertheless, the principal groups of diseases related to chemicals occurring in water may also be categorised in a similar way. However, there are a number of water-health associations that fall outside these categories. These include deficiency-related diseases and recreational uses of water. For recreational water use, the principal area of concern relating to faecal-oral disease transmission may be classified reasonably alongside other waterborne disease transmission. However, concern related to transmission of, for example, eye and ear infections does not readily fit into the classification system, nor does the increased transmission of diseases arising from the effect of immersion compromising natural defence systems.

Public health concern regarding cyanobacteria centres on the ability of many species and strains of these organisms to produce cyanotoxins. Cyanotoxins may fall into two of the four groups of water-related diseases. They may cause waterborne disease when ingested, and water contact disease primarily through recreational exposure. In hospitals and clinics, exposure through intravenous injection has led to human fatalities from cyanotoxins. These toxins pose a challenge for management. Unlike most toxic chemicals, cyanotoxins only sometimes occur dissolved in the water - they are usually contained within cyanobacterial cells. In contrast to pathogenic bacteria, these cells do not proliferate within the human body after uptake, only in the aquatic environment before uptake.

Cyanotoxins belong to rather diverse groups of chemical substances, each of which shows specific toxic mechanisms in vertebrates. Some cyanotoxins are strong neurotoxins (anatoxin-a, anatoxin-a(s), saxitoxins), others are primarily toxic to the liver (microcystins, nodularin and cylindrospermopsin), and yet others (such as the lipopolysaccharides) appear to cause health impairments (such as gastroenteritis) which are poorly understood. Microcystins are geographically most widely distributed in freshwaters. Recently, they have even been identified in marine environments as a cause of liver disease in net-pen reared salmon, although it is not clear which organism in marine environments contains these toxins. As with many cyanotoxins, microcystins were named after the first organism found to produce them, *Microcystis aeruginosa,* but later studies also showed their occurrence in other cyanobacterial genera.

The hazard to human health caused by cyanotoxins can be estimated from toxicological knowledge in combination with information on their occurrence. However, although the information clearly indicates hazards, there are few documented cases of human illness unequivocally attributed to cyanotoxins. In a number of cases, investigation of cyanobacteria and cyanotoxins was carried out only several days after patients had been exposed and had developed symptoms. This was because diagnosis moved on to considering cyanobacteria only after other potential causative agents had proved negative, or even years later when knowledge of cyanobacterial blooms in a water body was connected with the information on an outbreak of symptoms of unidentified cause.

The number of quantitative surveys on cyanotoxin occurrence is low, and the level of cyanotoxin exposure through drinking water or during recreational activities largely unknown. Surveys on cyanobacteria and cyanotoxins have been primarily ecological and biogeographical. Early surveys in a number of countries including Australia, Canada, Finland, Norway, South Africa, Sweden, the UK and the USA involved toxicity testing of scum samples by mouse bioassay. Surveys during the 1990s have tended to employ more sensitive and definitive methods for characterisation of the toxins, such as chromatographic or immunological methods. These studies provide an improving basis for estimating the range of concentrations to be expected in a given water body and season. However, monitoring cyanotoxin concentration is more difficult than many other waterborne disease agents, because variations in cyanobacterial quantities, in time and space, is substantial, particularly if scum-forming species are dominant. Wind-driven accumulations and distribution of surface scums can result in concentrations of the toxin by a factor of 1,000 or more and such situations can change within very short time periods, i.e. the range of hours. Therefore, discontinuous samples only provide a fragmentary insight into the potential cyanotoxin dose for occasional swimmers and into the amount entering drinking water intakes.

Very few studies of cyanotoxin removal by drinking water treatment processes have been published, although some water companies have carried out unpublished studies. Thus, a reliable basis for estimation of cyanotoxin exposure through drinking water is lacking. In regions using surface waters affected with cyanobacteria as a source for drinking water, actual toxin exposure will depend strongly on method of water abstraction and treatment.

In comparing the available indications of hazards from cyanotoxins with other waterrelated health hazards, it is conspicuous that cyanotoxins have caused numerous fatal poisonings of livestock and wildlife, but no human fatalities due to oral uptake have been documented. Human deaths have only been observed as a consequence of intravenous exposure through renal dialysis. Cyanotoxins are rarely likely to be ingested by humans in sufficient amounts for an acute lethal dose. Thus, cyanobacteria are less of a health hazard than pathogens such as *Vibrio cholerae* or *Salmonella typhi*. Nevertheless, dose estimates indicate that a fatal dose is possible for humans, if scum material is swallowed. However, swallowing such a repulsive material is likely to be avoided. The combination of available knowledge on chronic toxicity mechanisms with that on ambient concentrations occurring under some environmental conditions, shows that chronic human injury from some cyanotoxins is likely, particularly if exposure is frequent or prolonged at high concentrations.

Research into developing further understanding of the human health significance of cyanobacteria and individual cyanotoxins, and into practical means for assessing and controlling exposure to cyanobacteria and to cyanotoxins, is a priority. A major gap also lies in the synthesis and dissemination of the available information. Information concerning the efficiency of cyanotoxin removal in drinking water treatment systems is limited. Especially, simple, low-cost techniques for cyanobacterial cell removal, such as slow sand filtration, should be investigated and developed further. More information is also needed on the capability of simple disinfection techniques, such as chlorine, for oxidising microcystins and cylindrospermopsin. If this is found to be applicable, or if "conventional" treatments are found to be effective if properly operated, these approaches would provide a practical tool for removing cyanotoxins in many situations.

Whilst cyanobacterial blooms remain sporadic or occasional events, most emphasis is still placed upon the protection of drinking water supplies through the preparation of contingency plans and their activation when appropriate. Early warning systems and predictive models can facilitate this and should be based upon available information on the conditions leading to cyanobacterial bloom development and on occurrence, localisation and movement of scums.

Epidemiological evidence is of particular value in determining the true nature and severity of human health effects but is generally lacking in relation to human exposures to cyanobacteria. The limited studies undertaken to date in relation to recreational exposure require further substantiation. Opportunistic studies into exposures through drinking water may provide further valuable insights. Information from experimental toxicology also needs to be strengthened. In particular, long-term exposure studies should be carried out to assess the chronic toxicity of microcystins and cylindrospermopsins. Uptake routes require further investigation. Further systematic studies are also required into the suggested tumour-promoting effects of some cyanotoxins, particularly in the dose range of potential oral uptake with drinking or bathing water.

Lipopolysaccharide (LPS) endotoxins from cyanobacteria pose a potential health risk for humans, but knowledge of the occurrence of individual LPS components, their toxicology, and their removal in drinking water treatment plants, is so poor that guidelines cannot be set at present. Further bioactive cyanobacterial metabolites are also identified frequently and the health significance of these requires investigation. For management of cyanobacterial hazards to human health, a basic understanding of the properties, the behaviour in natural ecosystems, and the environmental conditions which support the growth of certain species is helpful.

NATURE AND DIVERSITY

Plants and animals possess consistent features by which they can be identified reliably and sorted into recognisably distinct groups. Biologists observe and compare what the organisms look like, how they grow and what they do. The results make it possible to construct systematic groupings based on multiple correlations of common characters and that reflect the greatest overall similarity. The basis for such groupings is the fact that all organisms are related to one another by way of evolutionary descent. Their biology and phylogenetic relationships makes the establishment of systematic groupings possible.

However, microbial systematics has long remained an enigma. Conceptual advances in microbiology during the twentieth century included the realisation that a discontinuity exists between those cellular organisms that are prokaryotic and those that are eukaryotic within the organisation of their cells. The microalgae investigated by phycologists under the International Code of Botanical Nomenclature (ICBN) included organisms of both eukaryotic and prokaryotic cell types. The bluegreen algae constituted the largest group of the latter category. The prokaryotic nature of these organisms and their fairly close relationship with eubacteria made work under provisions of the International Code of Nomenclature of Bacteria (ICNB) more appropriate. The prevailing systematic view is that comparative studies of the genetic constitution of the cyanobacteria will now contribute significantly to the revision of their taxonomy.

Relevant classification should reflect as closely as possible the phylogenetic relationships as, for example, encoded in 16S or 23S rRNA sequence data. The integration of phenotypic, genotypic and phylogenetic information render possible a consensus type of taxonomy known as polyphasic taxonomy.

The names "cyanobacteria" and "blue-green algae" (Cyanophyceae) are valid and compatible systematic terms. This group of micro-organisms comprises unicellular to multicellular prokaryotes that possess chlorophyll *a* and perform oxygenic photosynthesis associated with photosystems I and II.

The majority of cyanobacteria are aerobic photoautotrophs. Their life processes require only water, carbon dioxide, inorganic substances and light. Photosynthesis is their principal mode of energy metabolism. In the natural environment, however, it is known that some species are able to survive long periods in complete darkness. Furthermore, certain cyanobacteria show a distinct ability for heterotrophic nutrition.

Cyanobacteria are often the first plants to colonise bare areas of rock and soil. Adaptations, such as ultraviolet absorbing sheath pigments, increase their fitness in the relatively exposed land environment. Many species are capable of living in the soil and other terrestrial habitats, where they are important in the functional processes of ecosystems and the cycling of nutrient elements.

The prominent habitats of cyanobacteria are limnic and marine environments. They flourish in water that is salty, brackish or fresh, in cold and hot springs, and in environments where no other microalgae can exist. Most marine forms grow along the shore as benthic vegetation in the zone between the high and low tide marks. The cyanobacteria comprise a large component of marine plankton with global distribution. A number of freshwater species are also able to withstand relatively high concentrations of sodium chloride. It appears that many cyanobacteria isolated from coastal environments tolerate saline environments rather than require salinity. As frequent colonisers of euryhaline environments, cyanobacteria are found in salt works and salt marshes, and are capable of growth at combined salt concentrations as high as 3-4 molar mass. Freshwater localities with diverse trophic states are the prominent habitats for cyanobacteria.

Numerous species characteristically inhabit, and can occasionally dominate, both near-surface epilimnic and deep, euphotic, hypolimnic waters of lakes. Others colonise surfaces by attaching to rocks or sediments, sometimes forming mats that may tear loose and float to the surface. Cyanobacteria have an impressive ability to colonise infertile substrates such as volcanic ash, desert sand and rocks. They are extraordinary excavators, boring hollows into limestone and special types of sandstone. Another remarkable feature is their ability to survive extremely high and low temperatures. Cyanobacteria are inhabitants of hot springs, mountain streams, Arctic and Antarctic lakes and snow and ice. The cyanobacteria also include species that run through the entire range of water types, from polysaprobic zones to katharobic waters.

Cyanobacteria also form symbiotic associations with animals and plants. Symbiotic relations exist with, for example, fungi, bryophytes, pteridophytes, gymnosperms and angiosperms. The hypothesis for the endosymbiotic origin of chloroplasts and mitochondria should be mentioned in this context. The evolutionary formation of a photosynthetic eukaryote can be explained by a cyanobacteria being engulfed and codeveloped by a phagotrophic host.

Fossils of what were almost certainly prokaryotes are present in the 3,450 million year old Warrawoona sedimentary rock of north-western Australia. Cyanobacteria were among the pioneer organisms of the early earth. These photosynthetic micro-organisms were, at that time, probably the chief primary producers of organic matter, and the first organisms to release elemental oxygen into the primitive atmosphere. Sequencing of deoxyribonucleic acid (DNA) has given evidence that the earliest organisms were thermophilic and thus able to survive in oceans that were heated by volcanoes, hot springs and bolide impacts.

ORGANISATION, AND BEHAVIOUR

The structure and organisation of cyanobacteria are studied using light and electron microscopes. The basic morphology comprises unicellular, colonial and multicellular filamentous forms (Figure 1).

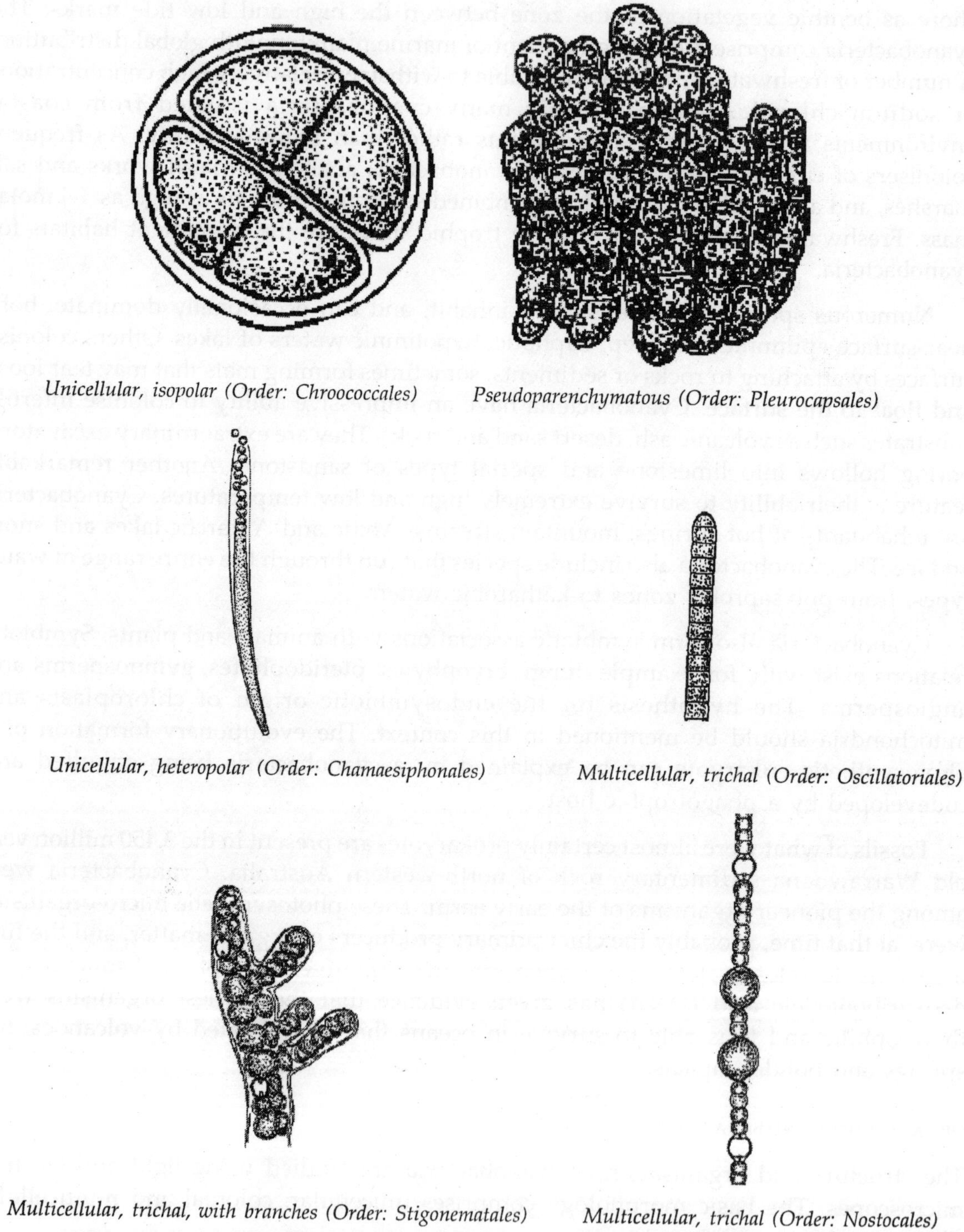

Unicellular, isopolar (Order: Chroococcales)

Pseudoparenchymatous (Order: Pleurocapsales)

Unicellular, heteropolar (Order: Chamaesiphonales)

Multicellular, trichal (Order: Oscillatoriales)

Multicellular, trichal, with branches (Order: Stigonematales)

Multicellular, trichal (Order: Nostocales)

Figure 1. Basic Morphology of Cyanobacteria

Unicellular forms, for example in the order Chroococcales, have spherical, ovoid or cylindrical cells. They occur singly when the daughter cells separate after reproduction by binary fission. The cells may aggregate in irregular colonies, being held together by the slimy matrix secreted during the growth of the colony. By means of a more or less regular series of cell division, combined with sheath secretions, more ordered colonies may be produced.

A particular mode of reproduction, which may supplement binary fission, distinguishes cyanobacteria in the order Chamaesiphonales and Pleurocapsales. In the Chamaesiphonales exospores are budded off from the upper ends of cells. In the second order, the principal mode of replication is by a series of successive binary fissions converting a single mother cell into many minute daughter cells (baeocytes or endospores).

Filamentous morphology is the result of repeated cell divisions occurring in a single plane at right angles to the main axis of the filament. The multicellular structure consisting of a chain of cells is called a trichome. The trichome may be straight or coiled. Cell size and shape show great variability among the filamentous cyanobacteria. Species in the order Oscillatoriales, with unseriated and unbranched trichomes, are composed of essentially identical cells. The other orders with a filamentous organisation are characterised with trichomes having a heterogeneous cellular composition. Vegetative cells may be differentiated into heterocysts and akinetes. In the order Stigonematales, the filaments are often multiseriated, with genuine branching. Both heterocysts and akinetes are present.

The only means of reproduction in cyanobacteria is asexual. Filamentous forms reproduce by trichome fragmentation, or by formation of special hormogonia. Hormogonia are distinct reproductive segments of the trichomes. They exhibit active gliding motion upon their liberation and gradually develop into new trichomes.

In contrast to eukaryotic microalgae, cyanobacteria do not possess membrane-bound sub-cellular organelles; they have no discrete membrane-bound nucleus; they possess a wall structure based upon a peptidoglycan layer; and they contain 70 S rather than 80 S ribosomes.

The photosynthetic pigments of cyanobacteria are located in thylakoids that lie free in the cytoplasm near the cell periphery. Cell colours vary from blue-green to violet-red. The green of chlorophyll *a* is usually masked by carotenoids and accessory pigments such as phycocyanin, allophycocyanin and phycoerythrin (phycobiliproteins). The pigments are embodied in phycobilisomes, which are found in rows on the outer surface of the thylakoids. All cyanobacteria contain chlorophyll *a* and phycocyanine.

The basic features of photosynthesis in cyanobacteria have been well described. Cyanobacteria are oxygenic phototrophs possessing two kinds of reaction centres, PS I and PS II, in their photosynthetic apparatus. With the accessory pigments mentioned above, they are able to use effectively that region of the light spectrum between the absorption peaks of chlorophyll *a* and the carotenoids. The ability for continuous photosynthetic growth in the presence of oxygen, together with having water as their electron donor for CO_2 reduction, enables cyanobacteria to colonise a wide range of ecological niches. Phycobiliprotein synthesis is particularly susceptible to environmental influences, especially light quality. Chromatic adaptation is largely attributable to a change in the ratio between phycocyanin and phycoerythrin in the phycobilisomes. Thus, cyanobacteria are able to produce the accessory pigment needed to absorb light most efficiently in the habitat in which they are present.

Cyanobacteria have a remarkable ability to store essential nutrients and metabolites within their cytoplasm. Prominent cytoplasmic inclusions for this purpose can be seen with the electron microscope. Reserve products are accumulated under conditions of an excess supply of particular nutrients. For example, when the synthesis of nitrogenous cell constituents is halted because of an absence of a usable nitrogen source, the primary products of photosynthesis are channelled towards the synthesis and accumulation of glycogen and lipids.

Dinitrogen fixation is a fundamental metabolic process of cyanobacteria, giving them the simplest nutritional requirements of all living organisms. By using the enzyme nitrogenase, they convert N_2 directly into ammonium (NH_4) and by using solar energy to drive their metabolic and biosynthetic machinery, only N_2, CO_2, water and mineral elements are needed for growth in the light. Nitrogen-fixing cyanobacteria are widespread among the filamentous, heterocyst forming genera (e.g. *Anabaena, Nostoc)*. However, there are also several well documented examples of dinitrogen fixation among cyanobacteria not forming heterocysts (e.g. *Trichodesmium)*. Under predominantly nitrogen limited conditions, but when other nutrients are available, nitrogen fixing cyanobacteria may be favoured and gain growth and reproductive success. Mass developments (often referred to as "blooms") of such species in limnic and marine environments (e.g. the Baltic Sea) are common phenomena world-wide.

Many species of cyanobacteria possess gas vesicles. These are cytoplasmic inclusions that enable buoyancy regulation and are gas-filled, cylindrical structures. Their function is to give planktonic species an ecologically important mechanism enabling them to adjust their vertical position in the water column. To optimise their position, and thus to find a suitable niche for survival and growth, cyanobacteria use different environmental stimuli (e.g. photic, gravitational, chemical, thermal) as clues. Gas vesicles become more abundant when light is reduced and the growth rate slows down. Increases in the turgor pressure of cells, as a result of the accumulation of photosynthate, cause a decrease in existing gas vesicles and therefore a reduction in buoyancy. Cyanobacteria can, by such buoyancy regulation, poise themselves within vertical gradients of physical and chemical factors. Other ecologically significant mechanisms of movement shown by some cyanobacteria are photomovement by slime secretion or surface undulations of cells.

The presence of very small cells of cyanobacteria (in the size range 0.2^{-2} μm) has been recognised as a potentially significant source of primary production in various freshwater and marine environments. These cyanobacteria constitute a component of the picoplankton in pelagic ecosystems. Cells can be recognised and estimates of their abundance made by using epifluorescence microscopy. The unicellular genus *Synechococcus* is one of the most studied, and geographically most widely distributed, cyanobacteria in the picoplankton. Toxigenic strains of *Synechococcus* have been reported.

BIOLOGICAL DIVERSITY

Although cyanobacteria probably evolved as a group of organisms about 2,000 million years before the advent of eukaryotes, they comprise fewer taxa than eukaryotic microalgae. The concept of species in the cyanobacteria has, however, no distinct boundaries. The situation is similar for most organisms, except for those that are sexually reproductive. Depending on the classification system used, the number of species recognised varies greatly. Based on the International Code of Botanical Nomenclature the class Cyanophyceae, for example, contains about 150 genera and 2,000 species.

Chemotaxonomic studies include the use of markers, such as lipid composition, polyamines, carotenoids and special biochemical features. The resulting data support the more traditional examinations of phenotypic and ecological characteristics. Physiological parameters are conveniently studied using laboratory cultures.

The diversity of cyanobacteria can be seen in the multitude of structural and functional aspects of cell morphology and in variations in metabolic strategies, motility, cell division, developmental biology, etc. The production of extracellular substances and cyanotoxins by cyanobacteria illustrates the diverse nature of their interactions with other organisms.

A molecular approach to the systematics of cyanobacteria may be most fruitful for inferring phylogenetic relationships. Macromolecules, such as nucleic acids and proteins, are copies or translations of genetic information. The methods applied involve direct studies of the relevant macromolecules by sequencing, or indirectly by electrophoresis, hybridisation, or immunological procedures. Nucleic acid technologies, especially the polymerase chain reaction (PCR), have advanced to the point that it is feasible to amplify and sequence genes and other conserved regions from a single cell. To date, 16S rRNA has given the most detailed information on the relationships within the cyanobacteria. However, the molecular results obtained should be integrated with other characteristics as the base for a polyphasic taxonomy. A considerable morphological, as well as a genotypical, polymorphy exists in the cyanobacteria, although as data from rRNA sequencing indicates they are correlated to a high degree.

The phylogenetic relationship of cyanobacteria is the rationale behind the meaningful systematic groupings. However, it is difficult to set up a system of classification that serves both the everyday need for practical identification, and offers an expression of the natural relationship between the organisms in question. Meanwhile, it will be necessary to use the available manuals and reference books to help in these investigations and with the proper identification of the cyanobacteria.

Vitamin B_{12} is the only growth factor that is known to be required by some species. Media must be supplemented with the essential nutrients needed to support cell growth, including sources of nitrogen, phosphorus, trace elements, etc. Toxigenic strains of cyanobacteria are deposited in international-type culture collections. Clonal cultures are distributed for research, taxonomic work and teaching purposes.

The cyanobacteria have both beneficial and detrimental properties when judged from a human perspective. Their extensive growth can create considerable nuisance for management of inland waters and they also release substances into the water which may be unpleasant or toxic. The water quality problems caused by dense populations of cyanobacteria are intricate, many and various and can have great health and economic impacts. As a consequence, the negative aspects of cyanobacteria have gained research attention and public concern.

The properties that make the cyanobacteria generally undesirable are also the qualifications for possible positive economic use. Blue-greens are the source of many valuable products and carry promising physiological processes, including light-induced hydrogen evolution by biophotolysis. Extensive research has taken place in the relevant fields of biotechnology. Cyanobacteria may be used for food or fodder because some strains have a very high content of proteins, vitamins and other essential growth factors and vital pigments of interest can also be produced. Cyanobacteria are also sources for substances of pharmaceutical interest. These examples are only a few of the possible applications of cyanobacteria for economic development and their exploitation is among the many challenges for biotechnology for the next millennium. Also in this context, their secondary metabolites and health relationships will become important.

FACTORS AFFECTING BLOOM FORMATION

Cyanobacteria have a number of special properties which determine their relative importance in phytoplankton communities. However, the behaviour of different cyanobacterial taxa in nature is not homogeneous because their ecophysiological properties differ. An understanding of their response to environmental factors is fundamental for setting water management targets. Because some cyanobacteria show similar ecological and ecophysiological characteristics, they can be grouped by their behaviour in planktonic ecosystems as "ecostrategists" typically inhabiting different niches of aquatic ecosystems. A number of properties and reactions to environmental conditions are discussed below in order to describe these ecostrategists and to aid the understanding of their specific behaviour.

Light Intensity

Like algae, cyanobacteria contain chlorophyll *a* as a major pigment for harvesting light and conducting photosynthesis. They also contain other pigments such as the phycobiliproteins which include allophycocyanin (blue), phycocyanin (blue) and sometimes phycoerythrine (red). These pigments harvest light in the green, yellow and orange part of the spectrum (500-650 nm) which is hardly used by other phytoplankton species. The phycobiliproteins, together with chlorophyll *a*, enable cyanobacteria to harvest light energy efficiently and to live in an environment with only green light.

Many cyanobacteria are sensitive to prolonged periods of high light intensities. The growth of *Planktothrix* (formerly *Oscillatoria) agardhii* is inhibited when exposed for extended periods to light intensities above 180 $\mu E\ m^{-2}\ s^{-1}$. Long exposures at light intensities of 320 $\mu E\ m^{-2}\ s^{-1}$ are lethal for many species. However, if exposed intermittently to this high light intensity, cyanobacteria grow at their approximate maximal rate. This light intensity amounts to less than half of the light intensity at the surface of a lake, which can reach 700-1,000 $\mu E\ m^{-2}\ s^{-1}$. Cyanobacteria which form surface blooms seem to have a higher tolerance for high light intensities. Paerl *et al.* related this to an increase in carotenoid production which protects the cells from photoinhibition.

Cyanobacteria are further characterised by a favourable energy balance. Their maintenance constant is low which means that they require little energy to maintain cell function and structure. As a result of this, the cyanobacteria can maintain a relatively higher growth rate than other phytoplankton organisms when light intensities are low. The cyanobacteria will therefore have a competitive advantage in lakes which are turbid due to dense growths of other phytoplankton. This was demonstrated in an investigation measuring growth of different species of phytoplankton at various depths in a eutrophic Norwegian lake. The results showed that the diatoms *Asterionella, Diatoma* and *Synedra* grew faster than the cyanobacterium *Planktothrix* at 1 m depth, while the growth rate was about the same for all these organisms at 2 m depth. At the very low light intensities below 3 m only *Planktothrix* grew. The ability of cyanobacteria to grow at low light intensities and to harvest certain specific light qualities, enables them to grow in the "shadow" of other phytoplankton. Van Liere and Mur demonstrated competition between cyanobacteria and other phytoplankton. Whereas the green alga *(Scenedesmus protuberans)* grew faster at high light intensities, growth of the cyanobacterium

(Planktothrix agardhii) was faster at low light intensities. If both organisms were grown in the same continuous culture at low light intensity, *Planktothrix* could out-compete *Scenedesmus*. At high light intensities, the biomass of the green alga increased rapidly, causing an increase in turbidity and a decrease in light availability. This increased the growth rate of the cyanobacterium, which then became dominant after 20 days. Although cyanobacteria cannot reach the maximum growth rates of green algae, at very low light intensities their growth rate is higher. Therefore, in waters with high turbidity they have better chances of out-competing other species. This can explain why cyanobacteria which can grow under very poor nutritional conditions often develop blooms in nutrient-rich eutrophic waters.

The light conditions in a given water body determine the extent to which the physiological properties of cyanobacteria will be of advantage in their competition against other phytoplankton organisms. The zone in which photosynthesis can occur is termed the euphotic zone (Z_{eu}). By definition, the euphotic zone extends from the surface to the depth at which 1 per cent of the surface light intensity can be detected. It can be estimated by measuring transparency with a Secchi disk and multiplying the Secchi depth reading by a factor of 2-3. The euphotic zone may be deeper or more shallow than the mixed, upper zone of a thermally stratified water body, the depth of which is termed the epilimnion (Z_m). Many species of planktonic algae and cyanobacteria have little, or only weak, means of active movement and are passively entrained in the water circulation within the epilimnion. Thus, they can be photosynthetically active only when the circulation maintains them in the euphotic zone. In eutrophic waters, phytoplankton biomass is frequently very high and causes substantial turbidity. In such situations, the euphotic zone is often more shallow than the epilimnion, i.e. the ratio Z_{eu}/Z_m is < 1, and phytoplankton spend part of the daylight period in the dark. Thus, the Z_{eu}/Z_m ratio is a reasonable approach to describing the light conditions encountered by the planktonic organisms.

Gas Vesicles

Many planktonic cyanobacteria contain gas vacuoles. These structures are aggregates of gas-filled vesicles, which are hollow chambers with a hydrophilic outer surface and a hydrophobic inner surface. A gas vesicle has a density of about one tenth that of water and thus gas vesicles can give cyanobacterial cells a lower density than water.

Growth Rate

The growth rate of cyanobacteria is usually much lower than that of many algal species. At 20 °C and light saturation, most common planktonic cyanobacteria achieve growth rates of 0.3-1.4 doublings per day, while diatoms reach 0.8-1.9 doublings per day and growth rates of up to 1.3-2.3 doublings per day have been observed for single-celled green algae. Slow growth rates require long water retention times to enable a bloom of cyanobacteria to form. Therefore cyanobacteria do not bloom in water with short retention times. A. comprehensive overview of mechanisms determining the growth rates of planktonic algae and cyanobacteria under different field conditions is available in Reynolds.

Phosphorus and Nitrogen

Because cyanobacterial blooms often develop in eutrophic lakes, it was originally assumed that they required high phosphorus and nitrogen concentrations. This assumption was maintained even though cyanobacterial blooms often occurred when concentrations of dissolved phosphate were lowest. Experimental data have shown that the affinity of many cyanobacteria for nitrogen or phosphorus is higher than for many other photosynthetic organisms. This means that they can out-compete other phytoplankton organisms under conditions of phosphorus or nitrogen limitation.

In addition to their high nutrient affinity, cyanobacteria have a substantial storage capacity for phosphorus. They can store enough phosphorus to perform two to four cell divisions, which corresponds to a 4-32 fold increase in biomass. However, if total phosphate rather than only dissolved phosphate is considered, high concentrations indirectly support cyanobacteria because they provide a high carrying capacity for phytoplankton. High phytoplankton density leads to high turbidity and low light availability, and cyanobacteria are the group of phytoplankton organisms which can grow best under these conditions.

A low ratio between nitrogen and phosphorus concentrations may favour the development of cyanobacterial blooms. A comparison between the optimum N:P ratios for eukaryotic algae (16-23 molecules N:1 molecule of P) with the optimum rates for bloom-forming cyanobacteria (10-16 molecules N: 1 molecule P), shows that the ratio is lower for cyanobacteria.

Population Stability

While many planktonic algae are grazed by copepods, daphnids and protozoa, cyanobacteria are not grazed to the same extent, and the impact of grazing by some specialised ciliates and rhizopod protozoans is usually not substantial. Cyanobacteria are attacked by viruses, bacteria and actino-mycetes, but the importance of these natural enemies for the breakdown of populations is not well understood. Because they have few natural enemies, and their capacity for buoyancy regulation prevents sedimentation, the loss rates of cyanobacterial populations are generally low. Thus, their slow growth rates are compensated by the high prevalence of populations once they have been established.

Temperature

Maximum growth rates are attained by most cyanobacteria at temperatures above 25 °C. These optimum temperatures are higher than for green algae and diatoms. This can explain why in temperate and boreal water bodies most cyanobacteria bloom during summer.

Cyanobacterial Ecostrategists

The physiological properties of cyanobacteria discussed above vary between different species. As a consequence, different "ecostrategists" inhabit different types of water bodies. A preliminary approach to describing these ecostrategists, based on ecophysiological laboratory work together with field observations, is described below. This information may be useful for management, because it helps to predict which cyanobacteria can be expected to occur under certain conditions. Further development of this approach will be possible as more data on occurrence of cyanobacteria under different growth conditions are collected from other continents.

Scum-forming Ecostrategists

During the vegetation period, a number of cyanobacteria develop large aggregates (colonies) of coccoid cells or filaments which are not homogeneously distributed over the water column. Important genera showing this development are *Microcystis, Anabaena* and *Aphanizomenon.* At the water surface the rate of photosynthesis of the colonies is high and the cells store large quantities of carbohydrates. Although the cells contain gas vesicles, the heavy carbohydrates acts as ballast and induce sinking within the colonies. According to Stoke's Law the sinking rate is dependent on the difference in density between the water and the cells, and on the square of the colony size (d^2). Large colonies sink faster than small ones, and single cells hardly show any vertical migration. By sinking, colonies move out of the euphotic zone into the deeper, dark water layers, where they use their carbohydrates during respiration and synthesise new gas-vesicles. They then become buoyant again and return to the euphotic zone. Buoyancy regulation enables the colonies to position themselves in light conditions which are optimal for their growth. A prerequisite is that the water body is not too turbulent. During the night, all colonies may become buoyant and some of the population may be accumulated on the water surface where they can be blown together by wind, forming stable scums along downwind shores. The frequency of vertical migration is dependent upon colony size.

In temperate regions, as temperatures decline in the autumn, photosynthesis becomes more rapid than respiration, and the carbohydrate "ballast" is not consumed. The colonies therefore sink to the bottom of the water body where they may survive the winter, gradually consuming their carbohydrate stores by respiration or fermentation. Cells which re-ascend from the bottom in the spring are unicellular or formed into very small colonies. During this period *Microcystis* spp. is difficult to recognise in plankton samples, and only becomes more conspicuous when the colonies increase in size during early summer.

Buoyancy regulation can be a substantial advantage in competition with other phytoplankton organisms. However, this type of regulation is only possible in water bodies with a shallow euphotic zone in relation to the depth of vertical mixing ($Z_{eu} < Z_m$). Therefore, in temperate climates, blooms of *Microcystis* spp. are found particularly in water bodies deeper than 3 m, because the euphotic zone is likely to be substantially more shallow than the mixed depth. However, even in shallow lakes, where they do not have the competitive advantage of vertical migration, *Microcystis* spp. may become dominant and form substantial blooms, as has been reported from Hungary, Australia, and particularly from subtropical and tropical regions. Reynolds characterises *Microcystis* spp. as notoriously and overwhelmingly dominant in some lakes of the lower latitudes that exhibit diel stratification.

Many cyanobacteria cannot survive high light intensities over longer periods. This may limit their distribution to more turbid, eutrophic ecosystems. However, *Microcystis* species are less sensitive to high light intensities because buoyancy regulation enables them to find light conditions that are optimal for their growth. This means that the presence of *Microcystis* cannot be related strictly to the level of eutrophication. The genus is therefore found in mesotrophic, eutrophic and in hypertrophic waters. However, the amount of biomass that this species can attain depends on the level of eutrophication. Most *Microcystis* blooms are found in lakes with an average summer chlorophyll *a* concentration of 20-50 $\mu g\ l^{-1}$ and a Secchi transparency of 1-2 m.

Homogeneously Dispersed Ecostrategists

This ecotype comprises filamentous species, such as *Planktothrix (Oscillatoria) agardhii* and *Limnothrix (Oscillatoria) redekei*. These species are extremely sensitive to high light intensities and do not form colonies. Because the filaments are quite small, vertical migration by buoyancy regulation is less pronounced than their passive entrainment by water circulation. Therefore, these species are homogeneously dispersed throughout the epilimnion.

This type of ecostrategist is found in eutrophic and hypertrophic shallow lakes. Many lakes with blooms of dispersed ecotypes have a depth of not more than 3 m and chlorophyll concentrations of 50 $\mu g\ l^{-1}$ and, in extreme cases, greater than 200 $\mu g\ l^{-1}$. The filaments are hardly grazed and do not sediment. Blooms of this type often lead to virtual monocultures which can prevail year-round for many years. Population dynamics in such lakes can be limited. In temperate regions, the autumn population can even survive under ice in winter. In such situations, the spring population starts growth with a relatively high density and thus has an advantage in competition with other species. By causing high turbidity, these cyanobacterial populations effectively suppress the growth of other phytoplankton species. Thus, the next summer population establishes itself almost without any seasonal succession between different species of phytoplankton. This high stability of the population precludes any redistribution of phosphorus and nitrogen to other components of the ecosystem and this can cause a resilience effect in lake restoration projects.

Stratifying Ecostrategists

Representatives of this ecotype develop stable summer populations in the intermediate zone of thermally stratified lakes and reservoirs known as the metalimnion see. The organisms contain the red pigment phycoerythrin to absorb the green light, which is the prevailing wave length at this depth. The most common of these species is *Planktothrix (Oscillatoria) rubescens,* but red varieties of other *Planktothrix* species can also form metalimnic populations. The single filaments of these species hardly show any vertical migration. However, in late autumn at the end of the growing season, the cells can become buoyant and then form red surface scums. The niche of this type of *Planktothrix* is very limited. It needs sufficient light in the metalimnetic zone, but may be inhibited by too much light. Most metalimnetic blooms are found at light intensities of 1-5 per cent of the surface irradiance and in a range of Z_{eu}/Z_m between 0.7 and 1.2.

Nitrogen Fixing Ecostrategists

The mass development of species capable of fixation of atmospheric nitrogen (species of the genera Anabaena, Aphanizomenon, Cylindrospermopsis, Nodularia, and Nostoc) can often be related to periodic nitrogen limitation. Examples are found in deep, as well as in shallow, systems. However, while these ecostrategists often dominate in ecosystems with low levels of inorganic dissolved nitrogen, the reverse does not necessarily apply. Numerous lakes with clear nitrogen limitation are not dominated by nitrogen-fixing cyanobacteria. Low light availability may be the reason for this, because nitrogen fixation requires high amounts of energy. In turbid lakes, insufficient light energy may be available for effective nitrogen fixation. A number of nitrogen fixing species can form colonies and possess gas vesicles. This means that they can regulate buoyancy, like Microcystis, and can form stable scums along downwind shores. Restoration measures which simultaneously reduce phosphate and nitrogen loading (sewage diversion, isolation) may strengthen prevailing nitrogen-limiting conditions and hence the probability of large populations of nitrogen-fixing cyanobacteria.

Small, Colony-forming Taxa

Cases of large populations of the small, colony-forming genus *Aphanothece* have been reported. Little information is available on buoyancy regulation and scum formation by the species involved. In several water bodies, *Aphanothece* dominance has occurred after a decrease of *Planktothrix rubescens* populations. Reynolds reported them as the only cyanobacteria present in the summer plankton of small, intermittently flushed lakes in England. The dominance of this group is not strictly related with phosphate or nitrogen limitation, and there are no obvious relationships that can explain the sudden dominance of these cyanobacteria. They seem to dominate in an intermediate state during lake recovery after restoration measures have been taken; their ecology is unknown.

Benthic Cyanobacteria

Besides the planktonic ecostrategists described above, cyanobacteria may grow on the bottom sediments of water bodies which are sufficiently clear to allow light penetration these surfaces. These benthic species may form coherent mats. Especially high rates photosynthesis by such mats sometimes leads to trapping of the photosynthetically produced oxygen as bubbles within the mats; parts of the mats may then become sufficiently buoyant to tear loose and rise to the surface. For monitoring and management of toxic cyanobacteria, awareness of these is important because cyanotoxin problems are usually not expected in clear, oligotrophic waters. However, toxic benthic cyanobacteria have caused animal deaths in Scotland, where beached mats along the shore of a clear loch were scavenged by dogs, and Switzerland where toxic benthic populations of *Oscillatoria limosa* were ingested by cattle drinking from pristine mountain lakes.

REFERENCES

Anagnostidis, K. and Komárek, J. 1985 Modem approach to the classification system of cyanophytes. 1 Introduction. *Arch. Hydrobiol. Suppl. 71, Algological Studies,* 38/39, 291-302.

Bryant, D.A. [Ed.] 1994 *The Molecular Biology of Cyanobacteria.* Kluwer Academic Publishers, Dordrecht, 879 pp.

Rai, A.N. 1990 *CRC Handbook of Symbiotic Cyanobacteria.* CRC Press, Boca Raton, 253 pp.

Wille, N. 1904 *Die Schizophyceen der Plankton Expedition. Ergebnisse der PlanktonExpedition der Humboldt-Stiftung.* Bd. II. M.f. Verlag von Lipsius & Tischer, Kiel, 88 pp.

Zevenboom, W. and Mur, L. 1980 N_2 fixing cyanobacteria, why they do not become dominant in dutch hypertrophic lakes. *Dev. Hydrobiol.,* 2,123-131.

3

Cyanobacterial Toxins

The cyanotoxins are a diverse group of natural toxins, both from the chemical and the toxicological points of view. In spite of their aquatic origin, most of the cyanotoxins that have been identified to date appear to be more hazardous to terrestrial mammals than to aquatic biota. Cyanobacteria produce a variety of unusual metabolites, the natural function of which is unclear, although some, perhaps only coincidentally, elicit effects upon other biota. Research has primarily focused on compounds that impact upon humans and livestock, either as toxins or as pharmaceutically useful substances. Further ranges of non-toxic products are also being found in cyanobacteria and the biochemical and pharmacological properties of these are totally unknown.

Studies on the occurrence, distribution and frequency of toxic cyanobacteria were conducted in a number of countries during the 1980s using mouse bioassay. Analytical methods suitable for quantitative toxin determination only became available in the late 1980s, but studies of specific cyanotoxins have been increasing since then. The results of both approaches indicate that neurotoxins are generally less common, except perhaps in some countries where they frequently cause lethal animal poisonings. In contrast, the cyclic peptide toxins which primarily cause liver injury are more widespread and are very likely to occur if certain taxa of cyanobacteria are present.

CLASSIFICATION

Mechanisms of cyanobacterial toxicity currently described and understood are very diverse and range from hepatotoxic, neurotoxic and dermatotoxic effects to general inhibition of protein synthesis. To assess the specific hazards of cyanobacterial toxins it is necessary to understand their chemical and physical properties, their occurrence in waters used by people, the regulation of their production, and their fate in the environment.

Hepatotoxic Cyclic Peptides

Globally the most frequently found cyanobacterial toxins in blooms from fresh and brackish waters are the cyclic peptide toxins of the microcystin and nodularin family. They pose a major challenge for the production of safe drinking water from surface waters containing cyanobacteria with these toxins. In mouse bioassays, which traditionally have been used to screen toxicity of field and laboratory samples, cyanobacterial hepatotoxins (liver toxins) cause death by liver haemorrhage within a few hours of the acute doses. Microcystins have been characterised from planktonic Anabaena, Microcystis, Oscillatoria (Planktothrix), Nostoc, and Anabaenopsis species, and from terrestrial *Hapalosiphon* genera. Nodularin has been characterised only from Nodularia spumigena.

The cyclic peptides are comparatively large natural products, molecular weight (MW) ≈ 800-1,100, although small compared with many other cell oligopeptides and polypeptides (proteins) (MW > 10,000). They contain either five (nodularins) or seven (microcystins) amino acids, with the two terminal amino acids of the linear peptide being condensed to form a cyclic compound. They are water soluble and, except perhaps for a few somewhat more hydrophobic microcystins, are unable to penetrate directly the lipid membranes of animal, plant and bacterial cells. Therefore, to elicit their toxic effect, uptake into cells occurs through membrane transporters which otherwise carry essential biochemicals or nutrients. This restricts the target organ range in mammals largely to the liver. In aquatic environments, these toxins usually remain contained within the cyanobacterial cells and are only released in substantial amounts on cell lysis. Along with their high chemical stability and their water solubility, this containment has important implications for their environmental persistence and exposure to humans in surface water bodies.

The first chemical structures of cyanobacterial cyclic peptide toxins were identified in the early 1980s and the number of fully characterised toxin variants has greatly increased during the 1990s. The first such compounds found in freshwater cyanobacteria were cyclic heptapeptides (that is they contain seven peptide-linked amino acids) with the general structure of:

cyclo-(D-alanine^1-X^2-D-MeAsp^3-Z^4-Adda^5-D-glutamate^6-Mdha^7) in which X and Z are variable L amino acids, D-MeAsp^3 is *D-erythro*-β-methylaspartic acid, and Mdha is *N*-methyldehydroalanine. The amino acid Adda, (2S,3S,8S,9S)-3-amino-9-methoxy-2,6,8-trimethyl-10-phenyldeca-4,6-dienoic acid, is the most unusual structure in this group of cyanobacterial cyclic peptide toxins.

These compounds were first isolated from the cyanobacterium *Microcystis aeruginosa* and therefore the toxins were named microcystins. Structural variations have been reported in all seven amino acids, but most frequently with substitution of L-amino acids at positions 2 and 4, and demethylation of amino acids at positions 3 and/or 7. About 60 structural variants of microcystins have been characterised so far from bloom samples and isolated strains of cyanobacteria.

In one species of brackish water cyanobacterium, an identically acting and structurally very similar, cyclic pentapeptide occurs. It has been named as nodularin after its producer, *Nodularia spumigena*. The chemical structure of nodularin is cyclo- in which Mdhb is 2-(methylamino)-2-dehydrobutyric acid. A few naturally occurring variations of nodularins have been found: two demethylated variants, one with D-Asp[1] instead of D-MeAsp[1], the other with DMAdda[3] instead of Adda[3]; and the non-toxic nodularin which has the 6Zstereoisomer of Adda[3]. The equivalent 6Z-Adda[3] stereoisomer of microcystins is also non-toxic. In the marine sponge, *Theonella swinhoei*, a nodularin analogue called motuporin has been found. It differs from nodularin only by one amino acid, having hydrophobic L-Val in place of the polar L-Arg in nodularin. The toxin might be cyanobacterial in origin because the sponge is known to harbour cyanobacterial symbionts.

The mammalian toxicity of microcystins and nodularins is mediated through their strong binding to key cellular enzymes called protein phosphatases. In solution, microcystins and nodularins adopt a chemical "shape" that is similar, especially in the Adda-glutamate part of the cyanotoxin molecule. Recent studies have shown that this region is crucial for interaction with the protein phosphatase protein molecule, and hence it is crucial for the toxicity of these cyanotoxins. Microcystins show an additional characteristic of forming a covalent bond between the Mdha residue and the protein phosphatase molecule.

Most of the structural variants of microcystin and nodularin are highly toxic within a comparatively narrow range. Only a few non-toxic variants have been identified. In general, any structural modifications to the Addaglutamate region of the toxin molecule, such as a change in isomerisation of the Addadiene (6(E) to 6(Z)) or acylation of the glutamate, renders microcystins and nodularins non-toxic. Linear microcystins and nodularin are more than 100 times less toxic than the equivalent cyclic compounds. The linear microcystins are thought to be microcystin precursors and/or bacterial breakdown products.

Microcystins and nodularin have been characterised from axenic cyanobacterial strains (i.e. strains free of contaminating bacteria) and thus the cyanobacterial origin of these compounds is clear. At the present time, it is known that microcystins are produced by bloom forming species of Microcystis, Anabaena, Oscillatoria (Planktothrix), and Nostoc, by a species of *Anabaenopsis* and by a soil isolate of *Haphalosiphon hibernicus*. Nodularins have been found, with the exception of the marine sponge *Theonella* already mentioned, only in *Nodularia spumigena*. Further species may yet be demonstrated to produce microcystin.

Neurotoxic Alkaloids

Mass occurrences of neurotoxic cyanobacteria have been reported from North America,

Europe and Australia, where they have caused animal poisonings. In mouse bioassays death by respiratory arrest occurs rapidly (within 2-30 minutes). Three families of cyanobacterial neurotoxins are known:

- anatoxin-a and homoanatoxin-a, which mimic the effect of acetyl choline,
- anatoxin-a(S), which is an anticholinesterase, and
- saxitoxins, also known as paralytic shellfish poisons (PSPs) in the marine literature, which block nerve cell sodium channels.

Anatoxin-a has been found in Anabaena, Oscillatoria and Aphanizomenon, homoanatoxin-a from *Oscillatoria,* anatoxin-a(S) from *Anabaena,* and saxitoxins from Aphanizomenon, Anabaena, Lyngbya and Cylindrospermopsis. Sixteen confirmed saxitoxins from cyanobacterial samples have been reported, some of which may be chemical breakdown products in some species.

The alkaloid toxins are diverse, both in their chemical structures and in their mammalian toxicities. Alkaloids, in general, are a broad group of heterocyclic nitrogenous compounds (i.e. they contain ring structures with at least one carbon-nitrogen bond) usually of low to moderate molecular weight (< 1,000). They are produced, in particular, by plants and by some bacteria, and are invariably bioactive and commonly toxic. The non-sulphated alkaloid toxins of freshwater cyanobacteria are all neurotoxins. The sulphated PSPs, C-toxins and gonyautoxins are also neurotoxins, but the sulphated alkaloid cylindrospermopsin blocks protein synthesis with a major impact on liver cells. Some marine cyanobacteria also contain alkaloids which are dermatoxins, but have also been associated with gastro-enteritis and more general symptoms such as fever.

Alkaloids have varying chemical stabilities, often undergoing spontaneous transformations to by-products which may have higher or lower potencies than the parent toxin. Some are also susceptible to direct photolytic degradation.

Anatoxin-a

Anatoxin-a is a low molecular weight alkaloid (MW = 165), a secondary amine, 2-acetyl-9-azabicyclo(4-2-1)non-2-ene. Anatoxin-a is produced by Anabaena flos-aquae, Anabaena spp, Anabaena planktonica, Oscillatoria, Aphanizomenon and Cylindrospermum. Homoanatoxin-a (MW = 179) is an anatoxin-a homologue isolated from an *Oscillatoria formosa (Phormidium formosum)* strain. It has a propionyl group at C-2 instead of the acetyl group in anatoxin-a. The LD50 of anatoxin-a and homoanatoxin-a are 200 - 250 $\mu g\ kg^{-1}$ bw.

Anatoxin-a(S)

Anatoxin-a(S) is a unique phosphate ester of a cyclic N-hydroxyguanine (MW = 252) produced by *Anabaena flos-aquae* strain NRC 525-17. It has more recently been identified in blooms and isolated strains of *Anabaena lemmermannii*. The LD50 of anatoxin-a(S) is 20 μg kg^{-1} bw. Structural variants of anatoxina (S) have not been detected.

Saxitoxins

Saxitoxins are a group of carbamate alkaloid neurotoxins which are either non-sulphated (saxitoxins - STX), singly sulphated (gonyautoxins - GTX) or doubly sulphated (C-toxins). In addition, decarbamoyl variants and several new toxins have been identified in some species.

Saxitoxins were originally isolated from shellfish where they are concentrated from marine dinoflagellates and have caused deaths in humans. Saxitoxins have been found in the cyanobacteria Aphanizomenon flos-aquae, Anabaena circinalis, Lyngbya wollei and *Cylindrospermopsis raciborskii*. The North American *Aphanizomenon flos-aquae* strains NH-1 and NH-5 contain mostly neosaxitoxin and less saxitoxin. *Anabaena circinalis* strains contain mostly C1 and C2 toxins, with lesser amounts of gonyautoxins 2 and 3. The freshwater cyanobacterium *Lyngbya wollei* produced three known and six new saxitoxin analogues. *Cylindrospermopsis raciborskii* in Brazil was found to contain mostly neosaxitoxin and a smaller amount of saxitoxin.In marine *Trichodesmium* blooms from the Virgin Islands, a neurotoxic factor has been reported which was not anatoxin-a or anatoxin-a(S) but remains to be characterised.

Cytotoxic Alkaloids

In tropical and subtropical waters of Australia, the alkaloid hepatotoxin cylindrospermopsin with a completely different mechanism of toxicity has caused health problems in drinking water supplies. It is a cyclic guanidine alkaloid *with a molecular weight of 415. It is produced by* Cylindrospermopsis raciborskii, Umezakia natans and Aphanizomenon ovalisporum. *In pure form, cylindrospermopsin mainly affects the liver, although crude extracts of* C. raciborskii *injected or given orally* to mice also induce pathological symptoms in the kidneys, spleen, thymus and heart. Pure cylindrospermopsin has an LD_{50} in mice of 2.1 mg kg l^{-1} bw at 24 h and 0.2 mg kg l^{-1} bw at 5-6 days. Recently, new structural variants of cylindrospermopsin have been isolated from an Australian strain of *C. raciborskii*, with one being identified as demethoxy-cylindrospermopsin.

Dermatotoxic Alkaloids

Benthic marine cyanobacteria such as *Lyngbya, Oscillatoria* and *Schizothrix* may produce toxins causing severe dermatitis among swimmers in contact with the filaments. The inflammatory activity of *Lyngbya* is caused by aplysiatoxins and debromoaplysiatoxin which are potent tumour promoters and protein kinase C activators. Another strain of

Lyngbya majuscula has caused. dermatitis and severe oral and gastrointestinal inflammation. It was found to contain lyngbyatoxin-a. Debromoaplysiatoxin along with other toxic compounds has also been isolated from other Oscillatoriaceae, such as *Schizothrix calcicola* and *Oscillatoria nigroviridis.*

Irritant Toxins

Weise *et al.* were the first to isolate LPS from the cyanobacterium *Anacystis nidulans* and numerous reports of endotoxins in cyanobacteria have followed. Lipopolysaccarides are generally found in the outer membrane of the cell wall of Gram negative bacteria, including cyanobacteria, where they form complexes with proteins and phospholipids. They are pyrogenic and toxic. Lipopolysaccarides, as the name implies, are condensed products of a sugar, usually a hexose, and a lipid, normally a hydroxy C_{14}-C_{18} fatty acid. The many structural variants of LPS are generally phylogenetically conserved, i.e. particular orders, genera and occasionally species, have identical or similar fatty acid and sugar components contained in their cell wall LPS. It is generally the fatty acid component of the LPS molecule that elicits an irritant of allergenic response in humans and mammals.

Lipopolysaccharides are an integral component of the cell wall of all Gram negative bacteria, including cyanobacteria, and can elicit irritant and allergenic responses in human and animal tissues that come in contact with the compounds. There is considerable diversity of LPS composition among the cyanobacteria, but differences are largely related to phylogeny. Thus, different genera typically have distinct LPS compositions that are largely conserved within that genus. Cyanobacterial LPS are considerably less potent than LPS from pathogenic gram-negative bacteria such as, for example, *Salmonella.* The chemical stability of cyanobacterial LPS in surface waters is unknown.

Structurally, LPS is a complex polymer composed of four regions. Region I, the Oantigen region, consists of repeating oligosaccharide units that may vary in structure, with numerous combinations of different sugar residues and associated glycosidic linkages. As suggested by its name, the O-antigen also exhibits several antigenic determinants that constitute the receptor sites for a number of lysogenic bacteriophages. Regions II and III are the outer core and backbone of a core polysaccharide. There is generally only minor variation in core structure between species. The backbone of the polysaccharide is connected to a glycolipid, lipid A, via a short link normally composed of 3-deoxy-D-mannoocmiosonic acid (KDO). Lipid A is a disaccharide of glucosamines highly substituted with phosphate, fatty acids and KDO, although the proportion of KDO is low or absent in cyanobacteria compared with other bacterial LPS. The lipid A component is also acetylated with amide and ester-linked hydroxy fatty acids.

Recent studies of the fatty acid composition of Australian species of cyanobacteria show a range of β-OH fatty acids ranging in size from C_{10} to C_{22}. Normal, saturated and branched chain acids have been detected. There was a stark predominance of straight chain 14:0 and 18:0 β-OH acids in *Microcystis* strains that was quite distinct from *Anabaena* and *Nodularia* strains where 16:0 β-OH predominated the LPS fatty acid fraction.

Although comparatively poorly studied, cell wall components, particularly LPS endotoxins from cyanobacteria may contribute to human health problems associated with exposure to mass occurrences of cyanobacteria. The few results available indicate that cyanobacterial LPS is less toxic than the LPS of other bacteria, such as *Salmonella*. More studies are needed to evaluate the chemical structures and health risks of cyanobacterial LPS.

Other Bioactive Compounds

Cyanobacteria are known to produce several other bioactive compounds, some of which are of medical interest, as well as compounds toxic to other cyanobacteria, bacteria, algae and zooplankton. Severe intoxication of fish embryos by crude extracts of *Planktothrix agardhii* has been reported by Oberemm *et al.* Skulberg *et al.* reported the presence of an unidentified "protracted toxic effect" in cyanobacterial samples that caused death within 4-24 hours in mice. Whether this effect was due to a specific cyanotoxin is unclear.

Cyanobacteria have been found to be a rich source of biomedically interesting compounds and therefore screening programmes for new bioactivities are underway. Cyanobacteria are known to produce antitumour, antiviral, antibiotic and antifungal compounds. Of the cyanobacterial extracts screened by a Hawaiian research group, 0.8 per cent showed solid tumour selective cytotoxicity. Depsipeptides called cryptophycins isolated from a cyanobacterium, *Nostoc* sp. strain GSV 224, are promising candidates for an anticancer drug. Recently, several new cyclic or linear peptides and depsipeptides from cyanobacteria have been characterised. Some are protease inhibitors, but the biological activity of the others remains to be characterised. Many of the cyanobacterial bioactive compounds possess structural similarities to natural products from marine invertebrates.

OCCURRENCE OF CYANOTOXINS

The toxicity of cyanobacterial mass occurrences (blooms) was originally brought to the attention of scientists through reports of animal poisonings by farmers and veterinarians, with the first well documented case being reported from Australia in 1878. In most, if not all, reported cases since that time, afflicted animals consumed water from water bodies where there was an obvious presence of a cyanobacterial scum on the water surface. More recent measurements of cyanobacterial toxins using sensitive modem analytical methods .have often revealed high frequencies of toxic blooms even when animal poisonings have not been reported. Throughout the world, it appears that liver-toxic freshwater blooms of cyanobacteria are more commonly found than neurotoxic blooms. Liver-toxic blooms have been reported from all continents and almost every part of the world where samples have been collected for analysis. Nevertheless, mass occurrences of neurotoxic cyanobacteria are common in some countries and these have been reported from North America, Europe and Australia, where they have caused several animal poisonings. Blooms which have caused both liver and kidney damage due to the toxin cylindrospermopsin have been reported in Australia, Japan, Israel and Hungary.

In recent years, surveys have been carried out in a number of countries in South America, Africa, Australasia, Asia and Europe. The conclusion that can be drawn from these surveys is that toxic cyanobacteria are internationally ubiquitous, and that as further surveys are carried out more toxic cyanobacterial blooms and new toxic species will be discovered. This is particularly true of tropical and subtropical regions that are currently under-represented in the literature. It seems likely that every country in the world will have water bodies which support blooms of toxic cyanobacteria at some time or another. It is also important to note that mass occurrences of toxic cyanobacteria are not always associated with human activities causing pollution or "cultural eutrophication". For example, massive blooms of toxic cyanobacteria have been reported in Australian reservoirs with pristine or near-pristine catchments (watersheds), and toxic benthic cyanobacteria have killed cattle drinking from oligotrophic, high-alpine waters in Switzerland.

Species Composition and Variation among Toxic Blooms

Cyanobacterial populations may be dominated by a single species or be composed of a variety of species, some of which may not be toxic. Even within a single-species bloom there may be a mixture of toxic and non-toxic strains. A strain is a specific genetic subgroup within a particular species, and each species may encompass tens or hundreds of strains, each with slightly different traits. Some strains are much more toxic than others, sometimes by more than three orders of magnitude. This can mean that one highly toxic strain, even when occurring in minor amounts amongst larger numbers of non-toxic strains, may render a bloom sample toxic.

Some of the studies shown in the table have been conducted over several years while others lasted only one season. The relative share of cyanobacteria in the samples varied; low frequency of cyanobacteria led to low percentages of toxic samples in some studies. In most of the studies the method used to detect toxicity is mouse bioassay, normally with a 4-hour time limit.

SDF may indicate low hepatotoxicity of samples or other unknown toxicity. Toxic and non-toxic strains from the same cyanobacterial species cannot be separated by microscopic identification. The use of molecular genetic methods, in particular the use of molecular probes and primers that target specific toxin production genes, will lead to the development of more precise identification methods for toxic cyanobacteria in the future. To confirm that a particular cyanobacterial strain is a toxin-producer, it is important to isolate a pure culture of that strain, preferably free of other bacteria; then to detect and quantify toxin concentrations in the pure culture; and, where possible, to purify and characterise fully the toxins. It is likely that the list of confirmed toxic species will increase in the future due to the isolation of new species and strains, and because of the use of improved isolation, culturing and analytical methods.

Microcystis sp., commonly *Microcystis aeruginosa,* are linked most frequently to hepatotoxic blooms world-wide. *Microcystis viridis* and *Microcystis botrys* strains also have been shown to produce microcystins. *Microcystis* is a non-nitrogen-fixing genus which is often dominant under nutrient-rich conditions, although it also forms blooms in less polluted waters. Microcystin-producing *Anabaena* sp. have been reported from Canada, Denmark, Finland, France and Norway. A recent study from Egypt revealed that 25 per cent of 75 *Anabaena* and *Nostoc* strains isolated from soil, rice fields and water bodies contained microcystins. *Planktothrix agardhii* and *Planktothrix rubescens* are common microcystin producers in the Northern Hemisphere; toxic strains of these have been isolated from blooms in Denmark, Finland and Norway. In addition, these species were frequently shown to be dominant in microcystin containing blooms in China, in Germany and in Sweden. In Swiss alpine lakes, *Oscillatoria limosa,* which is benthic, is a microcystin producer. In spite of the widespread occurrence of cyanobacterial blooms in Australia, *Planktothrix* blooms are rare there. This may be due to the higher temperature and tendency for elevated clay-derived turbidity in Australian water bodies.

The toxicity of the species listed in the table is in most cases verified by laboratory studies with isolated strains. A few bloom samples are also included from the new areas of occurrence where toxicity of the species is not verified by strain isolation but the toxins are determined in the bloom samples. The authors have suggested the listed species as the probable toxin producer but these reports should be treated as tentative until pure strains are studied.

Nostoc rivulare blooms in Texas, USA have caused poisoning of domestic and wild animals and, more recently, two unidentified *Nostoc* strains were shown to produce microcystins.

The hepatotoxin, cylindrospermopsin, has been found in *Cylindro-spermopsis raciborskii* in Australia and Hungary, in *Umezakia natans* in Japan, and in *Aphanizomenon ovalisporum* in Israel. In spite of their occurrence in Europe, it appears that cylindrospermopsin-producing genera most commonly form toxic blooms in subtropical, tropical or arid zone water bodies. However, there have been reports of increasing occurrences of *Cylindrospermopsis raciborskii in* Europe and the USA. The neurotoxin, anatoxin-a, was first shown to be produced by *Anabaena flos-aquae* strains originating from Canada, and later by Finnish strains of unidentified *Anabaena* species, and in individual species of *Oscillatoria, Aphanizomenon,* and *Cylindrospermum,* by benthic *Oscillatoria* from Scotland, and by *Anabaena* and *Oscillatoria* in Ireland. It also was present in *Anabaena planctonica* blooms in Sardinia, Italy, in *Anabaena* and *Aphanizomenon* blooms in Germany, and in minor amounts in some Japanese bloom samples, as well as in *Anabaena* strains. Homoanatoxin-a has been characterised from an *Oscillatoria formosum (Phormidium formosum)* strain from Norway. To date, anatoxin-a(S) has been found only from *Anabaena* species: *A. flos-aquae* in the USA and Scotland, and *A. lemmermannii* in Denmark.

Aphanizomenon flos-aquae blooms and strains were found to contain saxitoxins in the USA and this species was for a long time the only known saxitoxin producer amongst the cyanobacteria. More recently, saxitoxins have been shown to be common in Australian rivers and reservoirs and to be produced by *Anabaena circinalis.* In North America, a benthic freshwater *Lyngbya wollei* was found to produce saxitoxins, as was a strain of *Cylindrospermopsis raciborskii* in Brazil.

Cyanotoxin Patterns in Strains and Species of Cyanobacteria

Cyanobacteria may produce several toxins simultaneously. Among neurotoxic strains, several PSPs are found in the same strain, although there are considerable variations between species. Furthermore, simultaneous neurotoxin and hepatotoxin production has been noted; the best example studied being the *Anabaena flos-aquae* strain NRC 525-17 which produces anatoxin-a(S) and several microcystins.

Microcystin

Although many strains produce several microcystins simultaneously, usually only one or two of them are dominant in any single strain. Qualitative variation in the microcystins present is most frequently found among strains of *Anabaena* but also in *Microcystis*. Some taxa have a number of microcystins in common, such as planktonic *Anabaena, Microcystis* and *Planktothrix (Oscillatoria).* However, there is evidence of microcystin variants that are typical for certain cyanobacterial taxa. *Planktothrix* and some strains of *Anabaena* produce only demethylmicrocystins. *Planktothrix (Oscillatoria)* isolates from Finland seem to produce one of two types of microcystin.

German field samples dominated by *Planktothrix* have also shown these microcystins, with dominance of one major demethylated microcystin in populations of *P. rubescens,* and two or three of these variants in populations of *P. agardhii. Microcystis* strains from Japan appear to contain chiefly microcystin-LR, -RR and -YR, with some cultures showing all three variants, and some strains being dominated by one of them. These three microcystins are the only variants reported in several studies on *M. aeruginosa* and *M. viridis.*

In natural samples which usually contain many strains, or more than one toxin-producing species, different combinations of microcystins can be found. For example, in a *Microcystis* bloom from Homer Lake, USA, 19 different microcystins were characterised and in one Australian bloom of *Microcystis aeruginosa,* 23 microcystins were detected by high pressure liquid chromatography (HPLC), none of which were microcystin-LR.

Microcystin-LR is often mentioned as the most frequently occurring microcystin, although such observations may be biased by the fact that a chemical standard for the analysis of microcystin-LR was the earliest to be commercially available. It has been reported to be the major toxin in bloom and strain samples from Portugal, France, Canada and frequently co-occurring with microcystin-RR and -YR in Japan.

Geographical patterns are indicated by some survey results. Wide variation among the L-amino acids of microcystins has been reported for South Africa, frequent presence of demethylmicrocystins (RR and LR) has been seen in Finnish strains and microcystins in Danish samples show wide variation. In part, these patterns probably reflect regional differences in dominance of cyanobacterial species or strains. Water bodies with regular dominance of specific taxa are likely to exhibit characteristic patterns of microcystin variants.

Nodularin

The cyanotoxin, nodularin, is found in waters where *Nodularia spumigena* is present; the most prominent areas being the Baltic Sea and brackish water estuaries and coastal lakes of Australia and New Zealand. However, the best known *Nodularia spumigena* bloom location, Lake Alexandrina, Australia, has salinities which are only slightly elevated above normal river water and at levels still suitable for drinking water. The presence of variants of nodularin in environmental samples is usually rather insignificant. In the Baltic Sea, the collection of samples for several years has shown nodularin to be the major compound present. The same is true for the almost 90 hepatotoxic *Nodularia* strains isolated from the same source. Analyses of several strains isolated from blooms across Australia have revealed similar results, with nodularin variants being found rarely, and then only at low relative abundance.

Saxitoxins

There is much diversity of saxitoxin distribution in cyanobacteria from around the world. In addition, the lack of analytical standards for many saxitoxins has probably restricted findings in some countries. Saxitoxin-producing *Anabaena circinalis* blooms are widespread in Australian rivers and reservoirs, and the relative abundance of individual toxins is remarkably constant in toxin-producing strains, which is quite opposite to the microcystin-producing cyanobacteria. In all healthy *Anabaena circinalis* cultures examined in Australia, the relative composition of individual saxitoxins is very similar and dominated by C-toxins. As blooms and cultures age, the proportion of decarbamoyl-gonyautoxins breakdown products increases at the expense of the C-toxins. In North American *Aphanizomenon flos-aquae* and in Brazilian *Cylindrospermopsis raciborskii* samples, mostly neosaxitoxin and smaller amounts of saxitoxin have been detected. In the case of the *Aphanizomenon flos-aquae* from North America, only bloom samples and strains from New Hampshire, USA have been found to contain saxitoxins. Mat-forming *Lyngbya wollei* from Alabama, USA was found to produce decarbamoyl saxitoxin (dcSTX), decarbamoylgonyautoxin-2 and-3 (dcGTX2 and dcGTX3) and six new saxitoxins.

Concentrations of Cyanotoxins in Surface Waters

Information on the concentrations of cyanotoxins in surface waters has been appearing only recently in the international literature. In early studies, the toxicity of bloom samples was determined by mouse bioassays, but this method is unsuitable for measuring the low concentrations of cyanotoxins that usually prevail in cyanobacterial populations when they do not accumulate in scums. The development of better analytical methods, in the first instance HPLC and more recently enzyme linked immunosorbent assay (ELISA), has made the quantification of total and individual toxins possible.

The quantitative determination of toxin concentrations is mostly performed from lyophilised cultures, bloom samples or seston. Results are usually expressed as milligrams or micrograms of toxin per gram dry weight (dw). Whereas in cultures and bloom samples, the dry weight originates from cyanobacteria, it will encompass further particles in plankton samples taken outside of scum areas. The highest published concentrations of cyanotoxins from cyanobacterial bloom samples, measured by HPLC, are:

— microcystin - 7,300 $\mu g\ g^{-1}$ dw from China and Portugal,

— nodularin - 18,000 $\mu g\ g^{-1}$ dw from the Baltic Sea,

— cylindrospermopsin - 5,500 $\mu g\ g^{-1}$ dw from Australia,

— anatoxin-a - 4,400 $\mu g\ g^{-1}$ dw from Finland,

— saxitoxins - 3,400 $\mu g\ g^{-1}$ dw from Australia,

— anatoxin-a(S) - 3,300 $\mu g\ g^{-1}$ dw from the USA.

Higher concentrations have been reported in unpublished studies and may be expected under some circumstances.

For the purposes of water treatment and public health management, toxin concentration per litre of water is often a more relevant unit, for example for relating ambient concentrations to guideline levels. Concentration per litre usually refers to toxins contained within the cells as well as dissolved in water, and taken from a defined volume of water. Very high concentrations of microcystins per litre of water, have been reported up to 25,000 μg l^{-1} microcystin and up to 3,300 μg l-1 anatoxin-a(S). It should be noted, however, that these very high concentrations of toxins would be from scums or from very dense accumulations of cyanobacteria.

In one or two studies where many toxin concentration measurements have been made from more or less randomly collected plankton samples, the distribution of data revealed a small number of samples with high toxin concentrations, some samples with moderate concentrations, and many samples with low or zero concentrations. However, such surveys tend to include a large proportion of samples where the particle content is not dominated by cyanobacteria. If cyanobacteria constitute only a small fraction of the total dry weight, toxin concentration per gram dry weight will necessarily be low. In contrast, if mass developments of toxic cyanobacterial species are investigated, cyanobacteria will constitute a large fraction of the total dry weight. In these cases, toxin content is often high. Recent German data show that the relative frequency of high, moderate and low toxin concentrations per gram dry weight is dependent on the particular species that is dominant.

For management, the option of estimating toxin content from the biomass of a dominant cyanobacterial species can be helpful. Survey data from Germany show that microcystin toxin quotas differ between taxa, but variation within most of the samples dominated by the same taxon is only moderate. In contrast, an Australian investigation with other species of cyanobacteria showed substantial variation of toxin concentrations per dry weight, even in cases with dominance of the same cyanobacterial species. Further regional investigations are needed to clarify the toxins and species for which toxin quotas may be reasonably stable and, thus, predictable. The further development of quantitative, pre-concentration methods coupled with the highly sensitive analytical methods noted above has also enabled the measurement of very low concentrations of extracellular, dissolved toxins in water.

Seasonal Variations in Bloom Toxin Concentration

The timing and duration of the bloom season of cyanobacteria depends largely on the climatic conditions of the region. In temperate zones, mass occurrences of cyanobacteria are most prominent during the late summer and early autumn and may last 2-4 months. In regions with more Mediterranean or subtropical climates, the bloom season may start earlier and persist longer.

In France, four months is not uncommon, and in Japan, Portugal, Spain, South Africa and southern Australia blooms may occur for up to six months or longer. By contrast, in dry years, in tropical or subtropical areas of China, Brazil and Australia, cyanobacterial blooms may occur almost all year round, perhaps waning only briefly during reservoir overturn.

In shallow lakes, particularly in north-western Europe, populations of *Planktothrix agardhii (Oscillatoria agardhii)* may prevail perennially for many years. In deeper, thermally stratified lakes and reservoirs with moderate nutrient pollution, *Planktothrix rubescens (Oscillatoria rubescens)* may form blooms at the interface between the warmer upper and colder deeper layers of water during summer, but maintain high, evenly distributed density throughout the entire water body during winter. Both *Planktothrix* species may contain high amounts of microcystins. Blooms of cyanobacteria, especially *Planktothrix agardhii,* have been found in winter under ice in Scandinavian and German lakes and can thus be an all year round problem.

Although toxic cyanobacteria occur in a large number of lakes, reservoirs and rivers in the world, quantitative reports on seasonal variation of cyanobacterial species composition and toxin concentration are rare. Only a few studies on seasonal, spatial and diel (day to night) variations in lakes have been published. Carmichael and Gorham showed a high degree of spatial variation of bloom toxicity that was due mostly to variations in the relative amounts of toxic *Microcystis aeruginosa* throughout the lake, rather than to substantial variations in cell toxin content. Other measurements of toxin concentrations in lakes have revealed similar trends; samples taken at the same time from different parts of the lake may show wide divergence in cyanotoxin content. A study in Alberta, Canada, showed considerable variation in toxin concentrations among the three lakes studied, both within and between years, even though the lakes were located within the same climatic region.

In any year or season, individual water bodies have their own populations of cyanobacteria and algae, the dominance of which is dependent not only on the weather, but on the specific geochemical conditions of the lake. If there are no major changes in these conditions, toxic blooms are likely to recur annually in those lakes that have a history of toxic blooms. Certain species, including the highly toxic *Planktothrix agardhii* and *P. Rubescens,* are known to produce maximum mass occurrences deep in the water column and which may be overlooked by surface monitoring of waters. Such situations may also cause problems for water treatment.

Studies over prolonged periods usually show that toxin concentration per gram dry weight may vary substantially over a time scale of weeks to months, but rarely from day to day as is sometimes reported. The maximum toxin concentration per gram dry weight is usually reported in summer or autumn, when cyanobacterial biomass dominates dry matter. However, the time of toxin concentration maximum and biomass maximum are not necessarily coincident. Thus, there can be significant variation in the amount of toxin per mass of cyanobacteria over time, independently of changes in the size of the cyanobacterial population. The explanations for this are twofold. Firstly, there may be a waxing and waning of species or strains of quite different toxin quotas. Secondly, the toxin quotas may change up to five-fold in response to changes in environmental conditions. A study by Kotak *et al.* found substantially higher concentrations of microcystin in blooms during the day than at night, whereas a study from Australia found no variation in microcystin content when samples were incubated during 24 hours at different depths in a reservoir. Both findings need to be explored further.

High regional, seasonal, spatial, temporal and diel variations of toxin concentrations indicate that predicting or modelling the occurrence of toxin concentrations requires a comprehensive understanding of population development in different types of aquatic ecosystems, as well as of the variability of their toxin quotas. Data bases for such predictive models have yet to be compiled.

PRODUCTION AND REGULATION

Laboratory studies with pure strains of cyanobacteria have found that environmental factors can induce changes in toxicity or toxin concentration, but usually by a factor of no more than three or four. On a per cell basis, the changes in toxin content are probably even smaller. These environmentally-induced changes are far less than the range of more than three orders of magnitude in toxin content measured between individual strains grown in culture under identical conditions. This lends support to the assumption that much, if not most, of the variation in toxicity of "monospecific" natural blooms is the waxing and waning of strains of the same species, but with varying toxin quotas. The factors that control the growth and toxin content of individual strains are as yet unknown, but clearly the genetic regulation of cyanotoxin production is an important area for further study and understanding.

Both toxigenic and non-toxigenic strains exist within many species of cyanobacteria. When grown in the laboratory, particular strains always produce much greater amounts of toxins than others. Indeed, the difference may be as much as three orders of magnitude or more. Several attempts to differentiate toxin producing from non-toxic strains of the same species using microscopic methods have failed. The use of molecular biological methods to characterise toxic and non-toxic planktonic isolates has been initiated recently and will, in future, help to clarify the taxonomic status of these organisms, as well as the ecology of individual strain types.

Regulation by Chemical and Physical Factors

The production of toxin by a single cyanobacterial strain seems to be consistent and the spontaneous and permanent loss of toxin production has been seldom reported. The effects of several environmental factors on growth and toxin production by cyanobacteria have been studied in batch and continuous culture experiments. Culture age in batch cultures, and temperature, are the parameters most frequently examined, followed by light, nutrients, salinity, pH and micronutrient concentrations. Studies have been done with hepatotoxic Microcystis, Oscillatoria (Planktothrix), Anabaena and Nodularia;

anatoxin-a producing *Anabaena, Aphanizomenon* and *Planktothrix;* and saxitoxin producing *Aphanizomenon* and *Anabaena circinalis*. Microcystins and anatoxin-a are largely retained within cells when the conditions for the growth of the organism are favourable. The amount of microcystin in a culture increases during the logarithmic growth phase, being highest in the late logarithmic phase. The maximum anatoxin-a concentration is found during the logarithmic growth phase. While the variants of microcystins produced by a particular strain are rather constant, the ratios of individual microcystins may change with time, or under conditions of different temperatures and light. Environmental factors affect toxin content of cyanobacteria, but only within a range of less than an order of magnitude. The majority of studies indicate that cyanobacteria produce most toxins under conditions which are most favourable for their growth. For example, different cyanobacterial species have different light requirements: *Planktothrix* prefers low light intensities for growth, *Anabaena* moderate and *Aphanizomenon* high light intensities. All strains produce most toxin when grown under their optimum light conditions. Two- to threefold differences in toxin content have been reported in relation to light conditions.

Strains and species also differ slightly in their optimum growth temperatures. The toxin content in most studies was highest at temperatures between 18 °C and 25 °C, whereas low (10 °C) or very high temperatures (30 °C) decreased toxin content. Temperature gradients caused two- to threefold differences in toxin content. In a study using mouse bioassay to detect effects of pH on toxin production, cells were found to be more toxic when grown at high and low pH.

In high concentrations of phosphorus, hepatotoxic strains produced more toxins, but for anatoxin-a production phosphorus had no effect. The differences induced by low and high phosphorus concentrations vary between two- and fourfold. Similarly, in field studies, a positive correlation of total phosphorus with microcystin-LR concentration in cells of *Microcystis aeruginosa* or in bloom material of *Microcystis* spp. has been found. Non-nitrogen fixing species, like *Microcystis* and *Oscillatoria,* produce more toxins under nitrogen-rich conditions. Nitrogen fixing species are not dependent on the nitrogen in the media for their toxin production. In continuous cultures when the toxins were expressed in relation to cell protein rather than to dry weight, Utkilen and Gjølme found that nitrogen and phosphorus limited conditions had no effect on the toxin content of Microcystis aeruginosa.

Indications regarding the role of iron are contradictory. While studying the effect of trace metals on growth and on toxin content of *Microcystis aeruginosa,* Lukac and Aegerter found that in batch cultures only zinc was required for both optimal growth and toxin production. Orr and Jones have unified many of these rather disparate studies on the effect of growth conditions on cyanotoxin production. They showed that the rate of microcystin production by a cyanobacterial population in culture is directly proportional to its growth rate, no matter what environmental factor was limiting growth. Moreover, they showed that the amount of microcystin contained by a single cell of *Microcystis aeruginosa* is constant within a narrow range.

Biosynthesis

To understand how cyanotoxins are produced, it is necessary to study the biochemical and genetic basis of toxin production. Knowledge of the biosynthetic pathways of cyanotoxins is in its early stage and no complete biochemical pathways are known. Biosynthesis of several cyanotoxins has been studied by feeding labelled precursors to a cyanobacterial culture and following their incorporation into the carbon skeleton of the toxins. Shimizu *et al.* used an *Aphanizomenon flos-aquae* strain to study biosynthesis of saxitoxin analogues. They proposed a new pathway for neosaxitoxin biosynthesis, the key steps of which are the condensation of an acetate unit, or its derivative, to the amino group bearing an á-carbon of arginine or an equivalent, and a subsequent loss of the carboxyl carbon and imidazole ring formation on the adjacent carbonyl carbon. They established the origin of all the carbons in the toxin alkaloid ring system. The side-chain carbon was derived from methionine.

Anatoxin-a is related structurally to the tropane class of alkaloids found in higher plants. Based mainly on 14C-labelled precursors and enzymatic studies, Gallon *et al.* and Gallon *et al.* suggested the biosynthesis of anatoxin-a to be analogous to that of tropanes. Anatoxin-a was proposed to be formed from ornithine/arginine via putrescine, which is oxidised to pyrroline, a precursor of anatoxin-a. Labelling experiments using 13C NMR indicated that the carbon skeleton of anatoxin-a is derived from acetate and glutamate. The studies showed that C-1 of glutamic acid is retained during the transformation of anatoxin-a and not lost by decarboxylation, a finding incompatible with the tropane alkaloid theory. All of the carbons of anatoxin-a(S) are derived from amino acids. Three methyl carbons arise from L-methionine or other donors to the tetrahydrofolate C1 pool. L-arginine accounts for C-2, C-4, C-5 and C-6 carbons of the toxin. The intermediate in the biosynthesis of anatoxin-a(S) from L arginine is (2*S*,4*S*)-4 hydroxyarginine. The structure of the cylindrospermopsin suggests a polyketide origin for the toxin.

The origin of carbons in microcystin and in nodularin have been studied by following the incorporation of labelled precursors into the toxins by NMR. Carbons C1-C8 of Adda in nodularin are acetate derived and the remaining carbons presumably originate from phenylalanine, Methyl groups in carbons 2, 4, 6, 8, and the O-methyl group in the Adda unit, originated from methionine. The D-Glu and L-Arg carbons C4-C5 were acetate derived, with C1-C2 being from glutamate. Methyldehydrobutyrine was possibly formed from threonine, its methyl group coming from methionine. The α-methylaspartic acid was found to originate from condensation of pyruvic acid (C3-C4) and acetyl-CoA (C1-C2). The studies on the carbon skeleton of nodularin, with some minor differences, agree with work on microcystin-LR by Moore *et al.* In their study, L-Leu and DAla units in microcystin had acetate incorporation. The dehydroamino acid in microcystin has been proposed to be formed from serine rather than from threonine. Rinehart's group found linear nodularin, which was shown by culture experiments to be a precursor of cyclic nodularin. Three additional linear peptides were isolated from a bloom sample, one of them was possibly a precursor of cyclic microcystin-LR and the others possibly degradation product.

Genetic Regulation of Cyanotoxin Production

The genes and enzymes involved in cyanotoxin production are still mostly unknown. The first molecular biological studies on toxic cyanobacteria investigated the possible involvement of plasmids in toxin production. Four toxic strains of *Microcystis aeruginosa* contained plasmids, and no plasmid could be shown in one toxic and in several nontoxic strains. More recently, a similar study in Australia found no evidence for plasmid involvement in microcystin synthesis. Gallon *et al.* studied an anatoxin-a producing *Anabaena* strain NCR 44-1, which spontaneously became non-toxic. They found that the size of a plasmid was reduced in that non-toxic clone, but this work has not been repeated or confirmed.

Multi-enzyme complexes and peptide synthetase genes are involved in hepatotoxin production. Several cyclic and linear peptides, often with D-amino acids, are known to be produced, non-ribosomally, by multi-domain peptide synthetases via the so-called thiotemplate mechanism in bacteria and lower eukaryotes. The best characterised are the synthesis of gramicidin S and tyrocidin by *Bacillus*. Peptide synthetase genes have been detected and sequenced in *Microcystis aeruginosa* and in *Anabaena*. Analogous polymerase chain reaction (PCR) products to the peptide synthetase genes have been shown by using DNA from *Microcystis* and *Nodularia* as a template. Dittman *et al.* showed, in knockout experiments, that peptide synthetase genes are responsible for microcystin production. At least some strains which produce hepatotoxins also produce other small cyclic peptides which are likely to be produced by nonribosomal peptide synthesis.

FATE IN THE ENVIRONMENT

Partitioning between Cells and Water

It appears likely that cyanotoxins are produced and contained within the actively growing cyanobacterial cells. Release to the surrounding water, to form dissolved toxin, appears to occur mostly, if not exclusively, during cell senescence, death and lysis, rather than by continuous excretion.

In laboratory studies, where both intracellular and dissolved toxins have been measured, it is generally the case that in healthy log phase cultures, less than 10-20 per cent of the total toxin pool is extracellular. As cells enter stationary phase the increased rate of cell death may lead to an increase in the extracellular dissolved fraction. Even during log-phase cell growth in culture, a small percentage of cells in the population may be dying and lysing, even though there is an overall positive population growth. There are some indications that anatoxin-a may leak out of cells during growth especially in low light conditions. High concentrations of anatoxin-a, sometimes exceeding the intracellular pool of toxins, have been found in media in a batch culture study.

In the field, healthy bloom populations produce little extracellular toxin. The range of measured concentrations for dissolved cyanotoxins, in all cases except those where a major bloom is obviously breaking down, is 0.1-10 μg l^{-1} Cellbound concentrations are several orders of magnitude higher. In lakes or rivers, toxins liberated from cells are rapidly diluted by the large mass of water, especially if mixing of water by wind action or currents is vigorous. However, the concentration of dissolved toxins may be much higher in ageing or declining blooms. This is an important consideration for water treatment plant operators, because it means that removal of healthy cyanobacterial cells intact from the raw water supply may obviate or substantially reduce the need for additional adsorptive or oxidative toxin removal processes. The release of toxins from cells is enhanced by chemical treatments for the eradication of cyanobacteria, especially the use of algicides. Treatment of a bloom with copper sulphate, for example, may lead to complete lysis of the bloom population within three days and release of all the toxins into the surrounding water. The efficacy of copper sulphate treatment is, however, very much dependent on water chemistry, especially alkalinity, pH and dissolved organic content.

Chemical Breakdown

The four main groups of cyanotoxins: microcystins, anatoxins, PSPs and cylindrospermopsins, exhibit quite different chemical stabilities and biological activities in water.

Microcystins

Microcystins, being cyclic peptides, are extremely stable and resistant to chemical hydrolysis or oxidation at near neutral pH. Microcystins and nodularin remain potent even after boiling. In natural waters and in the dark, microcystins may persist for months or years. At high temperatures (40 °C) and at elevated or low pH, slow hydrolysis has been observed, with the times to achieve greater than 90 per cent breakdown being approximately 10 weeks at pH 1 and greater than 12 weeks at pH 9. Rapid chemical hydrolysis occurs only under conditions that are unlikely to be attained outside the laboratory, e.g. 6M HCl at high temperature. Microcystins can be oxidised by ozone and other strong oxidising agents, and degraded by intense ultra violet (UV) light.

In full sunlight, microcystins undergo slow photochemical breakdown and isomerisation, with the reaction rate being enhanced by the presence of water-soluble cell pigments, presumably phycobiliproteins. In the presence of pigments the photochemical breakdown of microcystin in full sunlight can take as little as two weeks for greater than 90 per cent breakdown, or longer than six weeks, depending on the concentration of pigment. A more rapid breakdown under sunlight has been reported in the presence of humic substances in field concentrations ranging from 2-16 mg l^{-1} dissolved organic carbon (DOC). Approximately 40 per cent of the microcystins was degraded per day under summer conditions of insolation. In deeper or muddy waters, the rate of breakdown is likely to be considerably slower.

Anatoxins

Anatoxin-a is relatively stable in the dark, but in pure solution in the absence of pigments it undergoes rapid photochemical degradation in sunlight. Breakdown is further accelerated by alkaline conditions. The half-life for photochemical breakdown is 1-2 hours. Under normal day and night light conditions at pH 8 or pH 10, and at low initial concentrations (10 μg l^{-1}), the half-life for anatoxin-a breakdown was found to be approximately 14 days. Anatoxina(S) decomposes rapidly in basic solutions but is relatively stable under neutral and acidic conditions.

Saxitoxins

In the dark at room temperature, saxitoxins undergo a series of slow chemical hydrolysis reactions. The C-toxins lose the N-sulphocarbamoyl group to form decarbamoyl gonyautoxins (dc-GTXs); while the dc-GTXs, GTXs and STXs slowly degrade to, as yet unidentified, non-toxic products. The half-lives for the breakdown reactions are in the order of 1-10 weeks, with more than three months often being required for greater than 90 per cent breakdown. Because dc-GTXs are much more toxic than C-toxins, a solution or water body containing a natural mixture of C-toxins and GTXs, for example from the lysis of an Australian bloom of *Anabaena circinalis*, will actually increase in toxicity over a period of up to three weeks, before toxicity begins to abate during the succeeding 2-3 months. Boiling an extract of *Anabaena* with predominant C-toxins may also substantially increase toxicity. Similar transformation reactions occur in living cells as they age in culture or in a natural bloom. No detailed studies have been carried out on saxitoxin breakdown in sunlight, either with or without pigments.

Cylindrospermopsins

Cylindrospermopsin is relatively stable in the dark, with slow breakdown occurring at elevated temperature (50 °C). In sunlight and in the presence of cell pigments, breakdown occurs quite rapidly being more than 90 per cent complete within 2-3 days. Pure cylindrospermopsin is relatively stable in sunlight.

REMOVAL ON NATURAL SEDIMENTS AND SOILS

Microcystins appear to be retained only weakly on natural suspended solids in rivers and reservoirs; usually no more than 20 per cent of the total microcystin concentration is adsorbed. In a laboratory experiment, some of the dissolved anatoxin-a and microcystins were reported by Rapala *et al.* to be adsorbed on lake sediments. Percolation through clay soils may provide some cyanotoxin removal, but this will depend greatly on the type of clay, surface charge, cation concentration of the water, etc. Cyanobacterial cells and microcystins were retained in soil columns, but less efficiently in sediment columns, in laboratory experiments simulating the fate of cyanobacterial toxins in artificial recharge of groundwater and bank filtration. No data are available for other cyanobacterial toxins, but some removal may be expected, again depending on the chemical conditions of soil and water. Sedimentation of living cells without lysis, for example through grazing by zooplankton and sinking of faecal pellets, may lead to accumulation and persistence of toxin material in sediments, although this process has received little scientific attention. As discussed in more detail below, microcystins retained in intact cells may persist for several months. Cells deposited in sediments may be subject to fairly rapid breakdown by sediment bacteria and protozoa, with the resultant release of toxins.

Biodegradation

Microcystins

In spite of their chemical stability and resistance to eucaryotic and many bacterial peptidases, microcystins are susceptible to breakdown by aquatic bacteria found naturally in rivers and reservoirs. These bacteria appear to be reasonably common and widespread. Degradative bacteria have been found in sewage effluent, lake water, lake sediment and river water.

Nonetheless, one Finnish study showed a complete lack of degradation of microcystin over a three-month period by an inoculum taken in winter from the Vantaanjoki River. There is usually an initial lag phase with little loss of microcystin and this period can be as short as two days or more than three weeks, depending on the water body, climatic conditions, the concentration of dissolved microcystin and in some cases, although not all, the previous bloom history of a lake. Once the biodegradation process commences, removal of microcystin can be more than 90 per cent complete within 2-10 days. This may vary depending on the water body, initial microcystin concentration and water temperature.

Jones *et al.* isolated a species of aquatic *Sphingomonas* that initiated ringopening of microcystin-LR to produce linear (acyclo-)microcystin-LR as a transient intermediat. This compound was nearly 200 times less toxic than the parent toxin. The products of complete bacterial degradation were non-toxic to mice at doses up to 500 µg kg^{-1} (compared with an LD_{50} for microcystin-LR of about 60 µg kg^{-1}). The same bacterium, however, did not degrade the closely related cyclic pentapeptide nodularin. In a strain of *Pseudomonas aeruginosa* from a Japanese lake, microcystin degradation appeared to proceed by attack on the Adda side chain of microcystin. Several bacteria were isolated from lake water and sediment in Finland capable of degradation of microcystins and some strains also degraded nodularin. One strain was identified as a *Sphingomonas* sp. and two of the strains belonged to the beta-subgroup of *Proteobacteria,* although the genera remains to be determined.

Other cyanobacterial toxins

Little work has been undertaken on the biodegradation of anatoxins, saxitoxins or cylindrospermopsin. Anatoxin-a may be readily degraded by bacteria associated with cyanobacterial filaments. Laboratory studies using non-axenic strains of cyanobacteria found low concentrations of dissolved anatoxin-a in the culture medium whereas high concentrations of anatoxin-a were found in the medium of a continuous culture using an axenic strain of the same species. A *Pseudomonas* sp. strain able to degrade anatoxin-a at a rate of 6-10 µg ml^{-1} per three days was isolated by Kiviranta *et al.* In the presence of lake sediment and natural bacteria, the half-life for breakdown of anatoxin-a in the laboratory was about five days. In a recent study by Jones and Negri no bacterially-mediated degradation of saxitoxins from *Anabaena circinalis* was observed in a range of surface water samples.

Bioaccumulation

Microcystins bioaccumulate in common aquatic vertebrates and invertebrates, including fish, mussels and zooplankton. In mussels, the highest microcystin concentrations are found in the hepatopancreas, and in vertebrates they are found in the liver. Williams *et al.*, have shown covalent binding and accumulation of microcystin LR in salmon liver and crab larvae. Whether the levels of microcystin accumulation are sufficient to pose a risk to humans is uncertain, and will depend on levels of consumption and the severity of toxic blooms in the area where fish or shellfish are caught or collected. Common advice given by water authorities is that the viscera of the fish should not be eaten, but caution should be taken in all cases where major toxic blooms occur. Saxitoxins from marine "red tide" dinoflagellates are well known for their propensity to bioaccumulate in marine vertebrates and invertebrates, often with disastrous consequences for animals and humans that consume them. Similarly, saxitoxins from the freshwater cyanobacterium *Anabaena circinalis* may bioaccumulate in an Australian species of freshwater mussel to concentrations exceeding international guidelines during as little as seven days exposure to a cell density of 100,000 cells per ml of a toxigenic strain. This cell density is commonly encountered in natural blooms of this species.

Persistence and stability in cells

Culture studies indicate that microcystins and nodularin degrade only very slowly, if at all, whilst contained within living cells. Similarly, scums of *Microcystis aeruginosa* that dry on the shores of lakes may contain high concentrations of microcystin for several months. These toxins are released back into the water body when re-immersed. Thus there is the potential for significant localised concentrations of dissolved microcystin even in the absence of living cells or a recently collapsed bloom.

In a lake study carried out over two summer - autumn periods, Lahti *et al.* found that dissolved microcystin was more persistent than particulate toxin, with 30 and 15 days respectively required for 90 per cent degradation to occur.

Impact on Aquatic Biota

Direct cyanobacterial poisoning of animals can occur by two routes: through consumption of cyanobacterial cells from the water, or indirectly through consumption of other animals that have themselves fed on cyanobacteria and accumulated cyanotoxins. Cyanotoxins are known to bioaccumulate in common aquatic vertebrates and invertebrates, including fish, mussels and zooplankton. Consequently, there is considerable potential for toxic effects to be magnified in aquatic food chains. Such toxicity biomagnification is well known for anthropogenic pollutants such as heavy metals and pesticides. There is no reason to suspect that the situation would be any different with natural cyanotoxins.

It is difficult to ascribe the deaths of natural populations of aquatic animals, especially fish, unequivocally to cyanotoxin poisoning. One of the main reasons for this is because the collapse of a large cyanobacterial bloom can lead to very low concentrations of oxygen in the water column as a consequence of bacterial metabolism; consequent fish deaths may be due to the anoxia. The best evidence for the potential for toxic effects on aquatic organisms comes from controlled laboratory trials with exposure of animals to toxic cyanobacteria or cell-free solutions of cyanotoxins.

Effects on aquatic bacteria

The influence of cyanobacterial toxins on bacteria is not fully understood and the scientific literature gives a number of contradictory statements. According to some authors neither an extract of *Microcystis aeruginosa* nor pure microcystin-LR have a biocidal effect on *Bacillus subtilis, Staphylococcus aureus, Escherichia coli* or *Pseudomonas hydrophila*. However, these limited tests should not be seen as general indicators of the potential impacts of cyanotoxins on aquatic bacteria. The majority of aquatic bacteria are yet to be cultured, and studies with common mammalian pathogens or "laboratory" bacteria should not be taken as all encompassing. It is quite possible that cyanotoxins impact on some species of aquatic bacteria and not others. Certainly, microcystins are not toxic to all bacteria because several species are known to degrade quite high concentrations of these toxins. It is even possible that the slow release of cyanotoxins from the cell surface or from senescent cells may stimulate associations of particular bacterial types which may even act as symbionts.

Attempts have been made to use bacterial toxicity tests to screen for the presence of cyanotoxins, especially microcystins. However it appears that the inhibition of bacterial phosphorescence is not related to the commonly known cyanotoxins. It has been suggested that the negative effect may be related to the presence of unidentified LPS endotoxins in the cell wall of the cyanobacterial cells.

Effects on zooplankton

Evidence of the potential effects of cyanotoxins on zooplankton from numerous studies, mostly in laboratory situations, is complex and inconsistent. The vast majority of published studies has been based on mouse bioassay data describing cyanobacterial toxicity, with only a few more recent studies having used analytical methods such as HPLC to quantify individual toxins. Overall, it appears that cyanobacteria may exhibit a deleterious effect on zooplankton, but the effect is highly variable between genera and species, and even between clones of individual zooplankton species. One of the main questions yet to be resolved is whether the observed inhibitory effects are due to the putative poor nutritional value of cyanobacteria, to the known cyanotoxins, or to other unidentified compounds. There is evidence in the literature to support all three effects as being significant, at least with particular species under experimental growth conditions.

A major difference in study design is whether organisms are exposed to cyanotoxins dissolved in water, or fed with toxic cyanobacteria. The latter is likely to lead to a substantially higher dose. Furthermore, Jungmann and Benndorf reported that exposure of *Daphnia* to dissolved microcystins showed effects only at concentrations several orders of magnitude above those found in field samples. They did, however, observe toxicity to *Daphnia* by unidentified metabolites other than microcystins from *Microcystis*. There is dramatic variation among zooplankton species in their response to toxic (and even non-toxic) cyanobacteria. For example, DeMott *et al.* showed that the four species of zooplankton differed in their sensitivity to hepatotoxins by almost two orders of magnitude, but toxicity was observed only at very high concentrations that are scarcely encountered in natural water bodies. Snell found that there was a genotype-dependent response of the rotifer *Asplanchna girodi* to toxic *Anabaena flos-aquae* and *Lyngbya* sp. Hietala *et al.* observed a variation in susceptibility of more than three orders of magnitude in the acute toxicity of *Microcystis aeruginosa* to 10 clones of *Daphnia pulex*. Both DeMott *et al.* and Laurén-Määttä *et al.* suggested that clone and species differences between zooplankton susceptibilities to toxic cyanobacteria may lead to selection pressures in favour of resistant strains or species in water bodies where toxic cyanobacteria occur frequently.

Benndorf and Henning found that the toxicity of a field population of *Microcystis* was increased by the feeding activity of *Daphnia galeata* over a period of a few months. A possible explanation for this phenomenon is offered by DeMott *et al.* who demonstrated that a number of zooplankton species will avoid grazing on toxic cyanobacteria, but continue to graze on non-toxic species. Similar results have also been shown for grazing by the phytoplanktivorous fish *Tilapia* and silver carp. Thus, grazing pressure from zooplankton and some fish may lead to the selective enrichment of toxic cyanobacterial strains over time.

It is likely that under natural conditions in water bodies, certain species and strains of zooplankton may be affected by cyanotoxins, whereas others will be unaffected. As such, cyanotoxins may influence the zooplankton community structure, especially during times when cyanobacteria are dominant within the phytoplankton.

Effects on fish

If fish are dosed with cyanotoxins by i.p. injections or by force-feeding, they develop similar symptoms of intoxication as laboratory mammals. The question relevant for field exposure is whether cyanotoxins enter healthy fish. For example, Tencalla *et al.* showed that gastrointestinal uptake by gavage (force-feeding) caused massive hepatic necrosis followed by fish deaths, whereas immersion of adults and juveniles in contaminated water did not cause toxic effects. Other reported evidence suggests that immersion in toxic cyanobacteria or cyanotoxins may be harmful to fish. Differences in sensitivity may be pronounced between species: goldfish were found to be nearly 30 times less susceptible to i.p. microcystin than mice. Release of toxic compounds from mass developments of cyanobacteria was considered to be the cause of fish kills by Penaloza *et al.* Histopathological investigations of fish deaths during cyanobacterial blooms in the UK, indicated that the cause of death was mostly due to damage of the gills, digestive tract and liver. The gill damage was probably caused by the high pH induced by cyanobacterial photosynthesis activity prior to the bloom collapse, together with the higher level of ammonia arising from the decomposition of the cyanobacteria. However, gill damage may have enhanced microcystin uptake and thus led to liver necrosis. Damage to gills by dissolved microcystin-LR has been shown experimentally in *Tilapia* and trout.

Other pathological symptoms ascribed to toxic cyanobacterial blooms include damage to the liver, heart, kidney, gills, skin and spleen. Garcia and Rodger *et al.* carried out experiments on trout, while Råbergh *et al.* experimented on carp. The latter study highlighted degenerative changes in kidney tubules and glomeruli. The effect of microcystins on European carp, *Cyprinus carpio,* under natural field conditions in

Australia has been described by Carbis *et al.* as atrophy of hepatocytes, gills with pinpoint necrosis, epithelial ballooning, folded lamellar tips, exfoliation of the lamellar epithelium, elevated asparate aminotransferase activity and serum bilirubin concentrations. Laboratory studies indicate that dissolved microcystins may affect fish embryos and behaviour of fish.

The most definitive effect of microcystin on fish concerns Atlantic Salmon reared in net pens in coastal waters of British Columbia and Washington State, USA. As yet unidentified microcystin-producing organisms produce a progressive degeneration of the liver in salmon smolts placed into open-water net pens. The disease, referred to as Net Pen Liver Disease (NPLD), has resulted in significant economic losses for the mariculture industry.

ANALYSIS OF CYANOBACTERIAL TOXICITY

Toxic cyanobacterial mass occurrences (blooms) are commonly found in fresh, brackish, and marine waters. Cyanobacteria growing in benthic environments have also been shown to contain toxins. The most common cyanobacterial toxins (cyanotoxins) are hepatotoxins, neurotoxins [anatoxin –a, anatoxin–a(S) and saxitoxins], cytotoxins, and dermatotoxins. Microcystins in freshwaters are most frequently produced by Microcystis, Planktothrix and Anabaena. Microcystin production has been also proven for *Nostoc*, terrestrial isolates of *Hapalosiphon* and *Phormidium*. In brackish waters such as the Baltic Sea or saline lakes and estuaries in Australia and New Zealand, nodularin producing cyanobacterium, *Nodularia spumigena*, frequently occur. Neurotoxins are commonly produced by *Anabaena* and less frequently by Aphanizomenon, Lyngbya and Oscillatoria. Cylindrospermum raciborskii, Aphanizomenon, Umezakia, Anabaena and *Radhidiopsis* produce cylindrospermopsins, whereas *Lyngbya, Schizothrix* and *Oscillatoria* are primarily implicated as main producers of dermatotoxins. Cyanobacterial toxins have caused hundreds of animal poisoning cases worldwide and are threat to human health. It has become evident that efficient methods to detect cyanobacterial toxins as well as toxic cyanobacteria are needed. This chapter will evaluate the potential of the current and emerging high throughput methods for analysis of cyanobacterial toxins and toxin producing organisms.

Detection of Toxins

To protect water users from poisoning and exposure to the toxins, it is important to know the identity and quantity of the toxins. Such cases include drinking water, dietary supplements, important areas for recreation, animal–poisoning cases, etc. Several papers in the current issue deal with toxin analyses in detail thus, only the potential for developing high throughput methods are considered here. The structures of the most wide spread cyanotoxins are known and that has made possible the development of high throughput analysis methods such as enzyme–linked immunosorbent assay (ELISA), protein phosphatase inhibition assay (PPIA), high–performance liquid chromatography (HPLC) and liquid chromatography – mass spectrometry (LC/MS). ELISA is available for cyanobacterial hepatotoxins and saxitoxins whereas, PPIA is used to detect microcystins and nodularins. Both methods have been evaluated to be sensitive and fast screening methods. Both of these analyses can be carried out in multi–well plate formats. The ELISA analyses can be fully automated by using new robotic technologies. The PPIA analysis involves the use of enzyme which readily inactivates without precautions. Both of these methods give total toxin concentrations but do not differentiate the compounds. Matrix–assisted laser desorption/ionisation time–of–flight mass spectrometry (MALDI–TOF) has been proven as a fast screening method for the detection of cyanobacterial peptide toxins and bioactive compounds.

To identify individual toxins, separation is needed and identification methods such as HPLC combined with UV, FT or mass detection are required. There are good applications for many toxins and they can be accurately determined. LC/MS is an excellent method to identify the individual toxins. Improvement of the instruments has also made the method very sensitive. All chromatographic methods are currently rather fast. The analysis times are in minutes and the equipment with auto samplers can be run continuously on nights and weekends. Thus, the limiting step still is sampling and sample preparation/extraction. The availability of standards of certain toxins is still a problem. New developments in this field leading to on–line measurements and fast and easy to use kit detection systems are likely to improve the rapid detection of cyanotoxins in the future.

Detection of Cyanobacteria and Genes involved in Toxin Production

In order to implement any rational mitigation scheme, further research is needed on the biology, ecology and proliferation of cyanobacteria producing toxins. The major toxin producers have been identified by isolating the organisms and showing their toxin production capability. This has created invaluable culture collections, which contain toxic and non–toxic planktonic cyanobacteria as well as benthic strains. Such culture collections have been valuable sources for physiological studies on toxin production, taxonomic/phylogenetic analyses of toxin producers as well as studies on the biosynthesis of toxins. We cannot identify toxin producers by microscopy since toxic and non–toxic strains of the same species are known to occur. The gene clusters involved in microcystin and nodularin production are known. Microcystins (nodularins) are produced non–ribosomally by the multi–enzyme complex consisting of polyketide synthases and peptide synthetases, which was verified by Dittmann et al. by gene knockout experiments. These biosynthetic gene clusters have been fully sequenced from *Microcystis, Planktothrix,* Anabaena and *Nodularia*. This has made possible to develop molecular methods to identify these toxin producers in samples. However, the biosyntheses of neurotoxins and other cyanobacterial toxins remains to be elucidated and verified by gene knockout experiments before molecular analysis can be developed with high certainty. Toxin analyses will not reveal the producers in most cases. The only exception is nodularin, which is produced thus far, only by Nodularia spumigena. Detection of toxic cyanobacteria by conventional or real–time PCR can serve as an early warning method since PCR detects the presence of toxin producers in very low concentrations. These and microarray technologies are the only methods currently available to study toxin producing cyanobacteria in situ.

Conventional PCR

Conventional PCR is a fast and cheap method to detect potentially toxin–producing strains in samples. This method requires only PCR and gel documentation facilities, which are common nowadays in laboratories. Like the methods for toxin analysis, the most time consuming task using molecular methods is the environmental sample collection, DNA isolation and purification. Conventional PCR can be used to detect microcystin/ nodularin producers since the gene clusters are now known and several primers are thus, available. These gene clusters are large and offer a wealth of information and possibilities for probe and primer design. In the recent years, it has become evident that cyanobacteria often produce plenty of non–ribosomal cyclic or linear peptides other than microcystins and nodularins and this should be taken into consideration in primer design.

Microcystis aeruginosa is found worldwide, the most frequently occurring microcystin–producing cyanobacterial species and its biosynthetic genes have been known for the longest time. For this reason, most studies have focused on the detection of toxic *Microcystis* strains but more recently extended to include other producers such as *Planktothrix* and *Anabaena*. Simultaneous occurrence of several potential microcystin producers in a lake is not uncommon. There are few general primers designed to detect several different producers of microcystins and to produce sequence information to design genus specific primers. In the development of genus–specific primers the testing with all possible microcystin–producing cyanobacteria is required in addition to *in silico* analyses. The other option, to detect specific microcystin producers, was presented by Hisbergues et al. PCR products generated by general primers were digested with restriction enzymes and producer genera were identified based on the differences in resulting bands. The gene region chosen in that study was *mcy*A, a region missing from the *Nodularia,* thus those primers were not suitable for detection of nodularin producers. Vaitomaa et al. and Rantala et al. successfully used *mcy*E primers to get PCR product from all known microcystin and nodularin producers. This gene region is involved in the construction of Adda and activation and condensation of glutamate. Adda and glutamic acid are present in both microcystins and nodularins and are shown to be the most important determinants for toxicity of these compounds. Structural variations of Adda and glutamic acid in microcystins and nodularins are less frequent than other parts of these molecules, which makes this gene region attractive for primer and probe design. The microcystin and nodularin synthetase sequence information has also been used to study the evolution of these genes. The microcystin synthetase genes were ancient and the non–toxic strains have lost these genes during evolution. This also implies that individual strains of cyanobacteria may have retained these genes and new toxin producers among the cyanobacteria currently regarded as non–producing species may still be found.

Conventional PCR is prone to typical PCR based method problems and requires some understanding of basic molecular biology. The primers are always designed and tested only with a subset of strains and thus, may not be readily applicable to new strains/ species without further analysis. Environmental samples may also contain PCR inhibitors and this should be controlled especially in the case of negative results. Multiplex–PCR when developed may further improve this method and make it faster.

DNA–based detection methods are only able to identify potential toxin producers. The analysis of high number of *Microcystis* strains and also environmental samples have found only few cases where microcystin synthetase genes were detected but the organism was unable to produce microcystins possibly due to mutations in the large gene cluster. The study of Kurmayer et al.showed that a rather high frequency of *Planktothrix* strains/filaments contained the genes but were not producing microcystins in Alpine lakes of Austria.

Molecular methods to detect producers of cylindrospermopsins have been developed. These detection methods are based on polyketide synthase (PKS) and peptide synthetase (PS) gene clusters present only in cylindrospermopsin producing strains, but verification of the involvement of these genes in cylindrospermopsin biosynthesis by knockout is yet to be done. Ferqusson and Saint developed multiplex PCR assay for *Cylindrospermopsis raciborskii, Aphanizomenon ovalisporum* and *Anabaena bergii* based on the PKS and PS genes which correlated well with the actual production of cylindrospermopsin. Lack of information on the biosynthesis and biosynthetic genes of other toxins such as the cyanobacterial neurotoxins limits the use of molecular methods to detect the producers of these toxins.

Quantitative Real–time PCR

The real–time PCR yields quantitative information and answers the question regarding which organism is a major producer of the toxins in a sample. The real–time PCR needs specific instrument that are more expensive than the standard PCR machines and have higher running costs. However, the benefits of this technology being quantitative are obvious compared with the conventional PCR. The current real–time PCR machines are designed for high throughput analyses since they use multi–well plates or carry cells with multiple capillaries. Currently, only few applications of this method for detection of cyanotoxin producers are found in literature. Two major methodologies, SyberGreen and TaqMan methods are available; the latter is regarded to be more specific. The optimisation of real–time PCR is often difficult and time consuming. The primers should preferably target amplification of short sequences (100 –200 bp) to be efficient. The prerequisite for this method is that the biosynthetic genes are known which restricts the current use of this method to microcystin and nodularin producers.

The quantitative real time PCR method is likely to produce valuable information on the toxin producers. Since it is quantitative, it can be associated with other environmental parameters possibly giving new insights to what factors promote the selection of toxic cyanobacteria and why especially toxic blooms occur. In case where several producers are found in environmental samples, it will reveal which organism is the major producer. It may have great importance when mitigation schemes are designed – *Microcystis* and nitrogen fixing cyanobacteria such as *Anabaena* or low light adapted *Planktothrix* or *Cylindrospermopsis* require different approaches.

DNA Chips

DNA chips (microarrays) offer new insights into cyanobacterial populations in natural environmental settings. This methodology is very new and has not been widely used in environmental analysis, yet. It is an attractive method for monitoring since large amount of data can be created fast and the method can be automated. Current monitoring of cyanobacteria in lakes is based on microscopic identification and cell/colony/filament counting. This microscopic method is prone to pitfalls: the identification of organism is subjective, it requires extensive training, and in addition, microscopic counting is very time consuming and tedious. DNA chip technology can identify all cyanobacteria that are present in a sample accurately and identify the toxin producers depending on the designed probes.

It should be emphasized that toxic cyanobacteria cannot be identified by microscopy, thus this new technology is superior when it comes to the identification of potentially toxic cyanobacteria in samples. At present, applications of DNA chip technology for the detection of cyanobacteria are scarce. Rudi et al. developed a microarray based on 16S rRNA genes for a few groups of cyanobacteria (*Phormidium, Microcystis, Planktothrix,* Anabaena, Aphanizomenon, Nostoc +Anab., Aph.), Eubacteria (+ chloroplasts). The array was based on membrane bound probes. The environmental DNA was isolated, amplified by PCR, labeled and hybridised with the complementary probes on the membrane. The method was used to detect cyanobacteria in eight Norwegian lakes. Another recent example of development of microarray to detect cyanobacteria comes from the European Union project MIDI–CHIP . The technology used in the MIDI–CHIP project was so called the universal microarray method. This method combines hybridisation with ligation detection reaction (LDR), which improves identification. In this method, two probes, a discrimination and a common probe, are needed. The universal array method avoids the limitation of hybridisation. It is very difficult to establish standard conditions for different polymorphic DNA targets to be analysed at the same time. The method consists of oligos (ZipCodes), which have similar thermodynamic behavior and are unrelated to probes. This makes array very flexible and new probes easy to add. The prototype microarray was based on 16S rRNA genes. Sequences from data bases (281) and strains isolated in MIDI–CHIP project were aligned and phylogenetic groupings were based on Neighbour joining three made in ARB–program. The arrays detected 19 major cyanobacteria groups and were tested with strains and environmental samples and found to be specific. These arrays have been developed further to include a few more important phylogenetic groups as well as the detection of microcystin/nodularin producing cyanobacteria.

Genome Projects

In recent years, the whole genome sequencing projects have also been extended to cyanobacteria. The first cyanobacterial genome sequenced was *Synechocystis* PCC 6803, which is a model organism to study photosynthesis, as well as stress or high light acclimation processes. This strain does not produce any of the known cyanobacterial toxins. Since then, tens of cyanobacterial genomes have been fully or partly sequenced. Most recently, toxin producing cyanobacteria genome sequencing projects have been started: two microcystin producing cyanobacteria *Microcystis aeruginosa* and *Anabaena* strain 90 genome as well as nodularin producing *Nodularia spumigena* from the Baltic Sea. Detailed annotation of these genomes is likely to reveal genes associated with toxin production and regulation. Comparison of these genomes will yield information on the metabolic versatility of these organisms and the differences between the various producers of toxins. Knowledge of genome sequences makes it possible to design expression arrays to study the gene expression of these organisms in various conditions as was carried out with *Synechocystis* PCC 6803 or *Anabaena* PCC 7120 strains.

Gene knockout experiments with planktonic cyanobacteria which will give important information about the function of the yet uncharacterised genes, have been difficult to accomplish. The whole genome sequence of an organism provides wealth of information, which can also be utilised in proteome research. The whole set of proteins of an organism can be separated by 2D gel electrophoresis and identified by mass spectrometry. Genome sequencing, as well as gene expression and proteome analyses are all high throughput methods. Combination of all these methodologies in case of toxin producing, bloom–forming cyanobacteria is likely to reveal key aspects of the biology of these important organisms. However, several ecologically relevant, mass–occurrence forming and toxin producing cyanobacteria such as *Planktothrix*, *Cylindrospermopsis* and all of the neurotoxin producers still wait for ushering into the genome sequencing programs.

REFERENCES

Matveev, V., Matveeva, L., and Jones, G.J. 1994a Study of the ability of *Daphnia carinata* King to control phytoplankton and resist cyanobacterial toxicity: implications for biomanipulation in Australia. *Aust. J. Mar. Freshwat. Res.*, 45, 889-904.

Moore, R.E., Ohtani, I., Moore, B.S., de Koning, C.B., Yoshida, W.Y., Runnegar, M.T. and Carmichael, W.W. 1993 Cyanobacterial toxins. *Gazzetta Chimica Italiana,* 123, 329--336.

Namikoshi, M., and Rinehart, K.L. 1996 Bioactive compounds produced by cyanobacteria. *J. Ind. Microbiol.,* 17, 373-384.

Ohtani, I., Moore, R.E. and Runnegar, M.T.C. 1992 Cylindrospermopsin, a potent hepatotoxin from the blue-green alga *Cylindrospermopsis raciborskii. J. Amer. Chem. Soc.,* 114, 7941-7942.

Raziuddin, S., Siegelman, H.W. and Tornabene, T.G. 1983 Lipopolysaccharides of the cyanobacterium *Microcystis aeruginosa. Eur. J. Biochem.,* 137, 333-336.

Rodger, H.D., Turnbull, T., Edwards, C. and Codd, G.A. 1994 Cyanobacterial bloom associated pathology in brown trout *Salmo trutta L.* in Loch Leven, Scotland. *J. Fish Dis.,* 17, 177-181.

Rudolph-Böhner, S., Mierke, D.F. and Moroder, L. 1994 Molecular structure of cyanobacterial tumour-promoting microcystins. *FEBS Letters,* 349, 319-323.

4

Antibacterial Activity of Freshwater Cyanobacteria

Cyanobacteria are a very old group of organisms and represent relics of the oldest photoautotrophic vegetation in the world that occur in freshwater, marine and terrestrial habitats. Cyanobacteria have drawn much attention as prospective and rich sources of biologically active constituents and have been identified as one of the most promising groups of organisms to be able of producing bioactive compounds. Cyanobacteria are known to produce metabolites with diverse biological activities such as antibacterial, antifungal, antiviral, anticancer, antiplasmodial, algicide, antiplatelet aggregation and immunosuppressive activities. Screening of cyanobacteria for antibiotics and other pharmacologically active compounds, has received ever-increasing interest as a potential source for new drugs. Cyanobacteria from local habitats seem to be a source of potential new active substances that could contribute to reduction of the number of bacteria, fungi, viruses and other microorganisms.

Cyanobacteria of Iran have not yet been studied for antimicrobial activity and little work has been done to screen cyanobacteria isolated from paddy-fields with regard to their production of bioactive compounds. In order to find the potential of cyanobacteria for production of antibacterial and antifungal compounds in rice-fields of north of Iran, 150 strains of heterocystous cyanobacteria were isolated and their potency were studied. The results are presented in this chapter.

ISOLATION OF CYANOBACTERIA

Soil samples were collected from different paddyfields in Golestan, Mazandaran and Gillan provinces. Soil samples in laboratory were cultured directly in Nfree BG-11 media, after colonisation, cyanobacteria were transferred to the same medium. Unialgal cultures were prepared using subculturing methods. Each isolated cyanobacterium was cultured in a 500 ml flask containing 150 ml of BG-11 medium without shaking, for 30 days. The incubation temperature was 28°C ± 2 and illumination at 3000 lux with a white continuous light.

Preparation of Supernatant and Cell Extracts

The cultures were harvested after 30 days by centrifugation at 5000 rpm for 15 min. The aqueous supernatant was collected and the algal pellet was extracted with 15 ml of methanol followed by 15 ml hexane, with shaking for 20 min. The culture supernatants and solvent extracts were dried under reduced pressure at 40°C and were stored at -10°C for further studies.

Antibacterial and Antifungal Bioassay

The following bacteria and fungi were used as test organisms: *Staphylococcus aureus* (PTCC 1112), *Staphylococcus epidermidis* (PTCC 1114), *Staphylococcus haemolyticus* (ATCC 29970), *Escherichia coli* (PTCC 1047), *Shigella sonnei* (PTCC 1235), *Proteus vulgaris* (PTCC 1312), *Candida krusei* (ATCC 44507), *Candida kefyr* (ATCC 38296), *Candida albicans* (ATCC 14053), *Candida guilliermondii* (IF 00838), *Cryptococcus neoformans* (PLM 589), *Aspergillus niger* (PLM 1140), *Aspergillus fumigatus* (ATCC 13073) and *Aspergillus fumigatus* (PLM 712).

Dried extracts and supernatants were dissolved in 4 ml of their extraction solvent, and antimicrobial activity was determined by the disc method. Filter paper discs (6.4 mm) were saturated with 50 μl of the test solution, dried under laminar air flow and placed on the Muller- Hinton agar plate for bacteria and Saubouraud's dextrose agar plate for fungi, which had been inoculated with a lawn of the test microorganisms. Plates were incubated at 37°C, for a period of 18-24 h for bacteria and at 25°C, for 24-48 h for fungi. Discs treated with 50 μl methanol or hexane were used as negative controls and gentamycine and nystatin discs were used (10 μg) as positive controls. The extracts and supernatants containing antibacterial and antifungal components produced distinct, clear, circular zones of inhibition around the discs and the diameters of clear zones were determined and used as an indication of antibacterial and antifungal activity.

The following formula was used for comparison of the antimicrobial activity of the sample with that of the standard (antimicrobial index):

$$\frac{\text{Inhibition zone of sample}}{\text{Inhibition zone of the standard}} \times 100$$

IDENTIFICATION OF CYANOBACTERIA

Identification was done using morphological variation studies and taxonomical approaches according to Anagnostidis and Komarek, Desikachary, Prescott and Anand. Semipermanent slides were prepared from each specimen and were coded and preserved in algal herbarium of Reaserch Institute of Applied Science, Shahid Beheshti University.

The results of culture supernatants and methanolic extracts of the isolated cyanobacteria that demonstrated antibacterial and antifungal activity are shown in Tables 1, 2, 3, and 4. Supernatant and methanolic extract of 21 strains from the 150 cyanobacteria strains, showed significant antibacterial activity against at least one Gram-positive or Gram-negative bacterium. Thirteen of them were identified as *Fischerella* species, where 5 *Stigonema* species, two *Nostoc* species and one *Hapalosiphon* species, were also among those showing antibacterial activity.

Supernatant and methanolic extract of 13 isolated strains showed antifungal activity against at least one of the fungal strains. Seven of them were *Fischerella* species, 4 were *Stigonema* species, and one strain of each of the *Nostoc* and *Hapalosiphon* species were also identified.

No antimicrobial activity was detected in the hexane extracts and none of the methanolic extracts or culture supernatants showed antibacterial activity against *E. coli.*

Results showed that antibacterial and antifungal activity were seen predominantly from the algal family Stigonemataceae and *Fischerella* and *Stigonema* genera.

The cyanobacteria such as Fischerella ambigua, Fischerella musciola, Nostoc commune, Scytonema hofmanni, Hapalosiphon fontinalis, Anabaena *spp.,* Nostoc spongiaeforme, Microcystis aeruginosa, Phormidium *sp., have been reported as the main cyanobacteria to produce antimicrobial substances.*

Screening efforts aimed to identify antimicrobial agents in cyanobacteria have revealed several promising lead compounds. Some of these substances identified including Nostocyclyne A, Nostofungicidine, Kawaguchipeptin B, Nostocin A, Ambigol A and B, Hapalindoles and Scytophycins. Most of the studies have only done as *in vitro* assays and, it is likely that most of these compounds will have little or no clinical application as they are either too toxic or inactive *in vivo*. They may however serve as useful lead compounds for synthesis of antibiotics or may find application in agriculture. For example Tjipanazoles isolated from the cyanobacterium, *Tolypothrix tjipanensis,* showed appreciable fungicidal activity against rice blast and leaf rust wheat infections.

A few studies have been done to screen cyanobacteria for production of antimicrobial substances from paddy-fields. Possibly the synthesis of highly active toxin is a defence option of cyanobacteria in these environments against other organisms like bacteria, fungi, viruses and eukaryotic microalgae.

In one study, the culture media of cyanobacteria belonging to Nostaceae, Microchaetaceae and Scytonematacaea isolated from the Argentinian paddyfields, were found to be active against *S. aureus* and Candida albicans.

In another study, it was shown that cyanobacteria from the paddy-fields of northern Thailand produce bioactive substances with antibiotic activity against Bacillus subtilis. Cyanobacteria of Iran have not yet been studied for antimicrobial activity and this screening program is among the first studies done for assessment of antibacterial and antifungal activity of Iranian paddyfield cyanobacteria.

In this investigation, out of 150 strains of cyanobacterial isolates, 21 showed significant *in vitro* antibacterial activity and 13 of them had antifungal effect. The proportion of the isolates with antibacterial and antifungal activities were approximately 14% and 9%, respectively, which is comparable with those published earlier in other screening programs: 11%, 7%, and 10%.

As depicted in Tables 1, 2, 3, and 4, *Fischerella, Stigonema, Nostoc* and *Hapalosiphon* species produce bioactive substances which may have potential for antibacterial and antifungal activity.

A variety of solvents with different polarities, were used for the extraction of algal bioactive materials. No antimicrobial activity was detected in the hexane extracts. This is probably because of polar nature of the active components. It shows that the chance of finding antimicrobial activity is higher in culture supernatants and in methanolic extracts. The results indicated that all of the supernatants and methanolic extracts of *Fischerella* and *Stigonema* species had high activity against Gram-positive bacteria.

Although some of the cyanobacteria produce active compounds against Gram-negative bacteria such as *S. sonnei* and *P. vulgaris* but no activity was found against *E. coli*. Different results have been reported by other authors in this case. Falch *et al.* and Hirata *et al.* reported active compounds against *E. coli* in the petroleum ether fraction of *Fischerella ambigua* and supernatant of *Nostoc spongiaeforme*, respectively.

Antifungal activity assays showed a good activity against C. krusei, C. kefyr and C. neoformans and the minimum against *Aspergillus* spp. Among the isolated cyanobacteria, *Hapalosiphon* species had the minimum activity against the test organisms.

Fischerella and Stigonema species from Stigonemataceae had the greatest frequency among the species that showing antibacterial and antifungal activity and exhibited the most prominent effect. The effect of antimicrobial activity of Stigonemataceae has been reported in other studies such as Patterson *et al.*, Falch *et al.*, and Smitka *et al.* Among all of the species studied in this investigation for antibacterial and antifungal activity, it seems *Stigonema* strains are being reported for the first time as producer of antibacterial substances. The results of this work indicate that this group of organisms displays a potential that warrants further investigations.

SCREENING FOR ANTIMICROBIAL ACTIVITY

Production of antimicrobial compounds seems to be a general phenomenon for most bacteria. An admirable array of microbial defense systems are produced, including broad-spectrum classical antibiotics, metabolic by-products such as organic acids, and lytic agents such as lysozyme. In addition, several types of protein exotoxins, and bacteriocins, which are biologically active peptide moieties with bactericidal mode of action, were described. This biological arsenal is remarkable in its diversity and natural abundance, since some substances are restricted to some bacterial groups while other are widespread produced.

The search for new antimicrobial agents is a field of utmost importance. The prevalence of antimicrobial resistance among key microbial pathogens is increasing at an alarming rate worldwide. Current solutions involve development of a more rationale approach to antibiotic use and discover of new antimicrobials, but the problem of antibiotic resistance is increasing globally and may render the current antimicrobial agents insufficient to control at least some bacterial infections.

Amazon basin is a source of enormous biological diversity, which is scarcelly studied. Particularly, few reports on microbial life are described. More recently, the characterisation of microorganisms with biotechnological interest has been reported, however, the production of antimicrobials by bacteria isolated in this region is poorly described. The aim of this work was to investigate the antimicrobial activity of bacteria isolated from aquatic environments neighboring Manaus, at the Brazilian Amazon Basin.

Bacterial Cultures

The strains used in this work included bacteria from aquatic environments of the Amazon region, near Manaus, Brazil (3º06'S, 60º01'W), and they were kindly provided by Dr. Spartaco Astolfi Filho. Indicator microorganisms for the characterisation of antimicrobial activity were strains from collections. Bacteria were maintained as stock cultures frozen at -21ºC in BHI broth supplemented with 20% glycerol. Strains were propagated twice before used in experiments. Morphological and biochemical characterisation was performed as described elsewhere.

Growth and Production of Antimicrobial Activity

Bacteria were grown in 250 ml Erlenmeyer flasks containing 100 ml of BHI broth incubated for 48 h at 31 ± 1ºC in shaker at 125 cycles/min. After growth, culture media were centrifuged at 10,000 g for 15 min, and the supernatants were filtered through 0.22 μm membranes. The resulting filtrates were used to evaluate antimicrobial activity. The pH of the filtrates was measured showing values always between 7.0 to 8.0.

Detection of Antimicrobial Activity

Antibacterial activity was determined essentially as described elsewhere. An aliquot of 20 μl cell-free culture supernatant was applied on cellulose disks (6 mm) on BHI agar plates previously inoculated with each indicator strain suspension, which corresponded to a 0.5 McFarland turbidity standard solution. Plates were incubated for 24 h at optimal temperature for the test organism. The inhibition zones around the disks were measured.

Effect of Heat and Enzymes

Proteolytic enzymes were tested on cell-free supernatant. Samples of 1 ml were treated at 37ºC for 1 h with 2 mg ml^{-1} of either trypsin or pronase E (Sigma). Samples were then boiled for 2 min to inactivate the enzyme. To analyse thermal stability, samples of the supernatants were exposed to temperatures ranging 40 to 100ºC for 15 min and 121ºC/ 105 kPa for 15 min before being tested for antimicrobial activity. Trichloroacetic acid (TCA) was added to the filtrates to reach a working concentration of 100 mg ml^{-1} and the samples were incubated for 2 h at 4ºC. After treatment with TCA, samples were centrifuged at 10,000 x *g* for 5 min and the supernatant was neutralised to pH 7.0 before testing for antimicrobial activity. After the treatments, the samples were tested for antimicrobial activity against *L. monocytogenes* ATCC 7644.

Direct Detection on Gels

Aliquots of 1 ml of culture filtrates were freeze-dried. Samples were suspended in 0.1 ml 125 mM tris pH 6.8 containing 0.1% SDS, 20% glycerol, and then applied to 14% polyacrylamide gels. Electrophoresis was carried out as described elsewhere using a Mighty Small II apparatus and 20 mA per gel. After running, the gels were washed with sterile distilled water for removal of SDS and then flooded on plates containing tempered BHI agar with 10^6 cfu/ml *L. monocytogenes*. Detection of antimicrobial activity was carried out as described by Naclerio *et al.*

Screening of the antimicrobial activity of 86 isolates was carried out against 7 indicator strains. Inhibitory activity against at least one indicator strain was detected for 59 isolates (68.6%). The majority of strains (64%) inhibited *B. cereus* growth. Forty three isolates (50%) inhibited the growth of *L. monocytogenes* and the strains P2 and P31A presented the highest inhibition zones (17 mm). *Staphylococcus aureus* was inhibited by only two isolates, P45B and P31A, while antimicrobial activity against *Escherichia coli* and *Salmonella* Enteritidis was detected only for isolates P10 and P30, respectively.

Those bacteria producing inhibition zones higher than 10 mm against at least two microorganisms were selected for further characterisation of their antimicrobial activity. Based on this criteria 19 bacterial isolates were selected. The isolates were straight Gram-positive rods with endospores, and were strongly catalase positive. Additional biochemical tests indicated that these isolates belonged to the genus *Bacillus*. The antimicrobial substances were partially resistant to proteolytic treatment, being only the antimicrobial substances produced by strains P8A and P34 sensitive to both trypsin and pronase. The activity produced by 11 strains, including P8A and P34, was completely lost after treatment with TCA. The antimicrobial activity of all filtrates was heat resistant for up to 100ºC for 15 min. After autoclaving, 14 filtrates still presented some residual activity, and 5 filtrates completely lost the activity after this treatment.

Direct detection of antimicrobial activity on polyacrylamide gels was carried out. All the substances showed inhibitory activity against *L. monocytogenes*, and presented a molecular weight lower than 14 kDa.

Screening for antimicrobial activity of bacteria isolated from Amazon has not been described yet. One study has demonstrated that production of antimicrobial substances is widespread among these bacterial strains. Almost 70% of the isolates exhibited antimicrobial activity against one or more indicator bacteria. This corresponds to a higher proportion than often described in other investigations. Among lactic acid bacteria isolated from raw milk, 82 of 298 strains (27.5%) displayed bacteriocin activity, while 8.7% of *S. aureus* strains isolated from cattle produced antimicrobial substances. De Vuyst *et al.* found that 122 of 426 *Enterococcus* strains (28.6%) of various origins produced enterocin.

Antimicrobial evaluation of cyanobacteria demonstrated that 16.3% was active against Gram-positive and 5.8% against Gramnegative bacteria, while 10.5% possessed antifungal activity. Bacilli isolated from the brittlestar *Amphipholis gracillima* exhibited a high range of complete inhibition of several test bacteria. The high proportion of antimicrobial producing strains may be associated with an ecological role, playing a defensive action to maintain their niche, or enabling the invasion of a strain into an established microbial community. The inhibitory effect was mostly against Gram-positive bacteria. Most strains inhibited *B. cereus*. The antimicrobial spectra, essentially restricted to Gram-positive bacteria, may suggest that the produced antimicrobial substances are related to a specific feature of Gram-positive bacteria. It is well known that activity of bacteriocins produced by Gram-positive bacteria is restricted to other Gram-positive bacteria.

The pH values of the crude antimicrobial substances indicate that the inhibitory effect was not due to production of organic acids. Most of these substances were partially or completely inactivated by proteases and TCA, suggesting that a protein moiety is involved in the activity. This may indicate that bacteriocin-like substances are implied in antimicrobial activity. These substances showed high thermal resistance and low molecular weight, which are characteristics of small hydrophobic peptides that constitute class II bacteriocins. Although these bacteria were not yet identified to the species level, morphological and biochemical characteristics indicate they belong to the genus *Bacillus*. A variety of antimicrobial compounds are produced by members of the genus *Bacillus*, many of these identified as peptides, lipopeptides and phenolic derivatives.

A wide range of antimicrobial substances produced by *Bacillus* spp. isolated from arthropods were recently described, including aromatic acids, acetylamino acids (amino acid analogs), and peptides. Bacteriocin-like substances have been related to *Bacillus* spp. isolated from soil and vegetal tissues. Isocoumarin antibiotics are produced by *Bacillus* spp., as demonstrated for *B. subtilis* strains from different habitats and geographic origins.

A siderophore with wide range of antibacterial spectrum is produced by *Bacillus* sp. NM12 from fish intestine and aminopolyol antibiotic by soil isolates of *B. cereus*. Although several antimicrobial substances described in this work appear to be peptides, other substances can not be ruled out since resistance to proteases and even to TCA was observed in some cases. The potential of these microorganism to produce antimicrobial compounds that can be useful for many applications is great and must be better explored.

REFERENCES

Mundt S. and Teuscher E. "Blue-green algae as a source ofpharmacologically-active compound". *Pharmazie.*, 43: 809-815 (1988).

Ostensvik O., Skulberg O.M., Underal B., and Hormazabal V. "Antibacterial properties of extracts from selected planktonic freshwater cyanobacteria- a comparative study of bacterial bioassays". *J. Appl. Microbiol.*, 84: 1117-24 (1998).

Papke U., Gross E.M., and Francke W. "Isolation, identification and detrmination of the absoluteconfiguration of Fischerellin B. A new algicide from the freshwater cyanobacterium *Fischerella muscicola*" (Thuret). *Tetrahedron Lett.*, 38: 379-382 (1997).

Patterson G.M.L. and Carmeli S. "Biological effects of tolytoxin (6-hydroxy-7-o-methylscytophycin b), a potent bioactive metabolite from cyanobacteria". *Arch. Microbiol.*, 157: 406-410 (1992).

5

Antibacterial Activity of Marine Cyanobacteria

Marine organisms are a rich source of structurally novel and biologically active metabolites. Secondary or primary metabolites produced by these organisms may be potential bioactive compounds of interest in the pharmaceutical industry. To date, many chemically unique compounds of marine origin with various biological activities have been isolated, and some of them are under investigation and are being used to develop new pharmaceuticals.

The cell extracts and active constituents of various algae have been shown to have antibacterial activity in vitro against Gram-positive and Gram-negative bacteria. A wide range of results of in vitro anti-fungal activities of extracts of green algae, diatoms, and dinoflagellates have also been reported. Similarly, to date, some micro-algae screened contain and/or excrete pharmacologically active compounds. For example, the dinoflagellates Gymnodinium sp. and Gonyaulax sp. produce an alkylguanidine compound that effects the central nervous system. In Rivularia firma, pharmacological activity is due to brominated bi-indoles. A red alga, Gracilaria lichenoides, excretes prostaglandins and the compound C_{20}.

The present chapter describes the details of a study conducted on the antimicrobial characteristics of methanol, acetone, diethyl ether, and ethanol extracts of some marine algae obtained from the coast of Urla.

Seaweeds were collected at a depth of 1-2 m from the coast of Urla, Izmir in May 2005 and were identified by the study team. Algae samples were cleaned of epiphytes and necrotic parts were removed. Then the samples were rinsed with sterile water to remove any associated debris. Half of these cleaned fresh materials were air-dried. As described by González del Val et al., 25 g of each fresh and air-dried algal sample were extracted in 50 ml of methanol, acetone, diethyl ether, and ethanol.

The strains of Candida sp., Enterococcus faecalis, Staphylococcus aureus, Streptococcus epidermidis, Pseudomonas aeruginosa, and Escherichia coli were obtained from the culture collection of the Basic and Industrial Microbiology Section of Ege

University, ‹zmir, Turkey, and were maintained on Brain Heart Infusion (BHI) agar medium at 4 °C until testing.

Antimicrobial activity was evaluated using the agar diffusion technique in petri dishes. Briefly, 25 μl of each extract was loaded on sterile filter paper discs 6 mm in diameter (E-760), and air-dried. Indicator microorganisms were spread on Mueller-Hinton agar plates with sterile effusion and the discs were placed on plates. After incubation for 24 h at 30 °C, a clear zone around a disc was evidence of antimicrobial activity. Diameters of the zones of inhibition were measured in millimeters. Each test was prepared in duplicate. Discs loaded with the extracting agents were tested as controls.

Extracts of 11 species of seaweed were tested against bacteria and yeast. The results of primary screening tests are summarised in Table 1, which show that the extracts of 6 algal species possessed antibacterial activity. For some species, the antimicrobial activity observed was similar to previous screening studies.

In the study, diethyl ether was the best solution for extracting the effective antimicrobial materials from the algae species used in the experiment, with the exception of *D. linearis*, for which ethanol was the most effective extraction solution. A significant difference in antimicrobial activity was not found between the acetone and methanol extracts of each alga. Antibacterial activity depends on both algal species and the efficiency of the extraction method. For instance, diethyl ether extracts of fresh C. mediterranea, E. linza, U. rigida, G. gracilis, and E. siliculosus showed effective results against all test organisms; however, the diethyl ether extracts of some species, such as P. pavonica, C. sniosa, D. linearis, D. membranacea, C. rubrum, and A. nojadiformis, gave different results. Although it is known that the diethyl ether extract of D. linearis was ineffective against microorganisms, the ethanol extract of the D. linearis indicated an antimicrobial activity against Gramnegative bacteria and Candida sp. This result could be related to the presence of bioactive metabolites in D. linearis, which are soluble in ethanol, but not in diethyl ether.

The diethyl ether extracts and ethanol extracts with the most effective antimicrobial activity were selected from 11 algal species for detailed antimicrobial producer tests. Regarding the obtained results, fresh-extracted and dry-extracted samples of the best halo-zone producers were assayed against 3 Gram (+) and 2 Gram (-) bacteria, and 1 yeast. Table 2 shows the effects of the different algae extracts collected from pathogenic microorganisms.

Dried extracts have less or no effects on bacteria in comparison to the fresh extracts. This result can be related to volatile antimicrobial compounds in the samples, such as hydrogen peroxide, terpenoid, and bromo-ether compounds (6,7). Another reason might be the loss of active materials that may be present in algae, like volatile fatty acids, during the drying process.

Ethanolic and diethyl ether fractions appear to be specific, particularly against the tested Gram-positive bacteria. In addition to their antibacterial activity, diethyl ether extracts of C. mediterranea, E. linza, U. rigida, G. gracilis, and E. siliculosus also inhibited the growth of yeasts.

Another significant result of the study was that the ethanol extracts of D. linearis (fw) showed antifungal and antibacterial activity against the Candida sp. and Gram-negative bacteria, respectively. Similarly, Moreau et al. reported the inhibitive effect of hexane extracts of D. dichotomaon fungal growth. Lima-Filho et al. found that the hexane extract of Gracilaria sp. inhibits only Bacillus subtilis. In contrast, the results showed that the diethyl ether extract of G. gracilis (fw) inhibited Candida sp., E. faecalis, S. epidermidis, S. aureus, E. coli, and P. aeruginosa. The activity against Gram-negative bacteria was less effective compared to Gram-positive bacteria.

Some studies concerning the effectiveness of extraction methods highlight that methanol extraction yields higher antimicrobial activity than n-hexane and ethyl acetate, whereas others report that chloroform is better than methanol and benzene. It is clear that using organic solvents always provides a higher efficiency in extracting compounds for antimicrobial activities compared to water-based methods.

Perez et al. found that the extract of Ulva lactuca had no antimicrobial activity. In contrast, the results showed that the diethyl ether extract of U. rigida inhibited all the test organisms. This difference may have been due to species variation.

In 2001, Gonzalez del Val wrote that methanol extracts of Padina pavonica show antibacterial activity only against *B. subtilis*; however, in the experiments, acetone, methanol, and diethyl ether extracts of *P. pavonica* had no antibacterial or antifungal activities, but the ethanol extract of *P. pavonica* showed weak activity against Candida, *E. faecalis, P. aeruginosa*, and *E. coli.* Gonzalez del Val also demonstrated the antibacterial or antifungal activities of the methanol extract of Enteromorpha compressa. Diethyl ether and ethanol extracts of *E. linza* showed high and low antimicrobial activities, respectively.

SCREENING OF ANTIBACTERIAL ACTIVITY

The production of antimicrobial activities was considered to be an indicator of the capacity of the seaweeds to synthesize bioactive secondary metabolites. There are numerous reports of compounds derived from macroalgae with a broad range of biological activities, such as antibacterial, antivirals, antitumorals, anticoagulant and antifouling. Numerous substances were identified as antimicrobial agents from algae: chlorellin derivatives, acrylic acid, halogenated aliphatic compounds, terpenes, sulphur containing heterocyclic compounds, phenolic inhibitors, etc. Antibacterial activity was found to vary with season.

The study was undertaken in order to examine the antibacterial effects of crude methanol extracts obtained from 32 marine macroalgae species harvested from the Atlantic and Mediterranean coast of Morocco against pathogenic bacteria *Escherichia coli* ATCC 25922, Staphylococcus. aureus ATCC 25923, E. faecalis ATCC 29212, Klebsiella pnomeuniae ATCC 700603 and Enterococcus faecalis ATCC 29213.

Seaweeds were collected by hand picking using Scuba diving or snorkeling and preserved on ice until further processing. The zones were investigated from 2003 to 2007 including 8 sampling stations of the Atlantic and Mediterranean coast of Morocco.

In the field, the algae were cleaned of epiphytes, washed thoroughly in sterile sea water. Then seaweeds were dried for 24 h under an artificial light in 24ºC and finally in a heater. The dry seaweeds were crushed in an electric mill until a fine powder was obtained. Every sample was preserved in a freezer until compound extraction.

The powder (5 g) of dried plant was extracted in soxhlet apparatus using methanol (200 ml) as solvent (8 h at 65ºC). The resulting extracts were concentrated to dryness in a rotary evaporator under reduced pressure. The extracts were diluted in 2 ml of pure.

Test was performed by the diffusion method, Sterile discs (BBLTM) de 6 mm were preparated by pipetting 25 l of extract to each disc, placed on the agar Mueller-Hinton (pH 7.4± 0.2 a 25°C) and incubated at 37°C during 24 h overnight. Inhibition results are expressed as width of the clear halo surrounding each disc on cultivated agar plats. Methanol solvent (100 %) without algal extract was also used as a negative control. All experiments were performed at least in triplicate. Representative halos were those measuring a diameter superior to 10 mm.

The strains used were *E. coli* ATCC 25922, *S. aureus* ATCC 25923, *E. faecalis* ATCC 29212, K. pnomeuniae ATCC 700603 and E. *faecalis* ATCC 29213. The organisms were obtained from the Department of Microbiology, Faculty of Pharmacy, University of Granada.

The antibacterial activity of crude methanol extracts of the 32 algal species is summarised in Figure 1 and Table 2. The antibacterial activity was classified from less active (+:10 mm Diameter of inhibition < 16 mm), to moderately active (++: 16 mm _ diameter of inhibition < 20 mm), to highly active (+++: diameter of inhibition 20 mm) and no active (- : diameter of inhibition < 10 mm) (Table 2). Twenty eight species showed antibacterial activity, no activity was observed from four methanolic extracts of *Fucus platycarpus*, *Phyllaria brevipes* (C. Agardh) E.C. Henry and G.R. South, *Codium bursa* (Olivi) C. Agardh and *Sargassum vulgare* C. Agardh. The majority of algal extracts were active against one or two microorganisms. Most of them, 28 (87.5%) showed activity against *S. aureus*, 22 (68.75%) showed activity against *E. coli* and 16 (50%), 9 (28.12%) and 3 (9.37%) presented inhibition activity against *K.* pnomeuniae, E. faecalis ATCC 29212 and E. faecalis ATCC 29213, respectively.

The methanol extracts of *Enteromorpha compressa* (Linnaeus) and *Cystoseira tamariscifolisa* (Hudson) Papenfuss showed activity against every bacterial strain tested. In all 13 species tested of Chlorophyceae, the extract of *Ulva lactuca* (Linnaeus) has a larger inhibition diameter against *E. coli* (16 mm), *S. aureus* (30 mm) and *E. faecalis* ATCC 29212 (13 mm). However, the extract of *E. compressa* (Linnaeus) Nees is the only one that showed activity against *E. faecalis* ATCC 29213 (14 mm).

Among the 19 *Phaeophyceae* tested, the extract of *Cystoseira mediterranea* sauvageau exhibited the highest activity against *E. coli* (16 mm) and *S. aureus* (33 mm), C. tamariscifolia (Hudson) Papenfuss, Dictyota linearis (C. Agardh) Greville and *Padina pavonica* (Linnaeus) Thivy acted against *E. faecalis* ATCC 29212 (13 mm), *K. pnomeuniae* (15 mm) and *E. faecalis* ATCC 29213 (11 mm).

S. aureus was more sensitive than all the stocks, with the largest inhibition diameter and *E. faecalis* ATCC 29213 was more resistant against all the methanol extracts of the marine algae tested. It was reported that the Gram-positive bacterial strains were more susceptible to seaweeds extract than Gram-negative bacterial strains.

The algal extracts such as *Enteromorpha ramulosa* (Smith) Carmichael and *Dictyopteris membranacea* (Stackhouse) Batters were active against Gram-positive and Gram-negative bacteria.

Lima-Filho et al., found that the hexane extract of *Ulva fasciata* Delile does not inhibit to *S. aureus* and *E. coli*. In the study, the methanol extract of *U. fasciata* Delile produced an inhibiting effect against *S. aureus* and *E. coli*.

Another significant result of the study was the non efficiency of the methanol extracts of *Sargassum vulgare* C. Agardh, which did not show antibacterial activity against *E. coli* and *S. aureus* growth. No growth inhibition of these two bacteria was also reported by Ballantine et al., when the hexane extracts of the same algae were tested.

Activity is classified according to the diameter of the inhibition zone around the point of application of the sample. Micro-organisms: Eco: *E. coli* ATCC 25922, Sta: *S. aureus* ATCC 25923, Ent: *E. faecalis* ATCC 29212, Kle: *K. pnomeuniae* ATCC 700603, Entf: *E. faecalis* ATCC 29213.

Moreau et al. reported that antibacterial activity is influenced by the reproductive state and seasonality. The results obtained in the study confirm this suggestion. The most active particularly against *S. aureus*, came from fertile plants of *U. lactuca* Linnaeus collected in August. Almost no activity was detected from vegetative plants collected in winter.

Similarly, *E. compressa* (Linnaeus) Nees, *Saccorhiza polyschides* (Lightfoot) Batters, *Cystoseira mediterranea* Sauvageau and *Cystoseira usneoides* (Linnaeus) M. Roberts were quite inactive in winter but strongly active in summer versus *S. aureus*. Macroalgae from the Moroccan coast are potential sources of bioactive compounds and should be investigated for natural antibiotics. The study has shown that the production of antibacterial substances by macroalgae is a regular occurrence among those found on the coast of Morocco. Biochemical analysis are currently undertaken to determine the structure and nature of these compounds.

REFERENCES

Borowitzka MA, Borowitzka LJ. "Vitamins and fine chemicals from micro algae". In: *Microalgal Biotechnology* 1992. Cambridge University Press. Great Britain p 179

Ely R, Supriya T, Naik CG. "Antimicrobial activity of marine organisms collected off the coast of South East India". *J. Exp. Biol. And Ecol.* 309: 121-127, 2004.

Febles CI, Arias A, Gil-Rodriguez MC et al. "In vitro study of antimicrobial activity in algae (Chlorophyta, Phaeophyta and Rhodophyta) collected from the coast of Tenerife (in Spanish)". *Anuario del Estudios Canarios* 34: 181-192, 1995.

Lima-Filho JVM, Carvalho AFFU, Freitas SM et al. "Antibacterial activity of extracts of six macroalgae from the Northeastern Brazilian Coast". *Brazilian Journal of Microbiology* 33: 311-313, 2002.

Moreau J, Pesando D, Bernad P et al. "Seasonal variations in the production of antifungal substances by some Dictyotales (brown algae) from French Mediterranean coast". *Hydrobiology* 162: 157- 162, 1988.

6

Effects of Cyanobacteria on Plankton

Cyanobacteria have a worldwide distribution, living in the sea, in brackish and freshwater areas, in damp soil, glaciers and deserts as well as in hot springs. They belong to the most archaic organisms on the Earth. The Proterozoic Era is sometimes called the 'Age of Cyanobacteria', because, at that time, between 2.5 and 0.5 billion years ago, cyanobacteria dominated the biota.Paleolimnological studies indicate that cyanobacterial blooms occurred during ancient conditions. These studies suggest that blooms are natural phenomena, which may be increased by anthropogenic eutrophication, but which are not dependent on this process. In the Baltic Sea, the cyanobacterial blooms are as old as today's brackish-water period, the 'Littorina Sea', which started about 7000 BP. The ancient blooms were most likely similar in size and magnitude to the present ones.

Mass-occurrences of cyanobacteria have, however, been reported to increase in frequency, as well as in intensity, due to eutrophication. In the Baltic Sea, annually recurrent blooms usually peak in July and August, at times of calm, warm weather. Different environmental conditions have been suggested to trigger cyanobacterial mass-occurrences. Increasing light and temperature, and decreasing turbulence, obviously stimulate the growth of Nodularia spumigena, though it can be difficult to identify which of these factors is the most significant in the field. Mixing of the water column, in association with upwelling of essential nutrients, has been shown to initiate the bloom of Aphanizomenon flos-aquae. Shallowing of the upper mixed layer due to solar heating has been suggested to initiate the bloom of Nodularia spumigena. Recently, Kahru et al. discussed that the frequency and magnitude of saltwater inflows into the Baltic Sea may have a crucial role in the onset of Nodularia spumigena blooms, due to their effect on phosphorus availability in the surface water layers.

A phytoplankton species can be considered harmful for several reasons. A harmful phytoplankton bloom may be an aesthetic nuisance, due to its dense occurrence in open water or on beaches. Decaying algal blooms may result in anoxia, due to high biological activity, and lead to fish mortality. Phytoplankton species with long spines, e.g. diatoms, may interfere mechanically with fish gills and cause damage. However, the most common reason for classifying algae as harmful is probably due to their toxin production. Cyanobacteria commonly produce toxins, and may also have other adverse effects on different species. It is therefore important to investigate how cyanobacteria influence other organisms and the structure and functioning of the aquatic ecosystem.

COMPETITIVENESS OF CYANOBACTERIA

Due to their early evolutionary history, cyanobacteria have developed an array of qualities that may favour them in competition with other organisms. Concerning temperature, it was stated in a review by Robarts & Zohary, that cyanobacterial dominance generally occurs during temperatures >20º C, but that temperature alone does not determine the occurrence of a species. Other factors include, among others, buoyancy, allelopathic compounds, grazing resistance, low total N : total P ratio, phosphorus and nitrogen storage capacity, effective competition for nitrogen via N_2-fixation and ability to grow at low light, as well as tolerance of high pH or low carbon dioxide concentration.

Buoyancy regulation allows the cyanobacteria to position themselves favourably within gradients of physical and chemical factors. Buoyancy regulation can be an important advantage in competition with other phytoplankton organisms, especially during non-turbulent conditions. Buoyancy of the two most common Baltic cyanobacteria, *Nodularia spumigena* and *Aphanizomenon flos-aquae*, is regulated by gas vesicles and changes in cellular concentrations of carbohydrates. The gas vesicles of these species survive mixing down to 60 m depth. *N. spumigena* seems to be more strongly buoyant than A. flos-aquae. Some species can migrate several metres per hour, and therefore, blooms may be formed in a short time. Furthermore, light and nitrogen are important factors controlling the buoyancy regulation.

Cyanobacteria produce a variety of secondary metabolites. Secondary metabolites are not used in the primary metabolism of an organism and include compounds that can operate as allelochemicals, antibiotics, hormones or toxins. Allelochemicals are substances produced by one organism that are toxic or inhibitory to the growth of another. Allelochemicals can suppress or stimulate the growth of other phytoplankton and consequently play an important role in the competition between species in the pelagic ecosystem. Furthermore, Kurmayer & Jüttner have suggested that chemical defences are more important than morphological characteristics in deterring grazers. Secondary metabolites of cyanobacteria have been studied extensively, and different allelopathic characteristics, such as antialgal, antibacterial, anticyanobacterial, and antifungal properties have been identified. Both dominant genera in the Baltic Sea, Aphanizomenon and Nodularia, have revealed antibacterial properties. In addition, *Nodularia* sp. showed antifungal activity and allelopathic effects towards other cyanobacteria and green algae.

Due to their ability to fix nitrogen, cyanobacteria are known to be favoured over other phytoplankton in nitrogen-poor waters. N_2-fixing genera were shown to be highly dependent on phosphorus input, and consequently on the N : P ratio of the water. N_2-fixation is directly related to phosphorus loading, especially in eutrophic aquatic ecosystems, but it may also be stimulated by iron. Furthermore, Lehtimäki et al. have shown that nitrogen fixation rates were highest during the exponential growth phase, and decreased in the stationary growth phase of the cyanobacteria. However, although N_2-fixation indicates an advantage to cyanobacteria, especially in nitrogen-poor waters, other sources of nitrogen are considered more important.

Cyanobacteria are superior competitors under low light conditions. Many species are also sensitive to long time periods of high light intensities and it has been shown that cyanobacterial blooms may be photo-inhibited. Species that form surface blooms, however, seem to possess a greater tolerance for light. Various mechanisms, such as elevated carotenoid concentration, can offer photo-protection for the cells in surface blooms. In the Baltic Sea, Aphanizomenon has been considered more susceptible to light than Nodularia, due to its avoidance of the surface.

Several of the hypothesised mechanisms probably contribute to the success of cyanobacteria simultaneously. This may explain why single relationships have failed to explain their dominance in various aquatic systems. Why do cyanobacteria produce toxins? Several hypotheses have been put forward to explain algal toxin production. Grazer repellence is probably the most studied hypothesis. Other hypotheses suggest that toxins are responsible for allelopathic activity, are functioning as a nitrogen source, or that toxins are the result of different metabolic processes.

According to Pajdak-Stós et al., cyanobacteria are highly resistant to grazing by protozoans, copepods and cladocerans, due to millions of years of co-evolution with these herbivores. It has, however, been pointed out that the occurrence of co-evolution is hard to assess in the marine environment. Notably, very few grazers are specialised on cyanobacteria, with the exception of several ciliate strains, e.g. *Nassula* sp. and *Pseudomicrothorax* sp., due to the grazing decreasing characteristics of cyanobacteria, other than toxins. Furthermore, to the best of my knowledge, it is not known when toxin production of cyanobacteria evolved. Steneck argues that the fossil record does not support the hypothesis that herbivory would generate strong selective pressures on the algal food species, although grazing is usually intense and universal in the marine environment. To sum up, the ecological purpose of the secondary metabolites produced by planktonic organisms, e.g. phytoplankton, is not known. Although a certain compound may affect copepod grazing adversely, it does not necessarily imply that it has evolved as a feeding deterrent.

CYANOBACTERIA—HARMFUL TO ZOOPLANKTON AND FISH

Cyanobacteria have commonly been considered as low quality food for zooplankton, due to their morphology, low nutritional value and toxin content. Many cyanobacteria have a filamentous or colonial morphology, and form aggregates, which could reduce feeding rates, or clog the feeding appendages of suspension feeding zooplankton. Unpalatability of a food species, i.e. bad taste or bad odour, is also considered to be a characteristic of poor food quality.

The low nutritional content of cyanobacteria, demonstrated as a low reproductive response, may be due to slow assimilation or lack of essential compounds, such as highly unsaturated fatty acids (HUFA). Saturated fatty acids (SAFA) are important due to their high calorific content and are mainly utilised for energy, whereas polyunsaturated fatty acids (PUFA) affect the production of eicosanoids, which are crucial for numerous physiological functions related to reproduction in invertebrates, e.g. egg production and egg hatching. Previously, the egg production and egg hatching success of copepods, provided with both toxic and non-toxic cyanobacterial monocultures, have been shown to be low. Deformed egg sacs have also been reported. Finally, zooplankton feeding may be inhibited, or zooplankton mortality increased, due to algal toxins or other secondary metabolites.

The hepatotoxins of cyanobacteria are considered extremely harmful to vertebrates. In the marine environment, cyanobacteria may affect fish physiologically or cyanobacterial secondary metabolites may deter juvenile fish from feeding. On the other hand, fish may learn to avoid harmful metabolites, though hunger limits this avoidance. Grazing studies with herbivorous fish have shown that feeding decreases when the percentage of toxic cyanobacteria cells increases. Cyanobacterial toxins have caused fish kills, been detected in fish livers and have been found to accumulate in fish tissues. Contrary to the accumulation hypothesis, Sahin et al. have shown that cyanobacterial toxins are excreted into the bile of the fish, rather than accumulating in the tissues. Experimental trials have also shown that some fish may be able to detoxicate cyanobacterial toxins.

Hansson suggested that herring larvae tend to avoid water containing dense Nodularia spumigena blooms in the Baltic Sea. Therefore the larvae are forced to stay in colder and deeper water, possibly reducing their rate of growth and increasing mortality. The mechanisms by which toxic cyanobacterial mass-occurrences affect fish behaviour and recruitment are still largely unknown.

POSITIVE EFFECTS ON PELAGIC COMMUNITY

Despite the above stated negative effects, evidence has recently started to accumulate that cyanobacteria may also have neutral or even positive effects on grazers. Meyer-Harms et al. studied grazing on cyanobacteria in the Baltic Sea and showed that copepods fed more actively on cyanobacteria when the blooms were dense and also during later growth stages. Rolff recognised strong signals of cyanobacteria in samples of stable isotopes dominated by rotifers and cladocerans in the Baltic proper. O'Neil & Roman suggested that pelagic harpacticoid copepods may be able to affect the distribution and species composition of Trichodesmium sp. by grazing.

A lack of evident effects on zooplankton, feeding on toxic algae, may be more general than potential adverse effects. Zooplankton can avoid harmful algae by selective feeding and vertical migration. Laboratory trials show that copepods may benefit from cyanobacteria when provided with appropriate mixtures of cyanobacteria and good quality food. Furthermore, the quality of cyanobacteria as food for zooplankton may change or even improve when the bloom is starting to age and decay. This may take place because the toxin concentration of blooms decreases during the senescent phase, and because a decaying Nodularia spumigena bloom is known to attract a diverse community of bacteria, flagellates, microzooplankton and crustaceans. Finally, Repka et al. have shown that detritus derived from cyanobacteria is good quality food.

GRAZING ON CYANOBACTERIA

Engström-Öst, J. et.al conducted a 2-week enclosure study in which they monitored a decaying cyanobacteria bloom for different parameters, organism abundances, toxins, fatty acids, protein and pigments. In the beginning of the study, when Nodularia was still in the growth phase, Acartia bifilosa fed mainly on ciliates and cyanobacteria (IV). Also, during the middle and at the end of the experiment, when N. spumigena was in its decay stage, A. bifilosa mainly selected ciliates. The fact that A. bifilosa fed on cyanobacteria is in contrast to their previous results, where A. bifilosa did not feed on any of the Nodularia strains, which were provided as sole food (I). The results suggest that grazing rates seem to differ considerably, depending on whether a food source is provided alone, or in a mixture with other food. Schmidt & Jónasdóttir made similar observations, where A. tonsa did not feed on Nodularia sp. when provided as sole food. In contrast, Meyer-Harms et al. found moderate grazing in situ by the copepod Acartia sp., on cyanobacteria, during the late phases of the bloom.

Grazing and foraging behaviour of two calanoid copepods, Acartia bifilosa and Eurytemora affinis, and the common mysid shrimp, Mysis mixta, in the presence of cyanobacteria were studied (I, II). Engström-Öst, J. et.al used cultures of different cyanobacteria strains in the experiments. Both juvenile and adult mysid shrimps, as well as the copepod *E. affinis*, fed more on the non-toxic Nodularia sphaerocarpa or Aphanizomenon flosaquae, than on the toxic N. spumigena, whereas grazing by *A. bifilosa* was non-detectable.

A herbivore can distinguish a toxic cell, either by recognising the toxin prior to ingestion of the cell, which would indicate the presence of an extracellular toxin in the water, or by learning, which would indicate the prior ingestion of a toxic cell, and subsequent avoidance due to its unpleasant taste or odour. According to DeMott et al., feeding inhibition is "either an adaptive behaviour to avoid eating toxic cells or a direct consequence of a weakened condition due to poisoning".

Eurytemora affinis is a suspension feeder and considered less selective than, e.g. species of the genus Acartia, which could be the reason for its feeding on the toxic strain. Cyanobacterial toxins are known to remain inside the cells during the exponential growth stage of the cyanobacteria, i.e. only 10-20% of the toxin in a log phase culture is extracellular. Extracellular toxin release is affected by several factors: temperature, light, salinity, growth stage and phosphorus concentration. Consequently, the experimental conditions should have promoted low nodularin release from the cyanobacteria, because Engström-Öst, J. et.al used growing cultures (I, II).

In the grazing experiments with mysid shrimps, adults fed less on the toxic strain of Nodularia spumigena than on the nontoxic strains of *N. sphaerocarpa* and *Aphanizomenon flos-aquae* (II). The juveniles also fed less on the toxic strain than on the non-toxic *A. flos-aquae*. Mysid shrimps are known to be omnivorous , and in the study they fed on cyanobacteria, probably because nothing else was available. In the field, the phytoplankton fraction of the stomach contents of Mysis mixta was dominated in September by dinoflagellates and also cyanobacteria to some degree. The results of the study suggest that, in nature, mysid shrimps ingest cyanobacteria, probably by suspension feeding, and in mixtures with other food.

Interference with feeding by cyanobacteria The mechanical interference of cyanobacterial filaments with feeding has been studied mostly in freshwater lakes dominated by Daphnia and other filter-feeding cladocerans. Engström-Öst, J. et.al studied the possible mechanical interference of filaments with copepods and mysid shrimps feeding on high quality food. One experiment was conducted with the calanoid copepod, Acartia bifilosa, grazing on the flagellate, *Brachiomonas submarina* (I), whereas in the other experiment, they provided mysid shrimps with *A. bifilosa* (II).

In the study with *A. bifilosa* as a grazer (I), no interference could be detected, i.e. no reduction in clearance rates was observed with increasing concentrations of cyanobacteria. The reason for this may be, however, that the concentration of high quality food was too high in this experiment. Consequently, the feeding may have been satiated and the potential interference effect of the cyanobacteria would have remained undetected. Another interpretation of the result is simply that cyanobacterial filaments did not interfere mechanically with the feeding of *A. bifilosa*. This is quite possible, because *Acartia* spp. are capable of feeding selectively, by suspension feeding on phytoplankton, as well as raptorially, on motile prey.

Effects of Cyanobacteria on Survival

Resistance to cyanobacterial toxins in zooplankton has been demonstrated a few times, mainly in freshwater systems or estuaries. Calanoid copepods from the Baltic Sea, Acartia bifilosa and Eurytemora affinis, seem to be very tolerant to cyanobacterial toxins as well. These observations were partially confirmed in the study. Survival of A. bifilosa was high during the middle and at the end of the mesocosm experiment but was somewhat lower at the beginning of the experiment (IV). This pattern cannot be due to nodularin-related toxicity, since nodularin concentrations remained high in the enclosures during the whole experiment.

Tolerance to nodularin may include advantages, e.g. the ability to derive nutrition from the filaments, or from the epifauna or flora associated with the filaments, during mass-occurrences of cyanobacteria. In the study, Daphnia-fed mysid shrimps showed high survival when exposed to toxic cyanobacteria (II). The mysid shrimps either avoided the filaments or fed on them together with Daphnia, which, in the latter case, suggests tolerance to nodularin. In any case, cyanobacteria probably have little short and medium-term (days to weeks) effects on mysids, whereas the true long-term (seasonal) effects on, e.g. reproduction could not be assessed in the study.

Effects of Cyanobacteria on Heterotrophic Community

Several hypotheses have been put forward concerning the fate of decaying cyanobacterial blooms. Grazing and grazing control have been considered of minor importance, due to the characteristics of cyanobacteria discussed above. Cyanobacterial blooms are commonly thought to decay within the water column. The cells are buoyant and remain on the surface as long as the gas vesicles remain intact. Sedimentation of both Aphanizomenon flos-aquae and Nodularia spumigena was concluded to be insignificant in the Baltic Sea and the Gulf of Finland, due to buoyancy and high mineralisation of senescent populations, whereas sedimentation is considered important by Kankaanpää et al.

In a study, Nodularia had proceeded well into the decay process by the middle of the 2-week experiment. Decay was indicated by the increasing particulate organic carbon (POC):chlorophyll a ratio, the decreasing polyunsaturated fatty acid (PUFA):total fatty acid ratio and decreasing filament length (III). These parameters give an estimate of growth conditions, presence of healthy cells and approximate growth rates. A further indication of decay was that total chlorophyll a and the two cyanobacterial pigments, echinenone and zeaxanthin, decreased towards the end of the experiment (III).

Hoppe reported high species diversity of bacteria and microzooplankton within decaying Nodularia aggregates, although he did not detect any feeding on the filaments themselves. Bacteria may also be strongly associated with actively growing blooms. Some bacteria are chemotactically attracted to the nitrogen-rich heterocysts. Bacterial epiphytisation, i.e. attachment, is associated with metabolically active, rather that senescent N_2-fixing species.

In the study, the number of filamentous bacteria (rods) increased towards the end of the mesocosm experiment (III). The observed size structure of the bacterial community can be a response to grazing pressure by bacterivores, since longer cells are probably more resistant to grazing than short ones. Also, the increasing mesozooplankton community probably predated strongly on the ciliates and thus, indirectly diminished the predation pressure on the bacterivores. A similar increase in the number of filamentous bacteria, due to strong grazing pressure, has previously been shown in a reservoir system and in freshwater enclosures. Furthermore, Christoffersen et al. showed that bacterial populations peaked shortly after the collapse of a cyanobacterial bloom, indicating that the bacteria were stimulated by lysis products released from the cyanobacteria. In contrast, the abundance of heterotrophic nanoflagellates decreased during a strong toxic Microcystis bloom.

Few studies have been performed to investigate the interactions between zooplankton and cyanobacteria in mesocosms. The disadvantages of enclosures increase with decreasing bag size. Different effects including 'wall effects' may result in low water circulation and unnatural mixing in the enclosures. In addition, species that thrive on substrates and surfaces may predominate. However, in situ enclosures can be considered more realistic than any 'bottle experiments' performed in the laboratory and are thus appropriate for studying community scale processes. In the study, the ciliate Euplotes sp. increased with time in the enclosures containing cyanobacteria (III). Although the 'wall effect' probably affected the abundance of the thigmotactic Euplotes sp. positively, it seemingly thrived better among the cyanobacterial filaments than in the control.

Engström-Öst, J. et al. were not able to detect any direct harmful effects attributable to the stable and high concentrations of nodularin on any of the studied organism groups in the shortterm experiment (III). The results show that the plankton community can exist in the presence of nodularin.

Food Quality Effects

In their study, Engström-Öst, J. et al. monitored the fatty acid composition of a decaying cyanobacterial bloom in enclosures (III). The main food quality indicators, fatty acids 20:5ω3 and 22:6ω3, were not associated with Nodularia spumigena in the mesocosm experiment (III). This shows that N. spumigena is deficient in highly unsaturated fatty acids (HUFA), which are important for grazers. Nevertheless, in the study, another fatty acid, 18:3ω3, was strongly associated with N. spumigena (III). 18:3ω3 is one of the main fatty acids in Nodularia sp. and other cyanobacteria, in addition to 16:0 and 18:2ω6. All herbivores, and probably also all omnivores, seem to be able to elongate 18:3ω3 to 20:5ω3 and 22:6ω3. Therefore, cyanobacteria may provide some elements that are useful or even essential for grazers. However, synthesis of HUFA is costly, thus animals grow best when provided with direct sources of 20:5ω3 and 22:6ω3.

Ageing cyanobacterial blooms attract numerous ciliates and other microzooplankton, due to bacteria attached to the filaments. They may thus serve as a significant food source for crustaceans, which are able to feed selectively. Ederington et al., however, showed that an algal diet supported higher egg production than a ciliate diet did in copepods, due to the sterols and PUFAs in the algae. On the other hand, bacterial fatty acids can be transferred from ciliates to copepods and their eggs. It has also been shown that protozoans may 'upgrade' low quality food by producing long-chain polyunsaturated fatty acids (PUFAs). The growth of copepods in turn improves when upgraded food is consumed.

In their study, Engström-Öst, J. et al. measured egg production of an abundant calanoid copepod, Acartia bifilosa (IV), during different phases of a cyanobacterial bloom. In contrast to previous findings, the results showed that both copepod species were able to feed, and A. bifilosa even produced eggs, during all phases of a Nodularia bloom (IV), though the results did not give full support to the hypothesis of improved food quality of the decaying cyanobacterial bloom. However, the cyanobacteria clearly did not have any harmful effects on copepod egg production. One reason for the relatively high egg production measured in the mesocosm experiment could be the availability of a diverse food environment. Generalist herbivores perform best when provided with a mixture of several plant species, or when their algal food is occasionally supplemented with protein-rich animal tissue. In conclusion, the results suggest that a cyanobacterial bloom, and its associated organisms, is a diverse and highly useful food source for the dominant copepods of the northern Baltic Sea, Acartia bifilosa and Eurytemora affinis.

Accumulation of Toxin in Higher Trophic Levels

An increase of cyanobacterial toxin concentration with time in the study organism, in comparison to the environment, i.e. by accumulation, has been recorded in different bivalves and crustacean zooplankton. Crustacean zooplankton may be an important vector for algal toxins from phytoplankton to planktivorous fish.

In their study, Engström-Öst, J. et.al conducted experiments with mysid shrimps and threespined sticklebacks, Gasterosteus aculeatus, in order to detect transfer and potential accumulation of nodularin, produced by Nodularia spumigena, to the planktivores, via cyanobacteria-fed copepods (V). Toxin samples were measured by two methods, in order to achieve more reliable results. The enzyme-linked immunosorbent assay, ELISA, measures the concentration of liver toxin in the sample, whereas protein phosphatase (PPase) inhibition assay measures inhibition of protein phosphatases by liver toxins, reflecting their toxicity. In the study, Engström-Öst, J. et al.detected accumulation in one trial, in the experiment using mysid shrimps and measured using ELISA. It is difficult to speculate whether a real accumulation had taken place or not, because the increase was detected only by ELISA and not by the other method, protein phosphatase (PPase) inhibition assay. However, it is difficult to imagine any other mechanism but true accumulation to explain the ELISA result.

The comparison between the two methods showed that the PPase inhibition assay gave higher values than ELISA (V). There may be several reasons for this. The methods can be classified according to their sensitivity and selectivity. In Engström-Öst, J. et al's experiments, ELISA had a lower detection limit, i.e. was more sensitive in detecting toxin, whereas the PPase inhibition assay gave somewhat higher values. Consequently, the PPase inhibition assay was probably less selective than the ELISA, as was also concluded by Harada *et al.* One fact that makes interpretation difficult is that the controls in the experiment were also positive. The reason for this is either that the predators had retained toxins in their tissues from the field, or that there was a bias in the measurements. The percentage of methanol in the toxin samples is a crucial factor, because false positives may arise if the concentration of methanol is too high. The final concentration of methanol in the samples was 10%, which should not be too high. In conclusion, Engström-Öst, J. et.al. were able to detect cyanobacterial toxin in the tissues of both of the planktivores used in the experiments and the results thus suggest that accumulation in higher trophic levels is possible.

REFERENCES

Kukkonen, J. 1999: *Comparison of cyanobacterial detection methods. – M. Sc. thesis, Department of Applied Chemistry and Microbiology,* University of Helsinki. 64 pp.

Lampert, W. & Taylor, B. E. 1985: "Zooplankton grazing in a eutrophic lake: implications of diel vertical migration". *Ecology,* 66: 68-82.

Maneiro, I., Frangópulos, M., Guisande, C., Fernández, M., Reguera, B. & Riveiro, I. 2000: "Zooplankton as a potential vector of diarrethic shellfish poisoning toxins through the food web". – *Mar. Ecol. Prog. Ser.* 201: 155-163.

Niemistö, L., Rinne, I., Melvasalo, T. & Niemi, Å. 1989: "Blue-green algae and their nitrogen fixation in the Baltic Sea in 1980, 1982 and 1984". – *Meri* 17: 1-59.

Paerl, H. W. 1990: "Physiological ecology and regulation of N2 fixation in natural waters". – *Adv. Microb. Ecol.* 11: 305-344.

7

Structural and Functional Features of Cyanobacteria

Cyanobacteria are a group of eubacteria capable of oxygen-evolving photosynthesis. This group of ancient autotrophs, more mundanely known as blue-green algae, have an oxygen-evolving photosynthetic apparatus similar to plants and they are considered to be the ancestors of plant chloroplasts. In as early as 1905 a Russian scientist, Konstantin Mereschowsky, suggested that chloroplasts might originally have been free cyanobacteria living symbiotically inside a eukaryotic cell. His idea was, however, ignored for decades until the development of new techniques allowed the comparison of chloroplasts and cyanobacterial cells on ultra-structural, molecular and genetic level. According to current estimates, the primary endosymbiosis event involving the engulfment of a cyanobacterium by a primitive eukaryotic cell occurred circa 1.5 billion years ago.

The origin of cyanobacteria has been traced back to the Archaean period. Without a protective ozone layer UV radiation razed all surfaces, making the environment hostile against the emergence of life forms. Despite the unwelcoming conditions, fossil records in sedimentary rocks provide evidence for the existence of photosynthetic microbial mats and colonies from 2.5 to 3.5 billion years ago. The cyanobacterial clade has been very successful over its existence. They are both abundant and widespread. Morphologically diverse species occupying a vast number of different niches are found from aquatic to terrestrial habitats. Some cyanobacteria are unicellular while others form filaments of adjacently attached cells. Furthermore, cells with specialized purposes, such as heterocysts, hormogonia or akinetes, can differentiate from vegetative cells. In addition cyanobacterial metabolism is flexible, some species being capable not only of photoautotrophy but also of photoheterotrophyand even heterotrophy in a light-activated manner. Cyanobacteria are considered to be among the smallest organisms to exhibit endogenous circadian rhythms in their physiology and gene expression. The circadian clock of *Synechococcus* sp. PCC 7942, is set to a roughly 24 h period by the products of the *kaiABC* gene cluster. While cyanobacteria are the only group with all three genes, homologues of *kaiA* and *kaiB* have been found in other bacteria as well.

Being the major primary producers in the oceans cyanobacteria are responsible for nearly one half of the net primary production on Earth. Some cyanobacteria, such as *Anabaena* sp. PCC 7120 and *Nostoc punctiforme,* form a crucial link in the nitrogen cycle because they are capable of fixing atmospheric nitrogen. Nitrogen-fixing cyanobacteria are used to fertilize fields in the cultivation of rice and thus have a role in global food production. Cyanobacteria are often environmentally hazardous. Serious health risks for humans and animals arise when toxic cyanobacteria are present in water used for consumption, agriculture or recreational purposes. Human activities promote harmful cyanobacterial growth because we contribute a great deal to the eutrophication of water bodies.

STRUCTURAL AND GENETIC FEATURES

Cyanobacteria are generally categorized among Gram-negative prokaryotes. This assignment is based on similarities in cell wall composition and structure. The cell is surrounded by an outer membrane consisting of lipopolysaccharides, lipids and transport proteins, a peptidoglycan layer and a selectively permeable plasma membrane. Peptidoglycan is a complex polymer consisting of cross-linked chains of carbohydrates and amino acids. Its role is to give the cell mechanical support. A surface layer of polysaccharides, the glycocalyx, serves to give additional protection from desiccation. Thylakoid membranes, located in the cytoplasm, are the sites for photosynthetic electron transfer reactions. Cyanobacterial thylakoids are loosely organized into networks and layers of stacked sheets instead of the grana structures present in plant chloroplasts. According to van de Meene et al. they are continuous with the plasma membrane in the model species *Synechocystis* sp. PCC 6803 although Liberton et al. and Schneider et al. claim the contrary. Other structural features of a cyanobacterial cell include hexagonal carboxysomes, pili, gas vesicles, phycobilisome antennae, ribosomes and a variety of storage granules consisting of glycogen, cyanophycin, lipids or polyphosphate.

Prokaryote genomes are small and compact. This applies to cyanobacteria as well, in which known genome sizes vary between 1.6 Mbp in *Prochlorococcus marinus* MIT 9301 and 9.01 Mbp in *Nostoc punctiforme* PCC 73102. A multitude of detailed genetic information on cyanobacteria is already available. As of August 2008 (August 18th) 34 cyanobacterial genomes had been completely sequenced and many sequencing projects are currently in progress. Cyanobacteria possess a single circular chromosome and may have one or more plasmids. The plasmids contain functional genes although the majority have unknown or hypothetical functions. In cyanobacteria the size of the genome reflects the number of genes. Gene density is high compared to eukaryotic genomes averaging at about 1 gene per 1100 bp in *Synechocystis* sp. PCC 6803. As is typical in bacteria, some genes in cyanobacterial genomes are arranged into operons. Even though the cyanobacterial clade is considered to be of monophyletic origin, the degree of similarity in gene organization is negligible.

Synechocystis sp. PCC 6803 – a Model Cyanobacterium

Synechocystis sp. PCC 6803 (hereafter *Synechocystis*) is a unicellular, non-toxic fresh-water cyanobacterium that is not capable of nitrogen fixation. This popular model organism was the first photosynthetic organism whose genome was completely sequenced and made publicly available. Work by Kaneko et al. revealed that the total size of the *Synechocystis* genome is circa 3,6 Mbp and that there are over 3000 open reading frames. The *Synechocystis* strains used in research descend from the one stored in the Pasteur Culture Collection (PCC). The primary strain was isolated from a Californian freshwater lake in 1968. Many research laboratories, however, use a glucose-tolerant strain because of its capacity to grow photoheterotrophically in the presence of glucose. An important characteristic of *Synechocystis* is its natural competence for transformation with exogenous DNA. Other cyanobacterial model species include *Synechococcus elongatus* sp. PCC 7942, the filamentous and nitrogen fixing *Anabaena* sp. PCC 7120, the thermophilic *Thermosynechococcus elongatus* and the marine Prochlorococcus marinus MED4.

PHOTOSYNTHESIS IN CYANOBACTERIA

Cyanobacteria are thought to be among the evolutionarily oldest organisms: putative microfossils have been found that are 3.5 billion years old and that are attributed to cyanobacteria. A main reason for the evolutionary hardiness of cyanobacteria is their successful combination of effective metabolic pathways. They are among the very few groups that can perform oxygenic photosynthesis and respiration simultaneously in the same compartment, and many cyanobacterial species are able to .x nitrogen. Therefore, they can survive and prosper under a wide range of environmental conditions.

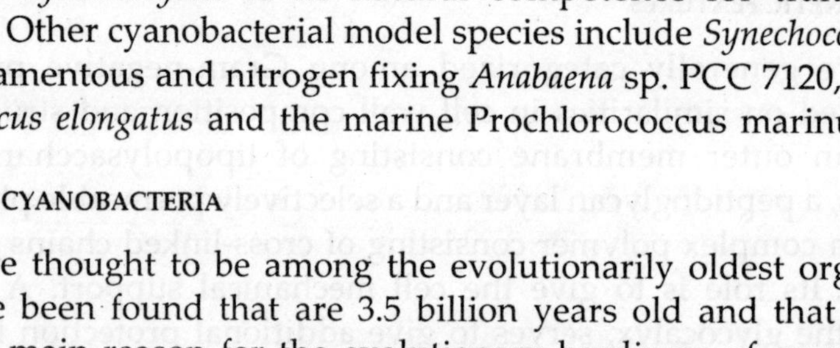

Figure 1: Outline of membranes and compartments in a cyanobacterial cell.

The combination of photosynthesis and respiration in a single compartment is thought to be quite unique, and the ramifications of this combination are reviewed.

Photosynthesis and respiration require electron transport pathways that to a large extent are catalysed by protein complexes in membranes. Figure 1 illustrates the compartmentalisation of the cyanobacterial cell. The thylakoid membrane, the internal membrane system that separates the cytoplasm from the lumen and that is present in virtually all cyanobacteria, contains both photosynthetic and respiratory electron transport chains. These electron transport chains intersect, and in part utilise the same components in the membrane. Note that oxygenic photosynthesis essentially is the reverse of respiration. The cytoplasmic membrane, separating the cytoplasm from the periplasm, contains a respiratory electron transport chain but not photosynthetic complexes in most cyanobacteria. Therefore, in most cyanobacteria, photosynthetic electron transport occurs solely in thylakoids, whereas respiratory electron flow takes place in both the thylakoid and cytoplasmic membrane systems. Overviews regarding aspects of cyanobacterial biochemistry and molecular biology related to photosynthesis and respiration are provided in Gantt and Schmetterer.

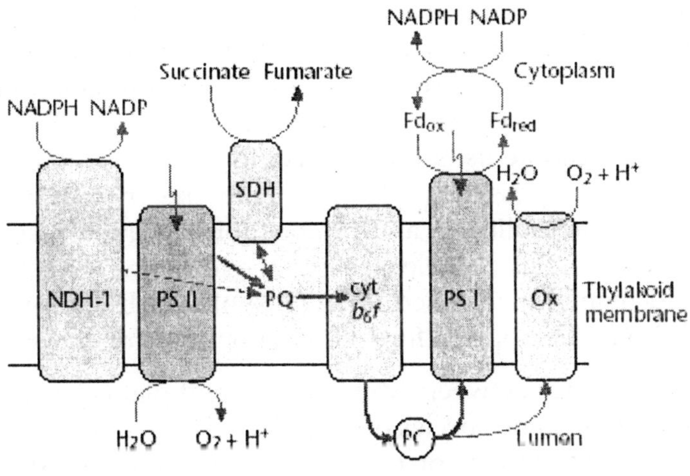

Figure 2. Schematic representation of the intersecting photosynthetic and respiratory electron transport pathways in thylakoid membranes of the cyanobacterium Synechocystis sp. PCC 6803.

The presence and simultaneous activity of photosynthetic and respiratory electron transport chains in the same membrane system is unusual. A schematic representation of the respiratory and photosynthetic electron transport chains in cyanobacterial thylakoid membranes is shown in Figure 2. As indicated, cyanobacteria utilise several redoxactive components in thylakoids for both photosynthesis and respiration, including the plastoquinone (PQ) pool, the cytochrome b_{6f} complex, and the soluble electron carriers in the lumen. In any case, the common use of components for essentially opposite metabolic processes that occur at the same time does not appear to be a significant impediment in view of the stunning longevity with which cyanobacteria have been a major presence in life on earth.

Photosynthetic Electron Flow

The photosynthetic electron transport chain in cyanobacteria is essentially identical to that in plants, even though some of the polypeptides that are part of electron transporting enzymes appear to be of different evolutionary origin in the two systems. Indeed, chloroplasts in plants are thought to have originated from cyanobacterial ancestors. The photosynthetic electron transport chain of oxygenic organisms has been reviewed extensively, so only a brief summary will be provided here. As indicated in Figure 2,photosystem II (PS II) uses light energy to split water and to reduce the PQ pool. Table 1 summarises the protein components involved with PS II and with other complexes mentioned in Figure 2. Electrons are transported from the PQ pool to the cytochrome b_{6f} complex and from there to a soluble electron carrier on the luminal side of the thylakoid membrane. In cyanobacteria this soluble carrier may be plastocyanin or cytochrome c_{553}, depending on the species and on the availability of copper. Either of these soluble one-electron carriers can reduce the oxidised PS I reaction centre chlorophyll, $P700^+$. This oxidised form of the reaction centre chlorophyll is formed by a light-induced transfer of an electron from PS I to ferredoxin (Fd) and eventually to NADP. Reduced NADP can be used for CO_2 fixation. Photosynthetic electron transfer leads to a proton gradient across the thylakoid membrane. In PS II, protons are released into the lumen upon water splitting, and protons formed upon plastoquinol oxidation by the cytochrome b6f complex are released into the lumen as well. The proton gradient across the thylakoid membrane is used for ATP synthesis by the ATP synthase in the thylakoid; this ATP may be applied for CO_2 fixation and for other cell processes.

One important difference between cyanobacteria and plants is that the stoichiometry of PS I and PS II in at least some species of cyanobacteria is much larger than 1, whereas in higher plants an equal amount of PS I and PS II is the rule. For example, in the cyanobacterium Synechocystis sp. PCC 6803 the PS I/PS II ratio is about 5. One possible explanation for this unusual stoichiometry in cyanobacteria is the involvement of PS I in a significant amount of cyclic electron flow around this photosystem, in which electrons would flow from PS I/Fd back to PQ and cytochrome b_{6f}, and from there to PS I again. This electron transport pathway should contribute to a proton gradient, which could be used for ATP synthesis, but would not lead to net NADP reduction. However, a high rate of cyclic electron flow around PS I in cyanobacteria cannot be demonstrated experimentally. Another and probably more plausible reason for the relatively large amount of PS I in cyanobacteria is the abundance of respiratory electron transfer pathways into the PQ pool, whereas the capacity of respiratory electron flow out of the PQ pool may be more limited. Therefore, an abundance of PS I may guarantee a rather oxidised PQ pool in the light, which is important to minimise photodamage. Moreover, the high amount of PS I may serve to compete effectively with cytochrome oxidase for electrons when light is available, thus maximising the number of electrons that can be used for CO_2 fixation.

Respiratory Electron Flow

The respiratory pathways in cyanobacteria have long been unclear or misunderstood. However, considerable insight has now been obtained in this important aspect of cyanobacterial physiology, primarily by making use of mutants lacking specific proteins. Much of this work has been performed on Synechocystis sp. PCC 6803, which is very amenable to targeted genetic modification and which was the first cyanobacterium for which a complete genomic sequence has become available. It has now been realised that the main respiratory electron transport activity into the PQ pool in cyanobacterial thylakoids involves succinate dehydrogenase (SDH) activity rather than electron flow from NAD(P)H; NADPH appears to be used preferentially for carbon fixation processes. The corollary of this new concept is that cyanobacteria are able to generate succinate as a respiratory intermediate, and therefore should have a citric acid cycle. According to the older literature, a citric acid cycle was presumed to be nonfunctional in cyanobacteria as a key enzyme, 2-oxoglutarate dehydrogenase, was absent. Indeed, this is true according to the genomic sequence of Synechocystis sp. PCC 6803, but this organism is able to convert 2-oxoglutarate to succinate in the absence of a traditional 2-oxoglutarate dehydrogenase complex, making use of an alternate pathway.

NAD(P)H Oxidation

The cyanobacterial equivalent of the mitochondrial and bacterial type-1 NADPH dehydrogenase (NDH-1) is an enzyme complex similar to the 14-subunit NDH-1complex from *Escherichia coli*, except that three subunits involved with substrate binding are not apparent from the cyanobacterial genome. Probably related to this observation, the preferred substrate of cyanobacterial NDH-1 is NADPH rather than NADH. Upon inactivation of ndhB, the gene for one of the essential NDH-1 subunits, mutants required high CO_2 for growth and showed a decreased respiration rate. The latter was taken to support the tacit assumption that NDH-1 was the major respiratory electron transport route into the PQ pool in cyanobacteria as it is in many other bacteria and in mitochondria from most eukaryotes. However, this view was challenged when mutants lacking succinate dehydrogenase activity showed a very slow rate of respiratory electron flow into the PQ pool. This suggests that NDH-1 activity in cyanobacteria in *vivo* is only very modest, and that most respiratory electrons enter the PQ pool via SDH. In cyanobacteria, NADPH is not used as a preferential respiratory substrate but rather is applied for the CO_2 fixation process. However, a large number of isogenes for two of the NDH-1 subunits are present in the Synechocystis genome; these isogenes may have different functions.

This suggests that NDH-1 may function in different capacities depending on the physiological state of the cell.

According to the genomic sequence of Synechocystis sp. PCC 6803, this cyanobacterium is capable of producing type-2 NDH (NDH-2), which is a single-subunit protein and that may not contribute to a proton gradient over the thylakoid membrane. Three genes for NDH-2s are present in the Synechocystis sp. PCC 6803 genome, but so far no evidence has been found that would suggest a large activity of NDH-2 in providing electrons to the PQ pool. Instead, NDH-2 in cyanobacteria may have a mostly regulatory function.

Succinate Dehydrogenase

Genes for soluble SDH subunits are easily identified in the Synechocystis sp. PCC 6803 genome, and upon deletion of such genes a phenotype consistent with a lack of SDH activity is found. Whereas fumarate is depleted in mutants lacking SDH, succinate accumulates. Moreover, in such mutants succinate accumulates to a higher degree when 2-oxoglutarate is added, signifying the capability of cyanobacteria to convert 2-oxoglutarate to succinate even though a traditional 2-oxoglutarate dehydrogenase is absent according to the genome sequence. This implies that cyanobacteria can utilise a modified citric acid cycle and that succinate and SDH are potentially important in cyanobacterial respiratory metabolism.

In the Synechocystis mutant lacking SDH activity, the rate of initial respiratory electron flow into the PQ pool was about an order of magnitude lower than in wild type, suggesting that essentially all respiratory electrons enter the PQ pool through SDH activity rather than through oxidation of NAD(P)H. Therefore, the large effect of inactivation of ndhB on respiration may be an indirect effect due to the absence of succinate in this mutant: oxidised NAD(P) is needed for running the citric acid cycle and other respiratory processes. This highlights the complexity of metabolism even in rather simple organisms, and cautions against one-sided interpretations. However, the biochemical analysis of Synechocystis mutants specifically lacking particular subunits provides an opportunity to critically test previous and current hypotheses and interpretations.

Terminal Oxidases

Using molecular-genetic approaches and the analysis of specific mutants, the nature and relative activity of terminal oxidases in Synechocystis sp. PCC 6803 has been clarified. According to biochemical data preceding the availability of the genomic sequence of this organism, a cytochrome aa_3-type cytochrome-c oxidase was found to be present along with another terminal oxidase. According to the genome sequence, apart from a cytochrome-c oxidase there may be a cytochrome *bd*-type quinol oxidase as well as a second quinol oxidase that most resembles a cytochrome *bo*-type oxidase. The cytochrome *bd*-type quinol oxidase could be demonstrated on the basis of inhibitor studies as well as by mutant analysis, but the cytochrome *bo*-type oxidase does not appear to be expressed under the conditions used thus far. Interestingly, under the conditions used, the cytochrome *bd*-type quinol oxidase appears to be located predominantly in the cytoplasmic membrane, whereas significant cytochrome oxidase activity is seen in thylakoids. Indeed, in mutants lacking PS I, rapid electron flow from PS II to cytochrome oxidase can be demonstrated, indicating a close functional relationship between PS II and thylakoidlocalised cytochrome oxidase.

The presence of cytochrome oxidase in the cyanobacterial thylakoid membrane raises the question of what the electron donor may be. If cytochrome c_{553}, the luminal electron carrier that can replace plastocyanin, is present, then this is the likely electron donor. However, electron transfer to the oxidase can also occur in the absence of this cytochrome. This electron transfer appears to involve plastocyanin, but also seems to require another cytochrome, cytochrome cM. This cytochrome c may serve as an electron transport intermediate and might be associated with the cytochrome oxidase complex. Limited sequence similarity exists between cytochrome cM and the cytochrome c-binding part of the cytochrome caa_3-containing cytochrome-c oxidase from bacteria with such a complex, and cytochrome cM appears to be required for electron flow out of the PQ pool under conditions that PS I and cytochrome c_{553} are absent.

Components of Photosynthesis and Respiration

Plastoquinone pool

In Synechocystis sp. PCC 6803, apart from PQ the only quinone found in thylakoids is vitamin K_1 associated with PS I. Therefore, essentially all electron transport through unbound quinones appears to proceed via PQ in cyanobacterial thylakoids. Plastoquinone in the thylakoid membrane shuttles electrons from both photosynthetic and respiratory electron transport chains to the cytochrome b_{6f} complex. PQ is reduced to the quinol form by accepting two reducing equivalents, but, when bound to protein, the semiquinone form can be stabilised. Therefore, PQ can function as an adapter converting one-electron redox reactions to two-electron reactions and vice versa.

Cytochrome b_{6f} Complex

A key complex in both photosynthetic and respiratory electron transfer in cyanobacteria is the cytochrome b_{6f} complex. Efforts to fully inactivate this complex in cyanobacteria have been unsuccessful thus far, indicating the central role this complex plays in electron transport. Indeed, all electron transfer in photosynthesis and respiration, with the exception of respiratory electron flow to quinol oxidases, needs to proceed through the cytochrome b_{6f} complex. However, as quinol oxidase may not be prevalent in thylakoids under many growth conditions and may occur predominantly in cytoplasmic membranes, essentially all plastoquinol oxidation in thylakoids occurs via the cytochrome b_{6f} complex in wild type under most growth conditions.

The redox-active components of the cytochrome b_{6f} complex are associated with the cytochrome b_6 apoprotein, the cytochrome *f* apoprotein, and the Rieske *FeS* centre. A fourth major component, subunit IV, does not carry redox-active cofactors. In mitochondria and Gramnegative bacteria containing a bc_1 complex, cytochrome b_6 and subunit IV are fused. Another interesting feature when comparing cytochrome b_{6f} and cytochrome bc_1 complexes is that cytochrome *f* and cytochrome c_1 do not have a common ancestor, in spite of their homologous function. In line with the different origins of the two cytochromes, in cytochrome *f* the N-terminal amino group of the processed protein forms a covalent bond with the haem group, whereas this is not the case for cytochrome c_1.

Perhaps the cyanobacterial mutant generated thus far with the lowest amount of cytochrome b_{6f} activity is one where *ccsB*, a gene coding for a protein involved in maturation of cytochrome *f*, has been impaired. This mutant contains an N-terminally modified *CcsB* that stabilises preapocytochrome *f* but that is inefficient in maturation of the cytochrome. Consequently, the *ccsB* mutant accumulates preapocytochrome *f* and has about a 5-fold reduced amount of mature, functional cytochrome b_{6f}. This strain grows only microaerobically, and partial restoration of *CcsB* function also restores the ability to grow aerobically. The role of oxygen remains to be elucidated, but it may involve regulation of gene expression.

An interesting feature of the cyanobacterial cytochrome b_{6f} complex is that in the *Synechocystis* sp. PCC 6803 genome three genes are found that may code for Rieske FeS centres, whereas only a single set of genes is present for the other components of the complex. One possible reason for this unexpected complexity is that the different Rieske FeS centres differ in their interaction of the cytochrome b_{6f} complex with soluble carriers, thus providing a mechanism for regulation of electron flow to PS I versus to the terminal cytochrome oxidase.

Soluble Electron Carriers

Both plastocyanin and cytochrome c553 can function in both photosynthesis and respiratory electron transport. In mutants lacking PS I, the gene for either soluble redox carrier can be deleted with only a moderate effect on the rate of electron flow from PS II via the quinone pool, the cytochrome b_{6f} complex and the remaining soluble carrier to the cytochrome aa_3-type cytochrome oxidase. In mutants lacking either plastocyanin or cytochrome c553, photosynthetic electron transport rates remain normal, indicating that either of the two soluble carriers is sufficient. Interestingly, efforts to delete the genes for both plastocyanin and cytochrome c_{553} in the same strain have been unsuccessful, indicating that one of the two carriers is required to shuttle electrons to PS I and cytochrome oxidase.

Table 1. Approximate capacity and probable source of rate limitation of photosynthetic and respiratory electron transfer in Synechocystis sp. PCC 6803.

Complex	*Capacity (μmol electrons h^{-1} per mg chlorophyll)*	*Source of rate limitation*
Photosystem II	1000	Light intensity, availability of oxidised PQ
Succinate dehydrogenase	200	Succinate concentration, PQ pool redox state
NADPH dehydrogenase	20	NADPH availability, PQ pool redox state
Cytochrome b_{6f} complex	1000	PQH_2 concentration
Photosystem I	3000	Light intensity, Fd/NADP availability, reduced cyt c_{553}/PC
Cytochrome-c oxidase	200	Reduced cyt c_{553}/PC availability

Regulation of Photosynthesis and Respiration

A central question that will now need to be addressed is how photosynthesis and respiration are regulated in a cyanobacterium. For regulation, an important issue is the capacity and rate of the various steps. The approximate capacities of electron transfer steps in *Synechocystis* sp. PCC 6803. The actual rate of the reactions may be significantly lower in vivo, depending on the conditions. It is clear that PS I activity is abundant relative to that of the cytochrome b6f complex, and this may explain in part the slow $P700^{+}$ rereduction kinetics that are generally observed after a period of illumination. If light is abundant, the photosynthetic electron transport chain has a much higher capacity of electron flow than has the respiratory chain, but at very low light intensity or in darkness respiratory rates are higher than those of photosynthesis. As the capacities of cytochrome oxidase and succinate dehydrogenase seem comparable, respiratory electron flow will not lead to major changes in the redox state of the PQ pool, leaving room for regulation of metabolic processes that are apparently mediated via the PQ redox state.

Parallels to the Situation in Chloroplasts

An important question is that of the relevance of the insights obtained thus far on Synechocystis for electron transport events in other organisms, such as plants. In photosynthetic eukaryotes, photosynthetic and respiratory electron flows have been separated as they occur in two different organelles. However, chloroplasts retain certain cyanobacterial features involved in respiratory electron flow. For example, genes for the NDH-1 complex are retained in the chloroplast genome, and a small amount of NDH-

1 activity can be demonstrated. However, there is little evidence for significant activity of SDH or of a terminal oxidase in chloroplasts under normal conditions. None the less, the case for respiratory activity in chloroplasts ('chlororespiration') is slowly building, even though it may not involve the same components that are prevalent in cyanobacteria. However, it is likely that comparisons with the cyanobacterial system will further guide the search for components involved in such activities in chloroplasts.

NITROGEN CONTROL IN CYANOBACTERIA

The oxygenic photosynthetic bacteria, the cyanobacteria, are autotrophs that fix CO_2 through the reductive pentose phosphate cycle. They make an important contribution to the Earth's nitrogen cycle by incorporating nitrogen into the biosphere through assimilatory processes. Organisms of this group frequently have the capacity to assimilate as sources of nitrogen a number of different simple N-containing compounds including ammonium, nitrate, nitrite and urea. Many cyanobacteria can also assimilate atmospheric nitrogen, and still some strains have the capability to assimilate some amino acids, particularly arginine and glutamine. The assimilation of most of these compounds provides intracellular ammonium, which itself is a preferred nitrogen source. Thus, in the presence of ammonium, the genes encoding permeases and enzymes for the assimilation of nitrogen sources alternative to ammonium are repressed, a process known as 'nitrogen control'.

Mechanisms of Assimilation of Combined Nitrogen

The incorporation into the cell of nitrogen-containing compounds, which are frequently found at low concentrations (e.g. below 1 μM) in the environment, takes place through permeases that are located in the cytoplasmic membrane. Multicomponent ABC (ATP-binding cassette)-type uptake transporters have been shown to be involved in the uptake of nitrate and nitrite or urea in a number of cyanobacteria. ABC-type permeases are also required for the transport of arginine and glutamine. These permeases use ATP to drive an active, concentrative transport of their substrates. On the other hand, a secondary transporter of the major facilitator superfamily has been identified as the nitrate–nitrite transporter in some marine cyanobacteria. The transport of ammonium is also mediated by secondary permeases, in this case of the Amt family. The Amt permeases can be probed with [^{14}C]methylammonium, which has been shown to be concentrated in cells of the unicellular cyanobacterium *Synechocystis* sp. strain PCC6803 to a level that suggests a membrane potential-driven transport.

Intracellular nitrate is sequentially reduced to nitrite and ammonium by nitrate reductase and nitrite reductase, which are the products of the *narB* and *nir* genes respectively. Cyanobacterial nitrite reductase is homologous with ferredoxin-dependent higher-plant nitrite reductase and contains a [4Fe-4S] cluster and sirohaem as prosthetic groups. Electrons from reduced ferredoxin are transferred to the iron–sulphur cluster and then to sirohaem, where nitrite is reduced to ammonium. Cyanobacterial nitrate reductase is homologous with Mo-containing bacterial oxidoreductases but is unique in that it uses ferredoxin as an electron donor, forming tight 1:1 complexes. The Mo cofactor is of the Mo-*bis*-molybdopterin guanine dinucleotide type, and the enzyme also contains a [4Fe-4S] cluster.

In this enzyme system, electrons flow from reduced ferredoxin to the iron–sulphur cluster and then to the Mo cofactor, where nitrate is reduced to nitrite. The *narB* and *nir* genes are clustered together with the nitrate/nitrite permease-encoding genes in numerous cyanobacteria forming an operonwith the structure *nir*-permease genes-*narB*. The high conservation of this gene arrangement, in which the expression level is higher for the upstream than for the downstream genes in the operon, suggests that it ensures the production of a balanced amount of the different proteins of the pathway. Regarding organic sources of nitrogen used by cyanobacteria, urea is degraded to ammonium and CO_2 by a standard bacterial Ni_2^+-dependent urease, whereas arginine is catabolised by an unusual pathway that combines the urea cycle and the arginase pathway rendering ammonium and glutamate as final products. Whatever the nitrogen source used for growth, intracellular ammonium is incorporated into carbon skeletons through the glutamine synthetase–glutamate synthase pathway.

In cyanobacteria, which lack 2-oxoglutarate dehydrogenase, the main metabolic role of 2-oxoglutarate is the incorporation of nitrogen. This metabolic arrangement makes 2-oxoglutarate an indicator of the C to N ratio of the cells. Figure 1 presents a scheme of the main nitrogen assimilation pathways that can be found in cyanobacteria. The scheme highlights the production of intracellular ammonium during the assimilation of different nitrogen sources and the role of 2-oxoglutarate as the C-skeleton for the incorporation of nitrogen into organic material. However, not all these pathways are present at the same time in a cyanobacterial cell, their expression being strictly regulated by the nitrogen source and also by the availability of carbon.

When the cyanobacterial cells are incubated in the presence of a limiting concentration of ammonium, but with an adequate supply of carbon, they sense a high C to N ratio that determines expression of genes encoding permeases and enzymes required for an efficient assimilation of ammonium or for the assimilation of alternative nitrogen sources. This activation of gene expression requires NtcA, a transcriptional regulator that belongs to the family of bacterial transcription factors whose best-known representative is CAP (catabolite gene activator protein) of *Escherichia coli*. Modelling of NtcA based on the CAP structure and preliminary X-ray structural data confirms that NtcA bears a structure similar to that of CAP, bearing a DNA-binding helix–turn–helix motif in its C-terminal domaim. NtcA binds to specific sites in the promoters of the regulated genes activating their transcription. The DNA-binding site of NtcA shows the sequence signature GTAN8TAC that is usually found 22 nt upstream from a promoter-

10 box. This promoter structure for NtcA-activated genes, which conforms to the structure of bacterial class II promoters, has been found in most cyanobacterial nitrogen-regulated genes investigated to date.

The *ntcA* gene is autoregulatory and the NtcA protein appears to be subjected to regulation being activated when the C to N ratio of the cell is high. Although the precise mechanism of modulation of NtcA is still not known, two elements that influence NtcA activity have recently been identified: 2-oxoglutarate and the PII protein. Binding of NtcA to some N-regulated promoters is stimulated by 2-oxoglutarate, suggesting that this metabolite can represent an allosteric effector for NtcA. The cyanobacterial signal transduction PII protein (*glnB* gene product) binds ATP and 2-oxoglutarate and is phosphorylated when the cells experience a high C to N ratio, it having been suggested that it operates as a 2-oxoglutarate sensor. The PII protein is required for the activation of NtcAdependent genes specifically under conditions of nitrogen stress, and it appears to promote gene expression more efficiently in its phosphorylated form. Therefore 2-oxoglutarate may affect NtcA activity directly and indirectly through PII. *In vivo* positive effects of 2-oxoglutarate on the expression of nitrogen-regulated genes have been observed in an engineered *Synechococcus* strain bearing the *E. coli* 2-oxoglutarate permease.

Some NtcA-dependent promoters appear to require additional regulators or transcription factors for proper operation. This is the case of the *Anabaena nir* operon promoter that requires binding of NtcB in addition to binding of NtcA. Use of this promoter is also dependent on CnaT, a putative glycosyl transferase that influences transcription through a mechanism that is yet to be worked out.

In addition to the regulation of gene expression, some elements of the nitrogen assimilation system are subjected to post-translational regulation in cyanobacteria. The nitrate/ nitrite permease is a target of regulation, being inhibited when the cells sense a high N to C ratio. The non-phosphorylated form of the P_{II} protein appears to effect specific inhibition of the permease when the cells are incubated under conditions that determine low 2-oxoglutarate levels. Additionally, in the unicellular cyanobacterium *Synechocystis* sp. strain PCC 6803, glutamine synthetase is subjected to ammonium-promoted inactivation by two small protein factors whose expression is repressed byNtcA when the cells are incubated in the absence of ammonium.

Heterocysts

Many filamentous cyanobacteria fix N_2 in specialised cells called heterocysts, which differentiate from some vegetative cells when the filaments are incubated in the absence of a source of combined nitrogen. The heterocysts bear a thick cell wall, which carries extra glycolipid and polysaccharide layers and shows a low permeability to gases. The heterocysts also exhibit a modified metabolism oriented towards keeping a microaerobic environment for nitrogenase expression and towards providing ATP and electrons for nitrogenase function.

NtcA is strictly required for heterocyst development, apparently linking differentiation to nitrogen deficiency, and is needed for the induction of the key heterocyst development regulator HetR. NtcA also participates in the activation of expression of genes whose products are required during the differentiation process or are active in the mature heterocyst. Early developmental genes such as the *hetC* regulatory gene and genes that are expressed later during the developmental process such as *devBCA*, encoding a glycolipid exporter, or the two *cox* operons, encoding cytochrome oxidases that are essential for N_2 fixation, are transcribed from NtcAdependent promoters. In the mature heterocyst, the *glnA* gene encoding glutamine synthetase, *petH* encoding ferredoxin:$NADP^+$ reductase and the *nifHDK* operon encoding the nitrogenase complex are also expressed from NtcA-activated promoters. Genes for the metabolism of cyanophycin, a nitrogen reserve made of aspartate and arginine that accumulates conspicuously in the heterocysts, are expressed from multiple promoters also including NtcA-dependent promoters.

Some of the heterocyst-related NtcA-dependent promoters do not present the structure of the standard, class II NtcAtype promoter, but rather show the structure of bacterial class I promoters with the NtcA-binding site being located approx. 90 nt upstream from the transcription start point. In summary, NtcA effects nitrogen control not only for triggering the developmental process but also for its progression and in the N_2-fixing heterocyst, and does so through the use of different types of NtcA-dependent promoters. The N_2-fixing heterocyst provides the vegetative cells of the filament with fixed nitrogen, probably in the form of amino acids. Glutamine is probably a transferred amino acid, but whether cyanophycin amino acids have a role in the nitrogen transfer from the heterocysts to the vegetative cells is currently not known. The mechanism of transfer is also not known, but mutants of some amino acid permeases of *Anabaena* sp. strain PCC 7120 have been shown to be impaired in diazotrophic growth, suggesting a role for such amino acid permeases in intercellular nitrogen transfer in the N_2-fixing filament.

Photosynthetic Reactions

Photosynthetic light reactions occur on the thylakoid membranes. Light reactions convert solar energy into chemical energy in the form of ATP and NADPH. Oxygen is released as a side product. Components of the photosynthetic light reactions include membrane-associated protein complexes: the phycobilisomes, photosystem I (PSI), cytochrome b6f, photosystem II (PSII) and ATP synthase as well as mobile components: plastoquinones, plastocyanins and cytochrome 552. Phycobilisomes are large complexes of proteins with covalently bound bilin-pigments. The phycobilisomes are mainly attached to PSIIs on the cytoplasmic side of the thylakoid membranes and may balance the distribution of captured energy depending on the prevailing light conditions. They serve as light harvesting antennae and absorb solar radiation mainly between 500 and 650 nm. Cyanobacteria can utilize most of the visible light spectrum because they have several pigments with different absorption characteristics. The chlorophyll a pigment has absorption maxima at the wavelengths 430–440 and 670 nm. In addition orange carotenoid pigments participate in light harvesting by absorbing mainly at 420–480 nm.

ATP and NADPH from the light reactions are used to fix atmospheric carbon dioxide into sugar compounds. The carboxylation reaction of the Calvin-Benson-Bassham cycle takes place in the cytoplasm. The Ribulose-1,5-bisphosphate carboxylase/oxygenase enzyme (Rubisco), which is packed into carboxysomes in the cytoplasm, catalyzes the fixation of CO_2 into ribulose 1,5-bisphosphate. The primary product of the Calvin-Benson-Bassham cycle is a three-carbon compound, glyceraldehyde-3-phosphate, which is used to make more complex sugars. In addition to carboxylation, the Rubisco enzyme can catalyze an energetically non-productive oxygenation reaction. In natural conditions the availability of inorganic carbon is often limiting for photosynthetic carbon fixation. Cyanobacteria have evolved carbon-concentrating mechanisms (CCMs) to concentrate inorganic carbon into the cell. Five active uptake systems for inorganic carbon have been identified in cyanobacteria, although not all are present in all species. BCT1, SbtA and BicA take up bicarbonate. Additionally there are two systems for CO_2 uptake: the low-CO_2-inducible $NDH1_3$ and the constitutively functioning $NDH1_4$, which have high and low affinities for dissolved CO_2, respectively. *Synechocystis* prefers to take up inorganic carbon in the form of bicarbonate.

Photoinhibition

Although light is an essential source of energy for photosynthesis, it also has destructive power. High-energy UV light is especially harmful to photosynthesis. In the light the PSII reaction centre is inactivated and the D1 protein is damaged. Inactivated PSII centres constantly undergo a repair cycle, which requires the removal and de novo synthesis of the D1 protein. When the rate of damage to the PSII reaction centre exceeds the rate of repair, oxygen evolution capacity decreases. This process is called photoinhibition. The rate constant of photoinhibition is directly proportional to light intensity in both plants and cyanobacteria. In the PSII repair cycle the damaged D1 protein is removed from the inactive PSII, degraded and replaced with a new copy after which the PSII is reassembled and activated. In *Synechocystis* the three-member *psbA* gene family codes for the D1 protein. The *psbA2* and *psbA3* genes are actively expressed and encode identical D1 proteins whereas the more divergent *psbA1* is virtually silent.

Different stages of PSII repair offer a wide range of regulatory possibilities in cyanobacteria at the levels of D1 degradation, synthesis and degradation of *psbA* mRNA encoding the D1 protein, translation of *psbA* mRNA, post-translational modifications of a pre-D1 form, assembly of the D1 protein into PSII and PSII activation. Different cyanobacteria have slightly different strategies for coping with photoinhibition. *Synechococcus* sp. PCC 7942, for example, switches between two forms of the D1 protein, D1:1 and D1:2. Under high light or UV light the proportion of the D1:2 form increases probably to facilitate acclimation to stress conditions. The underlying mechanism of photoinhibition itself has been under debate for a long time. Several models have been suggested. The four prevailing hypotheses are the acceptor-side, donor-side, low-light and manganese models.

REGULATION OF GENE EXPRESSION

In eubacteria the regulation of gene expression takes place on many levels: transcription initiation, transcription elongation and termination, transcript degradation, translation, posttranslational modifications and protein turnover. Transcription is initiated by the RNA polymerase (RNAP) holoenzyme, which consists of a catalytically active core and a sigma (σ) factor. The σ factor mediates promoter recognition. The RNAP holoenzyme binds to the promoter forming a closed complex. Next the DNA strands unwind, melt starting from the –10 region and form the open complex where the template strand is directed to the RNAP active site. Then follows a phase of abortive initiations where the RNAP repeatedly synthesizes short segments of RNA without éscaping the promoter. Promoter clearance occurs when the length of the transcript exceeds 12 bp and the holoenzyme proceeds to the transcription elongation phase.

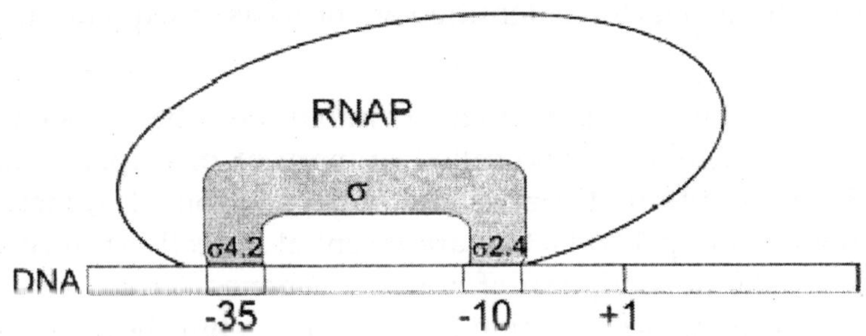

Figure 3. Diagram of the RNA polymerase holoenzyme attached to the DNA strand.

Promoter sequences and other *cis*-acting elements up- and downstream of the transcription start site regulate the expression level of that particular gene. Conserved sequence elements characterize bacterial promoters. The most important of these are recognized by the RNAP holoenzyme. In the model bacterium *Escherichia coli* the primary σ factor recognizes the hexameric –10 and –35 regions, whose consensus sequences are 5′-TATAAT-3′ and 5′-TTGACA-3′, respectively. The length of the spacer region between the –10 and –35 regions affects transcriptional efficiency and is optimally 17 bp. Some promoters lack the – 35 element but may have an additional extended –10 region.

Bacterial RNA Polymerase

Cyanobacteria, like other eubacteria, have only one type of RNAP, which is responsible for the transcription of mRNA, tRNA and rRNA. RNAPs are structurally and functionally conserved among prokaryotes and eukaryotes although cyanobacterial and chloroplast RNAPs have distinct features that characterize them as a group. The multi-subunit RNAP core, which is responsible for catalyzing RNA synthesis reactions, consists of 2 identical α subunits and the β, β' and ω subunits. Uniquely in cyanobacteria and their descendants, chloroplasts, the β' subunit is split into two: β' and γ. The structure of the RNAP has been determined with X-ray crystallography at 2.6 Å resolution in *Thermus thermophilus*.

While the RNAP core exhibits the catalytic activity in RNA synthesis, the σ subunit of RNAP is required for the recognition of specific promoter sequences. Burgess et al. first identified factors as proteins stimulating transcription in *Escherichia coli*. Although the RNAP core has unspecific affinity for DNA in general, it cannot locate promoter sites and initiate transcription without a σ factor. In general, bacterial genomes encode several σ factors and different σ factors recognize different promoters thus controlling the expression of distinct sets of genes. All σ factors compete for a limited amount of the RNAP core and the exchange of one σ factor to another is thought to be a major determinant in switching the gene expression pattern. Bacterial σ factors are divided into two families, σ^{54} and σ^{70}. The large 70 family derives its name from the 70 kDa-sized RpoD protein of *Escherichia coli* and its members have high homology even between species. All cyanobacterial σ factors, identified on the basis of sequence homology, belong to the σ^{70} family.

Lonetto et al. divided the σ^{70} type σ factors into three subgroups based on sequence similarity. Group 1 σ factors are also called primary or principal σ factors. They are very conserved between different species and only one group 1 σ factor is encoded by any bacterial genome. Group 1 σ factors are essential for cell viability and are thought to be responsible for the transcription of household genes. Group 2 σ factors are very similar to the group 1 σ factors in amino acid sequence, but they are nonessential. The alternative, or group 3, σ factors share less sequence homology with groups 1 and 2 and are known to have specialized stress- and development-related functions. A fourth group, the extracytoplasmic function σ factors, is sometimes separated from the alternative σ factors. These factors respond to signals from the periplasm. Four regions and their sub-regions, σ1.1- σ4.2, have been distinguished in σ^{70} type factors on the basis of amino acid sequence conservation. The extremely conserved domains σ2.4 and σ4.2 are critical for σ factor function because they recognize and bind the –10 and –35 promoter elements, respectively.

Synechocystis has nine σ factors. SigA is the primary σ factor, group 2 consists of SigB, SigC, SigD and SigE while SigF, SigG, SigH and SigI belong to group 3. In a phylogenetic analysis of amino acid sequences the *Synechocystis* SigA factor clusters together with the primary σ factors of *Escherichia coli* and *Bacillus subtilis*. The second cluster includes RpoS, the solitary group 2 σ factor from *Escherichia coli* and the closely related SigB, SigC, SigD and SigE factors of *Synechocystis*. SigB, the general stress-responsive σ factor in *Bacillus subtilis*, clusters together with the group 3 σ factors. The abundance of σ factors varies among different cyanobacterial species: *Synechococcus* sp. PCC 7002 has five, *Thermosynechococcus elongatus* BP-1 eight, *Prochlorococcus marinus* MED4 five, *Anabaena variabilis* ATCC 29423 at least ten and *Anabaena* sp. PCC 7120 twelve putative σ factor genes. Several σ^{70} type factors are found also in plant chloroplasts. Plant plastids have two types of RNAP, a nuclear-encoded RNAP and a plastid-encoded, bacterial type RNAP. Plant factors are nuclear-encoded but transported to plastids where they participate in transcription initiation in the same fashion as in cyanobacteria. Preliminary roles have already been suggested for all six (SIG1-SIG6) σ factors identified in *Arabidopsis thaliana*.

The function and activity of the σ factors themselves are under control. Anti-sigma factors are σ factor antagonists. Most information on the anti-sigma factors has been gathered from *Escherichia coli* and *Bacillus subtilis*. Guanosine-tetraphosphate, ppGpp is a global regulator involved in the bacterial stringent response. Artsimovitch et al. (2004) showed the binding of ppGpp to the transcription initiation complex in *Thermus thermophilus* and suggested that ppGpp might interfere with transcription initiation by destabilising protein-DNA interactions. On the other hand, in *Escherichia coli* transcription from σ^S regulated promoters requires the presence of ppGpp. ppGpp also interferes with the degradation of σ^S thus prolonging its lifetime. The double-stranded 6S RNA interacts with the RNA polymerase and is a regulator of σ^{70} in the *Escherichia coli* stationary phase. 6S RNA-mediated regulation can be either negative or positive depending on the target gene.

SENSING AND RESPONDING TO ENVIRONMENTAL CONDITIONS

Stress responses of bacteria can be divided into short- and long-term strategies: short-term strategies include motility or taxis, the reversible modification of enzyme activities and adjustment of photosynthetic light harvesting whereas long-term processes involve profound changes in the gene expression pattern, cell metabolism, structure and morphology. Stress triggers signalling cascades that lead to acclimation processes. Sensory proteins such as histidine kinases (Hik) first perceive a change in environmental conditions. Further downstream of the signal transduction chain are σ factors and specific transcription factors (figure 4). Their properties affect promoter selectivity and hence influence changes in target gene expression pattern.

Receptor-mediated mechanisms such as two-component systems form a central part of the regulatory network in bacteria. A classical two-component cognate pair consists of a sensory histidine kinase and a response regulator (Rre), which relays the signal onwards. Signal transduction by two component pairs can be initiated either on the cell membrane or in the cytoplasm. The *Synechocystis* genome has a total of 91 potential twocomponent system genes including 47 Hiks and 42 Rres. Mikami et al. and Murata and Suzuki used microarray analysis to study *Synechocystis* twocomponent systems. Their analysis of a large collection of *hik* and *rre* inactivation strains demonstrated that one Hik can sense signals originating form several types of stress: Hik33, for example, is involved in the perception of not only cold but also osmotic and redox stress signals. In addition to receptormediated systems, a variety of other sensing mechanisms function in cyanobacteria. Membrane fluidity has a significant role in the sensing of environmental signals, especially of temperature and osmotic changes. Also the synthesis of intracellular second messengers such as ppGpp, reactive oxygen species and cAMP can be rapidly enhanced by stress.

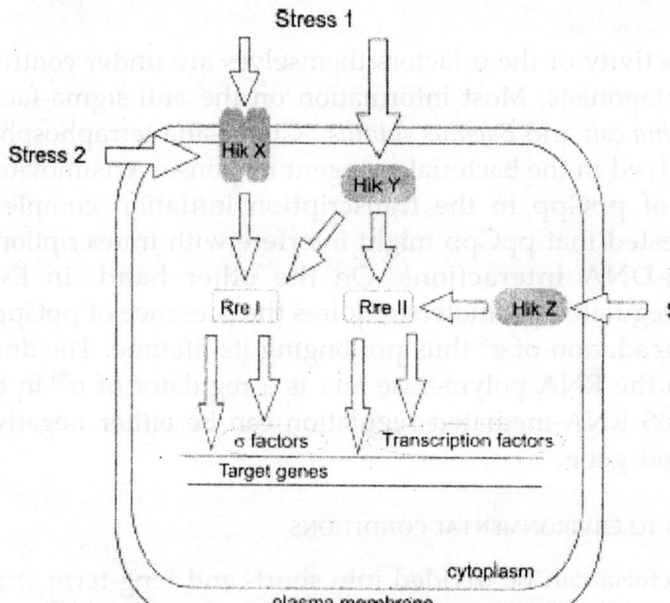

Figure 4. A generalized example of possible two-component signal transduction pathways in Synechocystis.

Also a variety of non-coding RNAs regulate gene expression in bacteria. Two types of 6S RNA have been reported from *Prochlorococcus* MED4. Moreover, Nakamura et al. and Voss et al. have identified a non-coding, 65-nucleotide-long RNA, Yfr1, from cyanobacteria. The analysis of an Yfr1 mutant strain revealed that it is involved in the regulation of stress-induced genes. In addition, Dühring et al. reported that an antisense RNA, IsrR, regulates the iron stress inducible *isiA* gene. σ factors lie at the end of the signal transduction chain together with specific transcription factors. The best-characterized transcription factor in *Synechocystis* is NtcA, which is together with the PII sensory protein involved in nitrogen metabolism.

Temperature

Synechocystis is a mesophilic cyanobacterium with an optimal growth temperature at 32 °C. *Synechocystis* cells can grow within the temperature range of 15 and 45 °C, but below 25 and above 43 °C it suffers from severe stress symptoms. Temperatures above 45 °C are considered to be lethal for *Synechocystis* although cells can acquire thermotolerance through pre-exposure to sub-lethal temperatures. During the first phase of the heat response, cells produce rapidly large amounts of heat shock proteins which act as molecular chaperones that enable the proper folding of proteins to their active state or solubilize aggregates of misfolded proteins. Furthermore, some heat shock proteins are proteases that degrade damaged proteins. The up-regulation of the heat shock proteins is a transient phenomenon and their levels decline to a steady state level. Cyanobacterial heat shock proteins include HspA, HtpG, ClpB, GroES and GroEL, DegP and DnaK. Inactivation of the *hspA* gene made the cells more sensitive to heat while over expression enhanced heat tolerance. Suzuki et al. found a total of 113 heat-inducible genes and 90 proteins in the *Synechocystis* genome. In their DNA microarray analysis most of the induced genes fell into the category "hypothetical and unknown", which means that it is not known how and why their activation affects heat acclimation. In *Synechocystis* light has a great influence on the heat stress response, as shown by Asadulghani et al.

High temperatures tend to make lipid bilayer membranes increasingly fluid and labile, which may affect the function of membrane-bound protein complexes. Hence another important aspect of the heat response is the stabilization of membranes by increasing the degree of fatty acid saturation. In *Synechocystis* the membrane-bound histidine kinase Hik33 relays signals of the state of the lipid bilayer to regulatory proteins and brings about changes in the gene expression pattern resulting in the rigidification of the membranes. Moreover, Suzuki et al. showed that Hik34 is a negative regulator of the heat shock protein genes *htpG* and *GroESL1*. Several bacterial species, including the model organisms *Escherichia coli* and *Bacillus subtilis*, are known to have specific factors that are responsible for the positive regulation of genes involved in high temperature stress. In *Escherichia coli* the heat shock response is mediated by the 32 and 24 factors. σ^{32} controls the expression of the "heat shock regulon", a set of genes involved specifically in the heat shock response. In contrast, no exclusive heat shock σ factors have been defined in cyanobacteria although some σ factors are known to respond to heat stress. The heat response can also be negatively regulated by a *cis*-acting DNA element, CIRCE, and the HrcA protein. CIRCE elements have been found on the promoters of several heat-inducible genes in *Synechocystis* and they are found in many other bacteria as well HrcA is a repressor that binds the CIRCE inverted repeat sequence thus preventing transcription initiation from that promoter. The expression of the *groESL* and *groEL2* genes is the best known examples of CIRCE-HrcA mediated negative regulation in *Synechocystis*.

Research on temperature stress in cyanobacteria has mainly focused on heat stress while less attention has been paid to low temperature. Nevertheless, some mechanisms of cold responses and acclimation have been identified. At low temperatures chaperones act to protect proteins that are susceptible to damage. Lower temperatures make cell membranes more rigid, and as a consequence the *desA* and *desB* genes encoding desaturase-enzymes in *Synechocystis* are activated. Desaturases increase the number of double bonds in fatty acid hydrocarbon chains thus increasing membrane fluidity. While its role in cellular sensing encompasses a variety of stresses, the membrane-bound Hik33 histidine kinase also perceives the degree of membrane fluidity and regulates the expression of the desaturases.

Light

Electromagnetic radiation from the sun ranges from very high-energy, short-wavelength ultraviolet-light to low-energy infrared radiation. In between the extremes is the spectrum of visible light. Cyanobacteria can sense both light quantity and quality. Proposed cyanobacterial photoreceptors include the photosynthetic pigments (phycobilins, chlorophylls and carotenoids), cryptochromes and phytochromes. Recently a novel green light receptor, CcaS, with a histidine kinase domain was identified in *Synechocystis*. In addition, green light can be sensed by the rhodopsins. Besides the actual photoreceptors, the redox-state of the photosynthetic electron transport chain might act as a light sensory system because it reflects the prevailing light conditions. Several mechanisms have been suggested to operate simultaneously to enable acclimation to different light conditions. State transitions of light-harvesting antennae may balance the distribution of light energy between photosystems I and II, chromatic adaptation involves adjustments in pigment composition and the PSII/PSI ratio may change. Some cyanobacterial strains may move closer to or further away from the light source by phototaxis. It has been proposed that cyclic adenosine monophosphate (cAMP), which mediates light signal transduction, is needed for phototaxis of a motile *Synechocystis* strain.

Salt and Osmotic Stress

Cyanobacteria perceive salt and hyperosmotic stress as different kinds of stress. Firstly, salt- and hyperosmotic stress induce distinct sets of genes whose products enable successful acclimation. Secondly, distinct physiological responses to salt- and sorbitol-induced osmotic stresses have been reported. Dissolved salts like sodium chloride (NaCl) cause simultaneously two distinct types of stress: ionic and hyperosmotic stress. Other solutes such as sorbitol and disaccharides cause non-ionic hyperosmotic stress. *Synechocystis* is moderately halotolerant and can withstand up to 1.2 M NaCl concentrations. High external salt concentrations result in dehydration due to the efflux of water from the cell. During the first minutes of salt shock, Na^+ ions enter the cell via K^+ channels. To balance ion concentrations, Na^+/H^+ antiporters on the plasma membrane remove excess Na+ ions from the cytoplasm and K^+ transporters import potassium cations to compensate for the extrusion of sodium.

In addition to the action of ion channels cyanobacteria produce compatible solutes. Glucosylglycerol, for example, is synthesized in large quantities in response to NaCl-induced osmotic stress. Inactivation of the gene encoding the glucosylglycerol-phosphate synthase enzyme disabled the production of glucosylglycerol. As a consequence, *Synechocystis* cells became sensitive to salt-induced osmotic stress and were not able to divide properly. Furthermore, sucrose, trehalose and, in *Synechococcus*, glycine betaine are synthesized or actively up-taken in order to protect cellular components.

Photosynthesis is vulnerable to salt stress. High internal salt concentrations disrupt photosynthesis by damaging the sensitive oxygen-evolving complex of PSII. As a result growth and productivity decrease. Osmoprotectants are known to stabilize photosynthetic membranes during osmotic stress. Chaperones have roles in osmotic stress and their responsibilities in salt- and sorbitolinduced stress are similar to those in heat stress: to facilitate and maintain the folding of proteins, to stabilize cellular structures and –in the case of proteases- to degrade irreversibly damaged proteins. The HspA heat shock protein, for example, participates in osmotic stress tolerance, as an *hspA* mutant strain is more sensitive to salt stress than a control strain. Microarray analysis of salt and osmotic stress responses in *Synechocystis* showed that although a large number of genes are saltinduced, their transient up-regulation lasted less than 24 h. This indicates that full acclimation to salt stress conditions can be achieved within 24 hours. Marin et al., showed that several histidine kinases, Hik16, Hik33, Hik34 and Hik41, are involved in sensing stress signals under salt and osmotic stress, and in transducing signals to the gene regulation machinery. However, these same histidine kinases are involved in sensing many other stresses as well, and thus cannot be considered to be exclusively salt specific.

pH

In the laboratory cyanobacteria are grown at a pH close to neutral but they are able to acclimate to a range of different pH conditions. If the growth medium is not buffered, pH tends to increase over growth time. In natural habitats cyanobacteria prefer alkaline environments. They did not tolerate pH below 5, whereas eukaryotic algae were present in as low pH as 1.9. Mapping of the transcriptome and proteome of acid stressed *Synechocystis* cells showed that the carbon concentrating mechanisms are intimately connected with acid stress tolerance. At a higher pH the concentration of bicarbonate is high and the availability of inorganic carbon is less likely to limit cyanobacterial growth. Sensitivity to low pH might thus be in part due to the effects of carbon limitation.

The most prominent changes in the proteome were restricted to the periplasm and the cytoplasmic fraction remained relatively untouched by pH stress. This implies that the cell surface moderates the effect of pH stress before intracellular components are severely affected.

Nutrient balance

A balance in nutrient availability is important because deprivation reduces productivity but on the other hand excesses may have toxic effects on cell metabolism. Homeostatic acclimation mechanisms maintain an internal balance amidst changes in nutrient availability. Non-nitrogen-fixing cyanobacteria uptake nitrogen preferably in the form of ammonium. Ammonia taken up by ammonium permeases is first incorporated into glutamate to produce glutamine. Glutamine and 2-oxoglutarate are then used to generate two glutamates. This pathway is the best known, although alternative assimilation pathways are known to exist. Many cyanobacterial species are capable of fixing atmospheric nitrogen. The nitrogenase enzyme is oxygen-sensitive and nitrogen fixation is separated from photosynthesis. In filamentous cyanobacteria nitrogen fixation takes place in heterocysts. In contrast, unicellular nitrogen-fixing species photosynthesize during the day and nitrogen fixation accompanied by vigorous respiration occurs in darkness. The intracellular nitrogen balance is sensed by the PII and NtcA regulators, which relay signals to the gene expression machinery. Phosphorus is another important macronutrient since it is the building block of for example membrane lipids, nucleic acids and ATP. A two-component signalling system that perceives phosphorus limitation is conserved among bacteria and homologs have been identified also in *Synechocystis*.

Strains and Standard Growth Conditions

Reecently, a study was conducted on the roles of *Synechocystis* group 2 factors by Maija Pollari. The details and findings of the study are described in the following pages. The glucose-tolerant strain of *Synechocystis* sp. PCC 6803 was used as the control strain (CS). *Synechocystis* cells were grown photoautotrophically in BG-11 medium supplemented with 20 mM Hepes-NaOH pH 7.5 at 32 °C. The continuous photosynthetic photon flux density (PPFD) was 40 mol m^{-2} s^{-1} and the cells grew in ambient CO_2 concentration. Liquid cultures were shaken 90 rpm. These will be referred to as standard growth conditions. BG-11 agar plates for maintaining the single inactivation strains were supplemented with kanamycin, plates for double inactivation strains with kanamycin, spectinomycin and streptomycin and plates for triple inactivation strains with chloramphenicol, kanamycin, spectinomycin and streptomycin.

Construction of σ Factor Inactivation Strains

All inactivation strains were made by interrupting the chosen σ factor gene or genes with antibiotic resistance cassettes. The *sig* genes in the single inactivation strains were interrupted with a kanamycin resistance cassette to obtain the strains ΔsigB, ΔsigC, ΔsigD and ΔsigE. The second *sig* gene was interrupted with the Ω fragment conferring resistance to spectinomycin and streptomycin. The *sigC* gene was inactivated in the ΔsigB strain to obtain ΔsigBC, the *sigD* gene was inactivated in the ΔsigB and ΔsigC strains to obtain ΔsigBD and ΔsigCD and the *sigE* gene was inactivated in the ΔsigB, ΔsigC and ΔsigD strains to obtain ΔsigBE, ΔsigCE and ΔsigDE, respectively. The triple inactivation strains were generated by interrupting a third *sig*-gene with a chloramphenicol resistance cassette. The *sigC* gene was inactivated in ΔsigBD to obtain ΔsigBCD and the *sigE* gene in ΔsigBC, ΔsigBD and ΔsigCD to obtain ΔsigBCE, ΔsigBDE and ΔsigCDE, respectively. *Synechocystis* cells were transformed according to Williams.

Structural Modelling of the Synechocystis RNA Polymerase

For structural modelling the amino acid sequences of the *Synechocystis* RNA polymerase subunits were obtained from CyanoBase. The sequences were aligned with their corresponding counterparts from the 2.6 Å *Thermus thermophilus* RNA polymerase structure using the MALIGN program and manual adjustment. Secondary structures were predicted with Predict Protein and final three-dimensional modeling was done with the MODELLER software. The computer-generated models were examined visually and their overall stereo-chemical quality was assessed with PROCHECK.

Physiological Characterization of the Inactivation Strains

Survival in high temperature

The survival of *Synechocystis* cells in heat stress was examined with a viability assay. Cell culture was incubated for 15 min at 48 °C under the PPFD of 5 µmol m^{-2} s^{-1} with or without a 1 h pre-treatment at 43 °C under 35 mol of photons m^{-2} s^{-1}. After the heat treatments the samples were diluted 1000 fold and 10 µl aliquots were spotted onto BG-11 agar plates. The cells were grown under standard growth conditions for one week after which colonies were counted. The percentage of surviving cells was calculated by dividing the number of colonies from treated samples by the number of colonies from untreated control samples.

Oxygen evolution measurements

Light saturated rate of photosynthetic oxygen evolution was measured *in vivo* from cells grown under standard growth conditions, at 80 mol of photons m^{-2} s^{-1} or after incubation at 48 °C. Oxygen evolution was measured with a Clark-type oxygen electrode at 32 °C in the presence of 10 mM $NaHCO_3$. In addition, photosynthetic activity was measured from cells grown for two days in standard growth light or at 80 mol of photons m^{-2} s^{-1} using the same light intensity as in the growth conditions.Light saturated rate of PSII electron transfer was measured in the presence of 0.5 mM 2,6-dichloro-*p*-bentsoquinone (DCBQ), an artificial electron acceptor. 0.7 mM ferricyanide was added to keep DCBQ in an oxidized form. Measurements of PSII capacity were done from cells grown under standard conditions or in BG-11 medium supplemented with salt or sorbitol. PSII capacity was also measured from cells grown at double light intensity and after 0 min, 15 min, 30 min and 45 min of illumination at 1500 mol of photons m^{-2} s^{-1} high light. The high light treatments were done also in the presence of lincomycin, a translation inhibitor.

Biophysical measurements

Flash-induced increase and subsequent decay of chlorophyll fluorescence was measured with a FL2000 fluorometer from cells grown under standard conditions. A 2 ml sample of cell suspension containing 10 μg chl ml^{-1} was dark adapted for 5 min in the fluorometer cuvette to ensure that the Q_A electron acceptor of PSII was in an oxidized state. Q_A was reduced with a short, strong flash and reoxidation was then followed with a series of weak probe flashes. Fluorescence relaxation was also measured in the presence of 10 μM DCMU (3-(3′,4′-dichlorophenyl)-1,1-dimethylurea), a quinone analogue, which inhibits the transfer of electrons from Q_A to Q_B.

77 K Fluorescence emission spectra were measured with an Ocean Optics S2000 spectrometer. Cells were grown for two days in liquid culture under standard growth conditions or at double light and concentrated to 40 g chl ml^{-1}. The sample was excited with blue or orange light. The spectra were normalized to the PSI emission peak value at 723 nm.Thermoluminescence was measured with a home-made luminometer. The A_{730} of the cell suspension was adjusted to 1.0, concentrated 300 times and mixed with glycerol to a final concentration of 30 %. To measure the Q-band 20 μM DCMU was added. Before measurement, the cell suspension was dark incubated. The temperature was lowered to -20 °C and a 4 μs Xenon flash was fired. Heating at the rate of 1 °C s^{-1} was started 30 s after the flash.

Gene Expression

Mild heat stress was induced by incubating the cells at 43 °C at 35 mol of photons m^{-2} s^{-1}. More severe heat stress was induced by short 15 min treatments at 48 °C under the PPFD of 5 mol m^{-2} s^{-1} with or without triggering acquired thermotolerance with a 1 h pre-treatment at 43 °C. Salt stress was induced by supplementing the BG-11 medium with 0.7 M NaCl. Hyperosmotic stress was induced by supplementing the growth medium with 0.5 M sorbitol. Cells were treated for 10 min, 30 min and 1 hour in otherwise standard conditions. High light stress was induced by illuminating cell culture under the PPFD of 1500 mol m^{-2} s^{-1}. After the treatments cells were harvested by centrifugation at 4 °C.

RNA analysis

RNA isolation and Northern blot. Total RNA from 15 ml samples of cell culture containing 10 μg chl ml^{-1} was isolated by extracting with acidic phenol according to. 10 μg of RNA was separated on 1.2% agarose-glyoxal gels and Northern blot hybridizations were performed as in. Gene specific probes were amplified by PCR from *Synechocystis* genomic DNA. The probes were labelled with 32P-dCTP or with digoxigenin-dUTP, which is less sensitive but non-radioactive. *Reverse-transcription PCR.* The expression of the *sigA* and *sigC* genes was studied with RTPCR. Isolated RNA was treated with RQ1-RNase-Free DNase (Promega). Gene specific primers and a total of 1 μg of RNA were used in cDNA synthesis with the You-Prime First-Strand Beads kit. PCR was performed and samples of the reactions were separated with agarose gel electrophoresis. RT-PCR is a semi-quantitative method, which gives an estimation of the relative amount of the template used in the reaction.

Quantitative real-time PCR. The *psbA2* and *psbA3* genes are two very similar members of the *psbA* gene family. The sensitive qRT-PCR technique was used to study the expression of these two genes in high intensity light. After isolation RNA samples were treated with Turbo DNase to eliminate contaminating genomic DNA prior to cDNA synthesis with the iScript kit. A total of 1 g of RNA was used for cDNA synthesis. Primers were designed for the *psbA2* and *psbA3* genes and two reference genes: *rrn16Sa* and *rnpB*. The same reverse primer was used for both *psbA* genes. Specific forward primers were designed to the less similar upstream regions of the *psbA2* and *psbA3* genes. All primers were first subjected to a temperature gradient and a melt curve analysis to determine optimal primer melting temperature oC and to confirm the specificity of the PCR product. Quantitative RT-PCR was performed on a BioRad iCycler. The PCR efficiencies of each reaction were estimated with the LineRegprogram and the expression of the *psbA* genes was calculated relative to the reference genes.

DNA microarray. Global gene expression profiling was performed as in Eisenhut et al. with an oligonucleotide DNA microarray. The used *Synechocystis* oligonucleotide microarray contained 8091 probes, each reading frame being represented from two to four times on the chip.

Protein analysis

Translation in high light stress was investigated by labelling the cells with radioactive methionine. 35[S]-methionine is incorporated into *de novo* synthesized proteins, which can be visualized by autoradiography. Membrane proteins were isolated and chlorophyll content was measured as in Tyystjärvi et al. The membrane proteins were separated on a commercially available sodium docedyl sulphate polyacrylamide gel and blotted onto Immobilon P-membrane. Equal loading was confirmed by staining the membranes with 0.1 % Ponceau S solution. The radioactive proteins were visualized by autoradiography.

SYNECHOCYSTIS RNA POLYMERASE

Comparative modelling is a useful tool for studying the three-dimensional structures of proteins for which X-ray crystal structures are not available. In the case of the bacterial RNAP holoenzyme, crystal structures have been determined for *Thermus thermophilus* and *Thermus aquaticus* at 2.6 Å and 3.3 Å resolutions, respectively. Crystallization of the RNAP has succeeded only in these two thermophilic bacteria. The structures of RNAP subunits are conserved in all bacteria and therefore it was possible to use the *Thermus thermophilus* structure as a template for the modelling of the *Synechocystis* RNAP holoenzyme.

All *Synechocystis* RNAP subunits except showed 40-50 % sequence identity with the template, which is high enough to ensure reliable modelling. The sequence identity of the ω subunit was only 20 % and therefore it was excluded from the model. *Synechocystis* RNAP has the split β′ subunit specific to cyanobacteria. The β′ subunit corresponds to the C-terminal and γ to the N-terminal sequence of the β′ subunits of other eubacteria. According to the model the first and last amino acids of the β′ and γ subunits, respectively, are located on the surface of the RNAP and therefore the split doesn't have a significant effect on overall structure. Another specific feature is that the *Synechocystis* β′ subunit contained a large, 635 amino acid insertion. The structure of the insertion could not be determined because of the lacking template. The accuracy of Maija et.al.'s model was examined by superimposing the model on the template structure. Ramachandran plots showed that circa 90 % of amino acid residues were in the most favourable regions and less than 1 % were in disallowed regions indicating good overall reliability. Previously bacterial RNAPs have been modelled in *Bacillus subtilis* and *Mycobacterium tuberculosis* using the *Thermus aquaticus* and *Thermus thermophilus* structures as templates, respectively.

Group 1 and group 2 σ factors in *Synechocystis*

Attempts to inactivate the *sigA* gene have failed indicating that SigA is essential for cell viability. The SigA protein is constitutively expressed in exponentially growing cells and *sigA* transcripts dominate the *sig* transcript pool under standard conditions. Furthermore, in phylogenetic analyses the *sigA* gene clusters together with the primary σ factors of other cyanobacteria. These results have led to the conclusion that SigA is the group 1 primary σ factor in *Synechocystis*. The primary σ factor has been identified and characterized also in other cyanobacterial species, including *Synechococcus* sp. PCC 7002, *Microcystis aeruginosa* K-81 and *Anabaena* sp. PCC 7120.

Amino acid alignment of the group 1 and group 2 σ factors showed that their sequences were similar. According to comparisons of three-dimensional models, also the threedimensional structure of SigA resembles closely the structures of the group 2 σ factors. High similarity in the regions binding to the –10 and –35 promoter elements suggests that the group 2 σ factors might have overlapping functions. The three-dimensional structures of the group 2 σ factors differed significantly only in the non-conserved domain (NCD) located between regions 1.2 and 2.1. SigA and SigC had the longest NCDs, 86 and 84 amino acids respectively, while the length of the domain was only 42 in SigB, 43 in SigD and 44 in SigE.

Although the length of the NCD was similar in SigA and SigC, their sequence identities were low and thus the structures of the domains cannot be considered similar. The only significant identity, 47 %, was found in the NCDs of SigB and SigD. It is tempting to speculate that the dissimilar NCD may be a potential candidate for the regulation of promoter selectivity. The NCD lies close to the 2.4 domain that recognizes the –10 promoter region. The position of the NCD suggests that it can possibly bind transcription factors that further specify promoter selection. Interaction with different proteins is known to regulate σ factor function and stability. An example of such regulation is provided by the AsiA-protein of the T4 bacteriophage, which binds the *Escherichia coli* 70 and specifically inhibits transcription form certain promoters. The specific roles of the group 2 σ factors were investigated using inactivation strains. Four σ factor single inactivation strains, six double inactivation strains and four triple inactivation strains were constructed and characterized.

Maija et.al. monitored growth in standard conditions and found that all inactivation strains grew as well as the control strain. The results indicated that one, two or three group 2 σ factors can be inactivated simultaneously in any possible combination and, apparently, *Synechocystis* can grow under standard conditions with only one group 2 σ factor present in the cell. The simultaneous inactivation of all four group 2 sigma factors has not been attempted yet. In addition to *Synechocystis*, single inactivation strains of group 2 factor genes have been constructed in *Anabaena* sp. PCC 7120 and *Synechococcus elongatus* sp. PCC 7942. All of these studies reported that the inactivation of group 2 σ factor genes did not affect growth under standard conditions. Also in agreement with the results, previously constructed sigBD, sigBE and sigDE strains in *Synechocystis* and a double inactivation strain in *Anabaena* sp. PCC 7120 grew as well as the control strain. In addition to the complete set of triple inactivation strains Summerfield and Sherman have recently constructed a sigBDE triple inactivation strain in *Synechocystis*. Similarly to the results, they found no differences in growth under standard conditions between the control and the inactivation strain.

Bhaya et al., Huckauf et al. and Imamura et al. have inactivated *Synechocystis* group 3 σ factors, which differ considerably from the group 1 and 2 factors. Bhaya et al. used a motile *Synechocystis* strain and inactivated the *sigF* gene. They found that the *sigF* inactivation strain did not form pili-structures and as a result had lost its capacity for movement. More recently, Asayama and Imamura showed that SigF specifically recognizes the promoter of the *pilA* gene, which is involved in establishing motility.

Inactivated *sigF, sigH* and *sigI* genes segregated completely, but the segregation of the *sigG* gene remained incomplete despite of a vigorous selection process. The group 3 σ factor gene *sigG* seems to be essential for cell viability. Huckauf et al. reported similar growth rates in the *sigF, sigH* and *sigI* inactivation strains compared to the control strain.

Physiological Roles of group 2σ factors

Photosynthesis

Photosynthetic and PSII capacities were measured with a Clark-type oxygen electrode from cells grown under standard growth conditions. Light-saturated rates of oxygen evolution in all inactivation strains were similar to that of the control strain. Clearly the inactivation of one, two or three group 2 σ factors did not affect photosynthetic oxygen evolution capacity nor cause defects in the function of the PSII electron transport chain.

Photosynthesis is fundamental process in cyanobacteria and one could assume that under standard conditions the primary housekeeping σ factor, SigA, is responsible for the transcription of photosynthetic genes. Shibato et al. and Imamura et al showed that the SigA factor may recognize *in vitro* the promoter of the *psbA2* gene encoding the D1 protein. Although previous studies of cyanobacterial group 2 σ factor inactivation strains have not presented data on photosynthetic oxygen evolution, the inactivation of the group 3 σ factor genes *sigF, sigH* and *sigI* did not significantly affect photosynthesis.

Growth phase transfer

Growth under standard conditions was similar to the growth of the control strain in all single, double and triple inactivation strains. The growth pattern of bacterial cell cultures is divided into three stages: the lag phase, the exponential phase and the stationary phase. According to Maija et al.'s study results, *Synechocystis* growth was exponential only during the first day, after which the cultures grew linearly for a few days before they reached a stationary phase of very slow growth. The *Synechocystis* growth pattern was not a typical bacterial one. In the experiments, samples were taken once every 24 h. With that sampling rate a lag phase was not detected and the exponential phase was very short. In the case of *Synechocystis,* a stage of linear growth is known to follow a short exponential phase.

When very dilute cell culture was spotted onto agar plates, the sigBD inactivation strain grew slower than the control strain but when denser cultures were spotted on the plates, the strains grew similarly. The same phenomenon was observed also in liquid cultures. If the initial A730 was set to 0.001, the sigBD strain had a lengthy lag phase that was not evident in the control strain. In contrast to the control strain, sigBD grew very slowly during the first day after which its growth rate increased significantly. After three days the differences in growth had disappeared as cell densities of the sigBD strain had reached those of the control strain. The results of Maija et.al's study show that the sigBD double inactivation strain is not able to transfer normally from the lag phase to the exponential growth phase. On the other hand, cells in very dilute cultures experience higher light intensities than in cells denser cultures. The lengthy lag phase might thus be an indirect result of light-sensitivity. The effect, however, persisted when a lower light intensity was used in the growth experiments. They tested if the effect might be caused by chemical signals that cells secrete into the growth medium.

Maija et al. grew sigBD cells in medium filtered from control strain cultures and vice versa. The sigBD strain always exhibited a very slow transfer from the lag phase to the exponential phase. Thus, chemical communication signals are unlikely to be responsible for the slow growth phase transfer in the sigBD strain. The slow transfer from the lag to the exponential phase was observed only in the sigBD double inactivation strain and not in the sigB or sigD strains. This indicates that the SigB and SigD factors, which are very similar in structure, may compensate for each other. A sigB strain has been studied earlier with regard to transfer from the exponential to the linear growth phase. Similarly to the results, no growth rate differences between the control and sigB strains were reported in the study although many differences were detected in transcription profiles. Foster et al., showed that fewer genes were differentially expressed in the sigB strain than in the control strain and that the expression of the *sigB* gene itself was up-regulated upon transition to the exponential phase.

Although all of the σ factor inactivation strains behaved like the control strain in the stationary phase under standard conditions, Asayama et al. showed that a sigC strain entered the stationary phase slightly earlier than the control strain in their standard conditions. In addition, they measured less viable cells in the sigC strain than in the control strain at the stationary phase. They propose that the SigC σ factor regulates the expression of the *glnB* gene encoding the PII transcription factor and as a result is important in the stationary phase. Differences in growth light, temperature, inorganic carbon concentration and other conditions may influence the role of the SigC factor in the stationary phase. Maija et.al. noticed that under high CO_2 and high temperature conditions sigC enters the stationary phase earlier than the control strain. The gradual slowing of growth may depend on the exhaustion of nutrient resources, cell density, decreased light availability due to shading or, most likely, a combination of many factors. The SigC factor may possibly have different roles depending on the cause of transfer into the stationary phase. A SigC homologue in *Synechococcus* sp. PCC 7002 seems to have a role in the stationary phase.

Acclimation to high temperature stress

Fluctuations in temperature are among the most important environmental factors affecting physiological processes in cyanobacteria. I studied the roles of the ΔSigB, ΔSigC and ΔSigD factors in both long- and short-term heat acclimation. Acclimation to long-term heat stress was studied by subjecting the cells to moderate heat stress at 43 °C for several days. At 43 °C, the ΔsigBD strain grew more slowly than then the control strain. Neither of the single inactivation strains ΔsigB or ΔsigD showed retarded growth, which strengthens the idea of partial functional redundancy of the ΔSigB and ΔSigD factors. Moreover, both the ΔsigC and ΔsigBC strains lacking the ΔSigC σ factor were not able to grow at 43 °C. The ΔsigCD strain behaves like the ΔsigC strain. These results indicate that the ΔSigC σ factor plays a key role in acclimation to long-term moderate heat stress. Under heat stress conditions the expression of the *sigA* gene decreased in general but declined even more in the ΔsigBD, ΔsigC and ΔsigBC strains compared to the control strain. Lower amounts of the sigA transcript may be one possible reason for slower growth at high temperature. It has been shown that lower amounts of the primary σ factor slow down the growth rate in *Escherichia coli*.

Viability experiments were conducted to investigate acclimation to short-term extreme heat stress. Cells were subjected to a 15-minute heat stress treatment at 48 °C and then spotted onto BG-11 plates. After one week of growth under standard conditions the number of colonies was counted and used to calculate survival percentage. 20 % of the control and ΔsigD cells survived, while only 2 % of the sigB and ΔsigBD and 4 % of ΔsigC cells did. Only 0,1 % of ΔsigBC cells survived which indicates that the double inactivation strain is more sensitive than either of the single inactivation strains. Photosynthesis is a highly heat-sensitive process, which is why I investigated the effect of high-temperature stress on the photosynthetic activity. Only 2 % of photosynthetic activity remained in the sigBC strain after 45 minutes at 48 °C whereas the control strain retained 21 % of its original activity. A similar, although slightly less dramatic effect was observed in the other inactivation strains as well. The ΔsigD strain maintained its photosynthetic activity at the same level as the control for the first five minutes but finally it, too, lost its activity faster than the control strain. Thus, both the SigB and SigC σ factors are required for short-term tolerance of extreme high temperatures whereas SigD has only a minor role.

Pre-treatment in sub-lethal heat stress conditions allows the cells to acclimate so that they can tolerate more severe stress. This phenomenon is known as acquired thermotolerance. Maija et.al. tested the capacity of the strains to acquire thermotolerance by incubating cells for 1 hour at 43 °C before subjecting them for 15 minutes to 48 °C. In this experiment nearly all of the control and ΔsigD strain cells survived, 45 % of ΔsigB and ΔsigBD cells, 60 % of ΔsigC but only 25 % of ΔsigBC cells survived. Based on these results, the SigC and SigB factors are required for full acquired thermotolerance. The abundance of heat shock gene transcripts was investigated under standard conditions and after 1 h treatment at 43 °C. Under standard conditions expression levels were similar and no differences were observed between the control and inactivation strains indicating that under standard conditions the inactivation strains are capable of producing normal amounts of heat shock gene transcripts..

The expression of the *sigB* gene is rapidly but only transiently up-regulated in response to heat stress. The expression pattern of the sigB gene under heat stress resembles that of *Synechocystis* heat shock genes. Since their products play central roles in the heat stress response, Maija et.al. analysed the expression of some *Synechocystis* heat shock genes. The ΔsigB and ΔsigBD strains produced less *hspA* transcripts after 1 hour of heat treatment at 43 °C and after a 15-minute treatment at 48 °C, indicating that the ΔSigB factor regulates positively the *hspA* gene in both long- and short-term heat stress. The *hspA* gene is highly induced at high temperatures and its product is involved in the protection of phycobilisomes and thylakoids during heat stress. Thus, lower amounts of the HspA protein may explain why photosynthesis in the ΔsigB and ΔsigBD strains is more susceptible to heat stress than in the control strain. The inactivation of the *hspA* gene causes similar effects as observed in the ΔSigB and ΔSigBD strains, although the phenotype is more dramatic. Furthermore, Singh et al. reported that in addition to *hspA*, many other heat shock genes, including *groEL1*, *groES* and *dnaK*, were less induced in a sigB strain than in the control strain at 45 °C. These results suggest that the normal activation of heat shock genes requires the SigB factor.

REFERENCES

Osanai, T., Ikeuchi, M. and Tanaka, K. (2008) "Group 2 sigma factors in cyanobacteria". *Physiol. Plant.* 133: 490–506.

Raivio, T.L. and Silhavy, T.J. (2001) "Periplasmic stress and ECF sigma factors". *Annu. Rev. Microbiol.* 55: 591–624.

Servant, P. and Mazodier, P. (2001) "Negative regulation of the heat shock response in Streptomyces. *Arch. Microbiol.* 176: 237–242.

Schagerl, M. and Müller, B. (2006) "Acclimation of chlorophyll a and carotenoid levels to different irradiances in four freshwater cyanobacteria". *J. Plant Physiol.* 163: 709–716.

Sudhir, P. and Murthy, S.D.S. (2004) "Effects of salt stress on basic processes of photosynthesis". *Photosynth.* 42: 482–486.

8

Analysis of Cyanobacterial Activity

Identification and quantification of cyanobacteria in water resources is the principal component of cyanotoxin monitoring programmes and can provide an effective early warning system for the development of potentially toxic blooms. Data on concentrations of total phosphate, nitrate and ammonia are valuable for assessing the potential for cyanobacteria to develop and whether or not nitrogen-fixing species are likely to occur. Whereas methods for these nutrients have been extensively reviewed and internationally harmonised by the International Organisation for Standardisation (ISO), approaches to the species determination and quantification of cyanobacteria are very variable and can be undertaken at different levels of sophistication.

Rapid and simple methods can be employed to analyse the composition of a sample at the level of differentiation by genera, which is often sufficient for a preliminary assessment of potential hazard and for initial management decisions. Further investigation may be necessary in order to address quantitative questions, such as whether cyanobacteria are present above a threshold density. Rapid quantitative counting methods can give useful estimates of cell numbers with a counting effort of less than one hour per sample, and the bulk method of biomass estimation by chlorophyll *a* determination can be very time-effective with only moderate equipment demands. More detailed taxonomic resolution and biomass analysis is required if population development or toxin content needs to be predicted. Distinction between these approaches is important because management must decide how available staff hours are most effectively invested. In many cases, the priority will be evaluation of a larger number of samples at a lower level of precision.

The choice of methods further requires informed consideration of sources of variability and error at each stage of the monitoring process, particularly with respect to sampling. Water bodies with substantial temporal and spatial variation of cyanobacterial cell density may show several-fold deviation in cell numbers between samples taken within a few minutes or within 100 m, and precise determination of biomass in one sample per week therefore will not produce a basis for assessment of population size. Much better information can be gained by investing the same effort into a less precise evaluation of a larger number of samples. Information return on working time investment can further be optimised by regular intralaboratory calibrations of methods and their quality control by comparing results with the rapid methods to results of elaborate and precise methods.

SAMPLE HANDLING AND STORAGE

Consideration of the type of information required and decisions regarding the type of analysis required should be made prior to sample collection. However, this is not always possible, particularly when a routine monitoring programme is not in place. Samples may therefore require immediate evaluation on arrival in the laboratory to determine if pretreatment is needed prior to appropriate sample storage. Samples that have been taken for microscopic enumeration should ideally be preserved with Lugol's iodine solution at the time of collection. These samples will be relatively stable and no special storage is required, although they should be protected from extreme temperatures and strong light. However, samples should be examined and counted as soon as practically possible because some types of phytoplankton are sensitive to prolonged storage, and Lugol's iodine solution disintegrates over extended storage periods.

Unpreserved samples for quantitative microscopic analysis require immediate attention either by addition of preservative or by following alternative counting methods which do not use preserved cells. Where unpreserved samples cannot be analysed immediately they should be stored in the dark with the temperature kept close to ambient field temperatures. Unpreserved samples are preferable for species identification because some characteristics cannot be recognised in preserved samples. For example, colonies of *Aphanizomenon* have a characteristic bundle structure which facilitates identification, but preservatives tend to disintegrate the colonies, and the single filaments are more difficult to distinguish from other genera. While samples for quantification must be preserved immediately or counted, samples for identification may be analysed within 24 hours because changes in numbers are less important.

Samples for the analysis of chlorophyll *a*, total and dissolved phosphate as well as nitrate and ammonia, should be filtered as soon after sampling as possible. Storage for a few hours in the dark in glass bottles is usually acceptable if temperatures do not exceed 20 °C. Filtration at the sampling site is recommended, particularly in warm climates, or filtration should occur immediately upon arrival in the laboratory. Filtered samples for nutrient analysis may be stored in the refrigerator for a few hours prior to analysis, or deep-frozen at -18 °C for several weeks. Although the suitability of storage of filters for chlorophyll *a* analysis at -18 °C is currently under debate, the method is employed by many laboratories if immediate extraction cannot be organised.

CYANOBACTERIAL IDENTIFICATION

Microscopic examination of a bloom sample is very useful even when accurate enumeration is not being carried out. The information obtained regarding the cyanobacteria detected can provide an instant alert that harmful cyanotoxins may be present. This information can determine the choice of bioassay or analytical technique appropriate for determining toxin levels.

Most cyanobacteria can be readily distinguished from other phytoplankton and particles under the microscope by their morphological features at a magnification of 200-1,000 times. Figure 1 shows the most frequently occurring of the species known to produce toxins. Cyanobacterial taxonomy, following the established botanical code, differentiates by genera and species. However, this differentiation is subject to some uncertainty, and organisms classified as belonging to the same species may nonetheless have substantial genetic differences, e.g. with respect to microcystin production. Genetically identical cells, obtained by isolation of one colony and cultivation of its daughter cells, are termed strains or genotypes, and field populations of one species consist of a number of genotypes which cannot be differentiated microscopically.

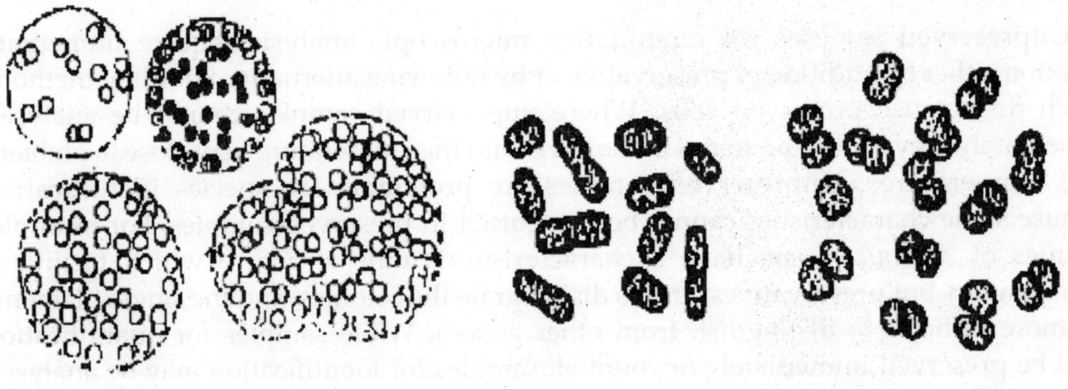

Coelosphaerium *Synechococcus* *Synechocystis*

Figure 1. The most frequently occurring species of cyanobacteria known to produce *toxins*.

Current understanding of the regulation of cyanotoxin production indicates that distinction of genera is very important for assessing potential toxicity, but that microcystin content varies extremely at the level of genotypes or strains, rather than at the level of species. This is one reason why identification to the taxonomic level of genera (e.g. *Microcystis, Planktothrix, Aphanizomenon, Anabaena)* is frequently sufficient. It is preferable to give only the genus name, especially if differentiation between species by microscopy is uncertain on the basis of current general taxonomic knowledge, a lack of locally available expertise, or lack of characteristic features of the specimens to be identified.

This must be emphasised because "good identification practice" has frequently been misunderstood to require determination down to the species level, and this has lead to numerous published misidentifications of species. Practitioners in health authorities with some experience in using a microscope can easily learn to recognise the major cyanobacterial genera and some prominent species which occur in the region they are monitoring. They should not be deterred by the pitfalls of current scientific work in cyanobacterial taxonomy which targets differentiation to the species level. Nevertheless, more precise identification of the dominant organisms down to species level may be useful for a more accurate estimate of toxin content. For example, *Planktothrix agardhii* and *Planktothrix rubescens* have both been shown to contain microcystins, but each species contains different analogues with different toxicity.

For establishing cyanobacterial identification in a laboratory, initial consultation and later occasional co-operation with experts on cyanobacterial identification is helpful. Training courses for beginners should focus on the genera and species relevant in the region to be monitored. Experts can assist in initially deriving a list of these taxa and the criteria for their identification. In the course of further monitoring, experienced experts should be consulted periodically for quality control and for updating such a list.

QUANTIFICATION

Rapid methods for frequent monitoring of large numbers of water bodies or sampling sites have been developed in some countries. These cannot be readily standardised and evaluated internationally, but require adaptation to regional or local conditions. Deciding on the appropriate classification of units to count depends upon variations such as whether prevalent taxa are filamentous or coccoid, whether they are colony-forming or occur as single filaments or cells, and whether populations are very diverse or largely mono-specific. One rapid method which is highly standardised and simplified involves counting a 0.5 ml sample of water for 2 minutes and determining the number of cyanobacterial units present. A table has been prepared which provides adjustment factors to take into account differences in the number of cells present in filaments or colonies.

Cyanobacterial biomass can also be determined using indirect methods, the most common being chlorophyll *a* quantification. This is a rapid and simple method but it is open to interference by chlorophyll *a* from other phytoplankton, hence it is best used when cyanobacteria are the main or dominant organisms present.

Enumeration

Microscopic enumeration of cyanobacterial cells, filaments or colonies has the advantage of directly assessing the presence of potentially toxic organisms. Little equipment in addition to a microscope is required. The method may be rather time consuming, ranging from a few minutes to several hours per sample, depending upon the accuracy required and the number of species to be differentiated.

Sample concentration by sedimentation or centrifugation

Direct counting of preserved cells is typically carried out by Utermöhl's counting techniques using a counting chamber and inverted microscope. This method is well suited for assessment of a large variation in cell type and is widely accepted as one of the most reliable. Counting chambers and sedimentation tubes are commercially available or can be constructed by the investigator. The most commonly used chambers have a diameter of 2.5 cm and a height of about 0.5-2 cm and thus can contain 2-10 ml of sample. These chambers can fit easily on the stage of an inverted microscope. If larger volumes of water need to be analysed, as is the case when cell density is low, then the height of the tube has to be increased. Such extended tubes, however, are too tall to fit on the inverted microscope stage and the light would have to pass through a considerable thickness of liquid before reaching the sedimented specimens. This problem can be overcome by using a tube in two sections which allows the supernatant to be removed without disturbing the sedimented cells on the bottom glass. The amount of sedimented water required depends on the density of cells, on the counting technique and on the magnification. If concentrations are high, even 2 ml may contain too many cells for enumeration, and the sample will need to be diluted.

Apparatus

— Inverted microscope with 10x and 40x objectives
— Counting chamber with sedimentation tube
— Cyanobacterial identification key
— Sample preserved in Lugol's iodine solution

Procedure

1. Allow the sample to equilibrate to room temperature. If cold samples are placed directly in the counting chamber, air-bubbles develop and prevent sedimentation.
2. Gently invert the bottle containing the sample several time to ensure even mixing of the phytoplankton.
3. Pour the sample into the sedimentation tube in place over the counting chamber.
4. Place the counting chamber on a horizontal surface where it will not be disturbed or exposed to direct sunlight.

5. Allow the sample to settle. Sedimentation times will vary depending on the height of the sedimentation tube. Allow at least 3-4 hours per cm height of liquid. Where neutralised formalin has been used as a preservative, double the time allowed for sedimentation. Note that buoyant cells may not settle and may require disruption of the gas vacuoles. However, this problem is frequently overcome by several days of storage with Lugol's solution, because uptake of iodine increases the specific weight of the cells.

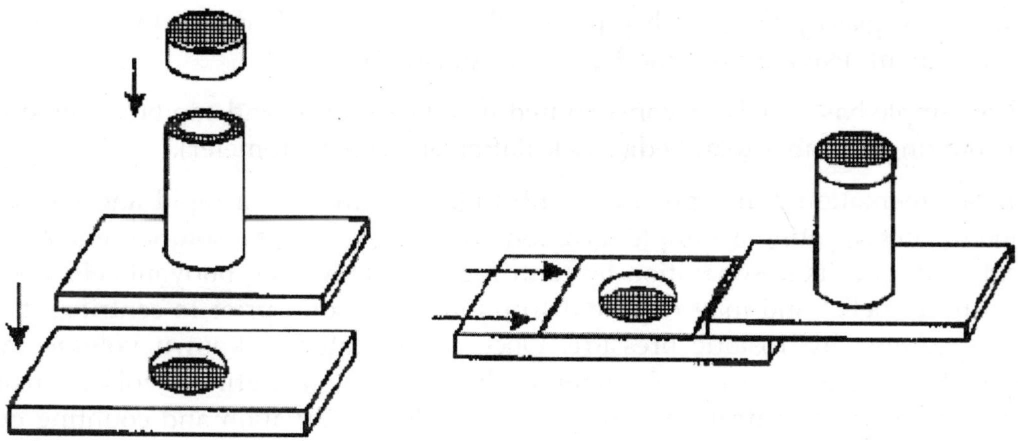

Figure 2. Counting chambers for use with inverted microscopes.

6. Phytoplankton density can now be determined by counting either the total number of organisms on the base of the chamber or by counting subsections.

If an inverted microscope is not available, and samples with low cyanobacterial density need to be counted, other techniques may be applied in order to concentrate samples sufficiently (e.g. sedimentation in a measuring cylinder, followed by careful removal of the supernatant).

Apparatus

- Glass measuring cylinder, 100 ml
- Glass pipette with pipette bulb or filler
- Standard laboratory microscope with 10x and 40x objectives
- Sample preserved in Lugol's iodine solution

Procedure

1. Allow the sample to equilibrate to room temperature.
2. Gently invert the bottle containing the sample several times to ensure even mixing of the phytoplankton.
3. Pour 100 ml of the sample into the measuring cylinder.
4. Allow the sample to sediment (3-4 hours per cm height of liquid) in a location where it will be out of direct sunlight and it will not be disturbed.
5. Using the glass pipette with pipette bulb or filler attached, carefully remove the supernatant, leaving only the last 5 ml undisturbed.
6. The sample has now been concentrated by a factor of 20 and can be counted using a counting chamber (e.g. Sedgewick-Rafter or haemocytometer).

Where sedimentation is not possible, centrifugation can offer a rapid and convenient method of concentrating a sample. Fixation with Lugol's iodine solution enhances the susceptibility of cells to separation by centrifugation. However, buoyant cells may still be difficult to pellet and may require disruption of vacuoles prior to centrifugation by applying sudden hydrostatic pressure. Once concentrated, a known volume can be quantified using a counting chamber or by counting a defined volume using a micropipette to place a drop on a microscope slide. Observation and counting can be done with a standard microscope.

Apparatus

— Centrifuge
— Centrifuge tube, 10-20 ml
— Syringe or bottle with cork, or plastic bottle with screw cap
— Standard laboratory microscope with 10x and 40x objectives

Reagents

— Aluminium potassium sulphate, 1.0 g $AlK(SO_4)_2.12H_2O$ in 100 ml distilled water

Procedure

1. Place 10-20 ml of sample in a centrifuge tube, seal with cap, and centrifuge at 360 × g for 15 minutes.
2. When pelleting needs to be enhanced, add 0.05 ml of aluminium potassium sulphate solution per 10 ml of sample. Mix and centrifuge as described.

3. Where problems occur with the pelleting of buoyant cells, try one of the following:
   i) Place sample in a plastic syringe, ensure the end is tightly sealed, then apply pressure to the plunger.
   ii) Place sample in a bottle with a tightly fitting cork then bang the cork suddenly.
   iii) Place sample in a well sealed plastic bottle and bring it down sharply onto a hard surface.

   These three approaches should be carried out with extreme care to avoid accidental exposure to toxic cyanobacteria. Once they have been subjected to this pressure shock, the gas vesicles should have been disrupted and cells should pellet when centrifuged.
4. Once cells have been centrifuged, carefully remove the supernatant and resuspend the pellet in a small known volume (e.g. 0.5 ml).
5. Samples concentrated by centrifugation can be counted using a counting grid or haemocytometer.

Counting cyanobacteria

Counting cyanobacteria involves defining the units to be counted. The majority of planktonic cyanobacteria are present as filamentous or colonial forms consisting of a large number of cells which are often difficult to distinguish. The accuracy of quantitative determination depends on the number of counted objects; the relative error is approximately indirectly proportional to the square root of the number of objects counted. The number of colonies, not the number of cells, is decisive for accurate enumeration. However, the number of colonies is often not very high, even in water containing a dense bloom where only 20-40 colonies may be present in a 100 ml sample. Both filaments and colonies can differ greatly in the number of cells present, hence results given as number of colonies give little information on the quantity of cyanobacteria present.

Typically, unicellular species are counted as cells per ml and filamentous species can either be counted as number of filaments, and quoting an average number of cells per filament. The cells per filament in the first 30 filaments encountered are often counted and averaged. Alternatively, the total filament length per ml may be assessed as the sum of the extension of each filament within a counting grid placed in the ocular of the microscope. The latter approach is more precise when filament length is highly variable. For colonial species, disintegration of the colonies and subsequent counting of the individual cells is preferable to counting colonies and estimating colony size. Disintegration of colonies sometimes occurs several days after fixation with Lugol's iodine solution. For more stable colonies, it can be achieved by alkaline hydrolysis or gentle ultrasonication. These methods often separate cells very effectively, and even where colonies are not totally broken down into single cells the colony size may be reduced sufficiently to allow single cells to be counted. If this approach is impossible, the geometric volume of individual colonies may be estimated. If colonies are relatively uniform in size, the average number of cells per colony may be determined and then the colonies can be counted. Generally, the use of published values for numbers per colony is not recommended because the size of colonies varies greatly.

There are several systematic methods for counting cyanobacteria. Most approaches aim at counting only a defined part of the sample and then calculating back to the volume of the entire sample. The most common methods are:

— *Total surface counting* which requires the counting of all cells settled within the chamber and may be very time consuming. This method is usually only appropriate for counting very large units at low magnification.

— *Counting cells in transects* from one edge of the chamber to the other. Transects should pass through the central point of the chamber. Some inverted microscopes are equipped with special oculars so that the transect width can be adjusted as required. However, in many cases, the horizontal or vertical sides of a simple counting grid can be used to indicate the margin of the transect. Back-calculating to a millilitre of sample requires measuring the area of the transects and of the chamber bottom as well as the volume of the counting chamber.

— Counting cyanobacteria occurring in randomly selected fields. It is recommended that the position of the chamber to find the next field should be changed without looking through the microscope in order to prevent a bias in the selection of fields. The Sichtfeld area covered by a counting grid is usually considered as one field. However, if no counting grid is available the total spherical Sichtfeld can be considered as a single field. Back-calculating to 1 ml of sample requires registration of the number of Sichtfelds counted, measuring the area of the Sichtfeld and of the chamber bottom, as well as knowing the volume of the counting chamber. The density of different species in one sample can vary and there can also be several orders of magnitude difference between the size of different species; hence it is necessary to select the counting method to suit the sample. Total chamber surface counting with low magnification (100x) is required for large species whereas transect or field counting with higher magnification (200x, 400x) is used for smaller or unicellular cyanobacteria. Accurate enumeration using transects or fields assumes on even distribution of cyanobacteria on the bottom of chamber surface after sedimentation. Due to inevitable convection currents, cells very rarely settle randomly on the surface of the bottom glass and are, almost always, more dense either in the middle or around the circumference of the chamber. Sometimes density also varies between opposite edges. The inaccurate estimate that arises from uneven distribution can be minimised by transect counting. Consequently, transect counting is the preferred method and counting four perpendicular diameters minimises the error. The relation of precision to counting time is very effective if about 100 counting units (cells, colonies, filaments) are settled in one transect. Samples are best diluted or concentrated so that the number of units of the important taxa lies within this range.

A note on filaments:

- Either count them just as cells (inaccurate if length is highly variable). In that case, the one extending out of the grid to the right would be counted, but not the one extending out to the left.
- Or estimate the length of filament within the boundaries of the grid. In this case: 3 boxes (bottom left) + 5 boxes (mid-right) = 8 boxes Specimens occurring exactly on the margin of the counting area (transect or field) present the common problem of whether to count them or not. When counting transects, those specimen that lie across the left margin are ignored while those that cross the right margin are included. When counting fields two predetermined sides of the grid are included and the other two are ignored.

There are different recommendations regarding the number of units per species that must be counted to obtain reliable data. It is particularly difficult to count each species with an acceptable error in each sample. Mass developments of cyanobacteria are characterised by dominance of one to three species. Even if total phytoplankton is to be counted, it is rare for more than six to eight species to contribute to the majority of the biomass. Therefore, for total phytoplankton counts, it is suggested that 400-800 specimens in each sample are counted, giving a maximum error for the total count as 7-10 per cent. In this situation there will be a 10-20 per cent error for the few dominant species, 20-60 per cent for the subdominant species and the rest of the species can be considered as insufficiently counted. If only cyanobacteria are to be counted, and only one or two species are present, counting up to the precision level of 20 per cent, by counting 400 individual units per species, can be accomplished within less than one hour.

The use of mechanical or electronic counters for recording cell counts can shorten counting time considerably, especially if only a few species are counted. Computer keyboards can also be used together with suitable programmes for recording cell counts. The use of an inverted microscope with counting chambers is generally the best approach for estimating cyanobacterial numbers. However, a standard microscope is sufficient for preconcentrated samples or for naturally dense samples from mass developments, provided the size of the water drop enumerated can be defined. Other counting chambers are available for use with a standard microscope. It can also be useful to monitor samples under high magnification with oil-immersion (1,000x) to check the sample for the presence of very small species which may be overlooked during normal counting. An alternative counting method which has been found to be useful is syringe filtration. This method is considerably less time consuming because it does not depend on lengthy sedimentation times and uses a standard laboratory microscope.

Apparatus

- Syringe, 10 ml
- Membrane filters, 13 mm diameter with 0.45 μm pore
- Membrane filter holder
- Glass microscope slides plus coverslips
- Standard laboratory microscope with 10x and 40x objectives

Reagents

- Immersion oil

Procedure

1. Mix water sample by inverting several times.
2. Take up 10 ml of the sample into the syringe.
3. Place filter holder with filter in place, on the end of the syringe.
4. Gently filter the sample through the filter by applying pressure to the syringe piston.
5. Once all the sample has passed through the filter, remove the filter from the holder and place it on a glass microscope slide with the captured cells uppermost.
6. Allow the filter to dry at room temperature then carefully add one or two drops of immersion oil to the filter. The will make the filter appear transparent and permit observation of the cyanobacterial cells trapped on its surface.
7. Finally, cover the filter surface with a glass coverslip and examine under the microscope.
8. The density of cyanobacteria can be easily calculated by counting the number of cells on the filter and dividing this by the volume of water filtered (i.e. number of cells per ml).

Determination of cyanobacterial biomass microscopically

Cell size can vary considerably within and between species, and toxin concentration relates more closely to the amount of dry matter in a sample than to the number of cells.

Hence, cell numbers are often not an ideal measure of population size or potential toxicity. This can be overcome by determining biomass. Two approaches are available, either estimation from cell counts and average cell volumes, or from chemical analysis of pigment content.

Cyanobacterial counts and cell volumes

Biovolume can be obtained from cell counts by determining the average cell volume for each species or unit counted and then multiplying this value by the cell number present in the sample. The result is the total volume of each species. Given a specific weight of almost 1 mg mm^{-3} for plankton cells, this biovolume corresponds quite closely to biomass. Average volumes are determined by assuming idealised geometric bodies for each species of each species, and calculating the corresponding mean volume of the respective geometric body.

Determination of Biomass using Chlorophyll a Analysis

The pigment chlorophyll *a* generally contributes 0.5-1 per cent of the ash-free dry weight of phytoplankton organisms. Although the pigment content may vary according to the physiological state of the organisms, chlorophyll a is a widely used and accepted measure of biomass. It is an especially useful measure during cyanobacterial blooms, when the phytoplankton chiefly consists of cyanobacteria, often of only one species. However, when chlorophyll *a* determination is used with mixed phytoplankton populations (cyanobacteria and other species), it gives an overestimation of cyanobacterial biomass. Rough microscopic estimations of the relative share of cyanobacterial cells among the total phytoplankton may be used to correct the overestimate.

Analysis of chlorophyll *a* requires relatively simple laboratory equipment, principally filtration apparatus, centrifuge and spectrophotometer. It is considerably less timeconsuming than microscopic biomass determination. Standard protocols have been described but preferred methods vary somewhat between laboratories. However, the main procedural steps in most methods are essentially the same: solvent extraction of chlorophyll *a*, determination of the concentration of the pigment by spectrophotometry, and adjustments to the result to reduce the interference by phaeophytin *a* which is a degradation product of chlorophyll *a*. A simple method following the ISO procedure for the determination of chlorophyll a in a lake water sample is outlined below.

Apparatus

— Spectrophotometer suitable for readings up to 750 nm, or photometer with discrete wavelengths at 665 and 750 nm

— Glass cuvettes, typically of 1 cm path length, or 5 cm for very low concentrations (e.g. from drinking water reservoirs at the beginning of population development)

- Centrifuge
- 15 ml centrifuge tubes, graduated and with screw caps
- Water bath at 75 °C or other heating device for boiling ethanol
- Glass fibre filters (GF/C), 47 mm diameter
- Filtration apparatus and vacuum pump
- Tissue grinder or ultrasonication device
- Pipette or similar for addition of acid

Reagents

- Ethanol (90 % aqueous)
- Hydrochloric acid, 1 mol l^{-1}

Procedure

Perform the following steps in low intensity of indirect light because light induces rapid degradation of chlorophyll.

1. After recording the initial volume of water, separate the cells from the water by filtration. Filter continuously and do not allow the filter to dry during filtration of a single sample. If extraction cannot be performed immediately, filters should be placed in individual, labelled bags (filters folded in half with cells innermost) or Petri dishes and stored at -20 °C in the dark. This step can be carried out at the sampling site and the samples are readily transported in this form. In preference to freezing, samples may be stored in the extraction medium for up to 4 days in the refrigerator.
2. Place the filter in a tissue grinder, add 2-3 ml of boiling ethanol, and grind until the filter fibres are separated. Ultrasonication can also be used. Pour the ethanol and ground filter into a centrifuge tube, rinse out the grinding tube with another 2 ml ethanol and add this to the centrifuge tube. Make up to a total of 10 ml in the centrifuge tube with ethanol. Place cap on the tube, label and store in darkness at approximately 20 °C for 24-48 hours.
3. Centrifuge for 15 minutes at 3,000-5,000 g to clarify samples. Decant the clear supernatant into a clean vessel and record the volume.
4. Blank Spectrophotometer with 90 per cent ethanol solution at each wavelength.
5. Place centrifuged sample in the cuvette and record absorbance at 750 nm and 665 nm. Readings at 665 nm should range between 0.1 and 0.8 units.

6. Add 0.01 ml of 1 mol l^{-1} HCl to sample in cuvette (adjust volume to suit the volume of cuvette being used, calculating approximately 0.003 ml of 1 mol l^{-1} HCl per ml of ethanol solution) and agitate gently for 1 minute. Record absorbance at 750 nm and 665 nm (750b and 665b).

Determination of Nutrient Concentrations

The capacity for development of a cyanobacterial bloom depends on the available concentrations of elements that the cells are composed of. These elements are needed in the ratio in which they occur in living cells. Hydrogen and oxygen are available in unlimited supply in an aqueous environment, and sulphur is usually also present in surplus concentrations. Carbon has been investigated as a potentially limiting factor, but has rarely been found to be relevant. Most often, phosphorus concentrations limit the amount of biomass that can form in a given water body but sometimes, nitrogen is limiting. The chief sources of nitrogen are nitrate and ammonia, but to some extent their lack can be compensated by some cyanobacteria through fixation of atmospheric nitrogen. Thus, even if phosphate is clearly the factor limiting carrying capacity, knowledge of nitrogen availability helps to predict whether nitrogen-fixing species are likely to grow.

Cyanobacterial cells appear to have little means of storing excess nitrogen, but can store enough phosphate for up to four cell divisions, which implies that one cell can grow into 16 without needing to take up dissolved phosphate. Information on dissolved phosphate concentrations, therefore, only demonstrates that if it can be detected, the phytoplankton population is not currently limited by phosphate. In order to assess the capacity of the water body to carry a cyanobacterial population, total phosphate must be determined, which can then be compared with the total concentration of nitrogen salts and organic nitrogen. However, in order to assess whether nitrogen may be limiting, analysis of dissolved components is sufficient. Among the methods available, the procedure of Koroleff for determining total phosphate has proved to be most reliable and is the basis of an ISO protocol.

Analysis of phosphorus according to ISO 6878

Phosphorus in various types of waters can be determined spectrometrically by digestion of organic phosphorus compounds to orthophosphate and reaction under acidic conditions to an antimony-phosphormolybdate complex which is then reduced to a strongly coloured blue molybdenum complex. The internationally harmonised method described by ISO/FDIS 6878 is applicable to many types of waters in a concentration range of 0.005 to 0.8 mg l^{-1}. Differentiation by the following fractions is possible through filtration procedures:

Option	Fraction	Filtration/procedure
1	Soluble reactive phosphorus	Filtered sample
2	Dissolved organic phosphate	Digested filtered sample
3	Particulate phosphorus	Option 4 minus option 2
4	Total phosphorus	Digested unfiltered sample

Digestion or mineralisation of organophosphorus compounds to orthophosphate is performed in tightly sealed screw-cap vessels with persulphate, under pressure and heat in an autoclave, or simply by gentle boiling. Polyphosphates and some organophosphorus compounds may also be hydrolysed with sulphuric acid to molybdate-reactive orthophosphate.

ANALYSIS OF CYANOTOXINS

There is a diverse range of laboratory methods used to detect and identify cyanotoxins in water and cyanobacterial cells. These methods can vary greatly in their degree of sophistication and the information they provide. Relatively simple low cost methods can be employed which evaluate rapidly the potential hazard and allow management decisions to be taken. In contrast, highly sophisticated analytical techniques can be employed which determine precisely the identity and quantity of cyanotoxins. Techniques can be selected depending on the facilities and expertise available, coupled with the type of information required. For example, analysis of water for recreational purposes may only require detection of potentially harmful cyanobacteria, whereas ensuring potable water does not exceed guideline levels for cyanotoxins may require highly specialised equipment and expertise. Information obtained from simple, rapid screening methods such as microscopic examination can be used to make an informed decision on the type of bioassay or physicochemical technique which should be employed. It is important to remember that, currently, there is no single method which can be adopted which will provide adequate monitoring for all cyanotoxins in the increasing range of sample types which have to be evaluated. Selectivity and sensitivity are important criteria for the selection of methods.

Sample Handling and Storage

When samples arrive in the laboratory, the type of analysis that will be carried out should have been anticipated previously. Full consideration should be given to the type of information required prior to sample collection. However, this is not always possible, particularly when a routine monitoring programme is not in place. Information obtained from the microscopic examination of the cyanobacterial taxa may also influence the choice of analysis but this information is not usually available until samples have the laboratory. Many samples will therefore require immediate evaluation on arrival in the laboratory to determine if any pre-treatment is needed prior to appropriate sample storage.

Samples for cyanotoxin analysis should be refrigerated in the dark to prevent toxin degradation but it is essential that storage be kept to a minimum. Where prolonged storage is required, samples can be frozen, although this will release toxins from the cells and only the total amount of toxin in the sample can then be determined. It is often desirable, especially for water treatment purposes, to estimate the amount of toxin contained within cells as well as that which is dissolved in the water. Where this information is required, samples should be gently filtered as soon as practically possible and if necessary the filters and filtrate can be refrigerated or frozen. A concentration step for cyanobacterial cells can be useful especially for less sensitive cyanotoxin assay methods. This can be achieved at the sampling location with the use of a plankton net or by using lakeside filtration apparatus. In the laboratory, concentration can be achieved by allowing buoyant cells to accumulate in the upper part of a separating funnel which enables excess water to be removed. This method can concentrate cells by at least ten fold but is dependent on cells being vacuolate. It may also lead to a bias if most of the cells do not accumulate at the surface because floating cells may have different toxin quotas than cells which do not. Centrifugation is useful but is often limited by the relatively small volume of laboratory centrifuge tubes and, furthermore, problems can be encountered when trying to pellet vacuolate cells. Filtration is increasingly popular because it can allow concentration of cells by several orders of magnitude and enables the weight of the cell mass to be determined. The following filtration method can be employed.

Apparatus

- Oven/incubator set at 45 °C or freeze-drier
- Vacuum dessicator
- Balance accurate to 0.0001 g
- Glass fibre filters 70 mm, GF/C, 1.2 μm will retain most cyanobacteria but the smaller pore size of GF/F (0.7 μm) is required for picoplankton cyanobacterial cells.
- Filtration cup and vacuum pump
- Measuring cylinder
- Plastic Petri dishes

Procedure

1. Place filters in dessicator under vacuum, then remove at intervals and weigh. Filters

are ready to use once a constant weight is obtained. It is useful to place each weighed filter in a separate Petri dish with the weight recorded on the dish. This makes samples easy to handle and prevents confusion.

2. Mix the water sample by inverting the bottle several times, then measure a known volume using the measuring cylinder. This volume will differ depending on the concentration of cells present, because only small volumes of water with a high concentration of cells can be filtered before the filter becomes blocked.
3. Using one of the pre-weighed filters, filter the water sample and then return the filter to the labelled Petri dish.
4. Keeping the filter in the Petri dish place it either in the drying oven or freeze-drier. If using on oven, the temperature must be kept below 50°C.
5. Once dry, return filters to the dessicator and weigh to constant weight. Calculate the dry weight of cells collected by subtraction of the initial weight of the filter.

By following this procedure it is possible to relate cyanotoxin content to the dry weight of particulate matter extracted as well as to the volume of water filtered. Sometimes dry weights are affected by the presence of large particles in the sample. This can easily be overcome by carrying out pre-screening using a 1-2 mm sieve if necessary. Furthermore, particulate matter may consist of plankton organisms in addition to cyanobacteria. A brief microscopic check of the fresh sample or a preserved sub-sample will reveal whether or not the dry weight can be attributed mainly to cyanobacteria. In dense blooms and scums this is usually the case but, in samples from more homogenously dispersed situations, other components may dominate and the relating of the toxins to dry weight will therefore underestimate the toxin content of the cells. If an accurate balance is not available, the filtration method can be used but the cyanotoxin content can be related only to the volume of water filtered. Dense cyanobacterial scums or samples concentrated using a plankton net are often freeze-dried. This provides a dry powder which can easily be weighed prior to extraction. However, great care must be taken with such dry powders because they can easily become airborne and may present a health hazard through inhalation. Most sample handling and storage methods for cyanotoxin analysis have been evaluated primarily for microcystins, hence the stability of other cyano-toxins may not be fully understood. It is therefore important to evaluate the chosen method if other cyanotoxins are being monitored.

Sample Preparation for Cyanotoxin Determination and Bioassays

Extraction

For cyanotoxin detection, samples may be extracted from cells or biological matrices such as animal tissue, or water samples. Cells have successfully been extracted in a number of different liquid phases, and some of the most popular are 5 per cent acetic acid, methanol, acidified methanol, aqueous methanol, and butanol:methanol:water. The efficiency of these methods depends on the sample and the cyanotoxins present. It has been found that the more polar extraction media, such as 5 per cent acetic acid, provide reasonable extraction efficiency of the more polar microcystins but give very poor recovery of hydrophobic microcystins. Methanol has been advocated as the most suitable solvent because it gives good extraction efficiency and has the added advantage of allowing rapid sample concentration through evaporation. It has since been suggested that 100 per cent methanol may give poor recovery of more polar microcystins; but the addition of a small percentage of water overcomes this. Fastner *et al.* showed that, especially for lyophilised field samples dominated by *Microcystis* spp., extraction with 75 per cent methanol and 25 per cent water was most effective. One approach which has been used routinely, with success, for microcystins is as follows:

Apparatus

- Rotary evaporation equipment
- Glass beakers, 50 ml
- Measuring cylinder, 20 ml
- Rotary evaporation flasks, 50 ml, pear-shaped
- Pipette, 0.25 ml
- Glass vials or microcentrifuge tubes, ~1 ml.

Reagents

- Methanol, high purity if possible.
- Aqueous methanol (75 per cent, v/v)

Procedure

1. Place each filter containing cells into a glass beaker with 20 ml of either pure methanol or 75 per cent aqueous methanol. Filters may be cut into small pieces with scissors but gloves must be worn and care must be taken not to cross contaminate samples or to lose material sticking to the scissors.
2. Allow filters to extract for 1 hour. If time is limiting, extract for 30 minutes because the shorter time has been found to reduce only slightly the recovery of microcystins.
3. Decant extract into rotary evaporation flask and dry *in vacuo* at 45 °C.
4. Add a further 20 ml of extraction solvent to the filter in the beaker and allow to extract as before.

5. Repeat this process a total of three times, each time decanting the extract into the same rotary evaporation flask.
6. Add 0.25 ml of methanol to the dry extract in the rotary evaporation flask, mix and remove the resuspended extract to a glass vial or microcentrifuge tube. Repeat this with a second 0.25 ml of methanol, placing both aliquots in the same vial or tube.
7. Consideration must be given to the type of analysis which is going to follow extraction because organic solvents are toxic to bioassay organisms. This can often be overcome by using a volatile medium, followed by evaporation and resuspension in a medium which is compatible with the assay method. With very toxic samples and/or sensitive bioassays, dilution with the assay medium is sufficient to obtain non-toxic solvent concentrations.

Two other rapid extraction procedures have also been successfully applied:

— After freezing and thawing to disrupt cells, toxins may be extracted in the filtration device by passing aqueous methanol or water, followed by methanol, through the filters by suction (use a total of 5-25 ml of solvent). Extracts can be used directly for analysis or further concentrated if enhanced sensitivity is needed. This method requires little equipment other than a filtration device and deepfreezer.

— Membrane filters with approximately 20 mg of freeze-dried material may be extracted in 2 ml microcentrifuge tubes by adding 1.5 ml solvent, sonicating, shaking for 30 minutes and centrifuging. This step extracts a large share of the microcystins, and extraction is more complete if the pellet is re-extracted twice and the supernatants are pooled. Saxitoxins are often extracted in acidified media including acetic acid, hydrochloric acid (HCl) and acidified methanol. A recent study found methanol acidified with TFA was the most efficient solvent when extracting saxitoxin and neosaxitoxin from cyanobacterial cells. Anatoxin-a has been successfully extracted with water, acidified water, acidified methanol, chloroform followed by hydrochloric acid or by dichloromethane after an acidification and neutralisation step. To store extracts prior to analysis or bioassays, blow to dryness and deepfreeze (-18 °C). Samples may then be re-dissolved in the solvent at a concentration adequate for the subsequent chemical analysis or bioassay.

Sample cleanup

The purpose of cleanup is to eliminate impurities by a simple operation without loss of analyte and where the concentration of cyanotoxins is low, it also enables enrichment of the analyte. Whether or not cleanup is necessary depends largely on the precision of the toxin determination and quantification required. Without cleanup, small toxin peaks in chromatograms may be missed due to masking by other matrix substances eluting simultaneously, and toxin concentrations may be overestimated if peaks are not clearly separated from matrix substances. However, for many screening and monitoring purposes, particularly of bloom samples containing little material other than cyanobacteria, the carrying out of bioassays or toxin analysis without sample cleanup has given satisfactory results. The establishment of a versatile cleanup method would give an additional advantage to the analysis of microcystins in other matrixes such as biological samples, fish and shellfish. Furthermore, it would give useful information for preparative separation of microcystins and their degradation products. Octadecyl silanised (ODS) silica gel has been employed extensively to facilitate sample cleanup and trace enrichment of microcystins and nodularin because it retains the toxins and allows interference compounds to pass through.

Concentration and cleanup method for microcystins and nodularins in water samples

Apparatus

- Porcelain filter funnel, 110 mm diameter and Buchner flask, 1 litre or similar
- Filter disks, GP/C, 110 mm
- Measuring cylinder, 500 ml
- Pipettes, various
- Water vacuum pump
- Glass bottles, 500 ml
- Vacuum manifold
- Solid phase extraction cartridges, 1 g trifunctional, end-capped C18
- PTFE tubing and cartridge adapters
- Glass sample tubes
- Drying hot block (45 °C) with blow-down nitrogen
- Microcentrifuge tube (1.5 ml)

Reagents

- Sodium thiosulphate solution, 1 g $Na_2O_3.5H_2O$ made up to 100 ml with distilled water
- Aqueous methanol, solutions containing 10, 20 and 30 per cent methanol in water
- Trifluoroacetic acid solution, 10 per cent TFA v/v in water
- Trifluoroacetic acid solution, 0.1 per cent TFA v/v in methanol
- Methanol

— Water

Procedure

1. Mix the water sample by inverting the container several times, then measure a 500 ml portion of the sample and filter it gently through a GF/C filter disc. The filter can be retained and extracted as described above to determine the particulate microcystin concentration.
2. Add 0.1 ml sodium thiosulphate solution to eliminate free residual chlorine. Shake the water sample vigorously and let it stand for a few minutes, then add 5 ml of the 10 per cent TFA and mix before passing the sample through a GF/C filter disc. 3. Place the sample in a 500 ml glass bottle, add 5 ml methanol and mix. The sample is now ready for solid phase extraction (SPE).
4. Solid phase extraction cartridges are prepared by attaching them to the vacuum manifold system and then conditioning them with 10 ml methanol followed by 10 ml water, ensuring that the cartridge does not become dry at any time. The methanol and water eluates are discarded.
5. Using PTFE tubing and adapter, attach a tube from the bottle containing the water sample to the top of the SPE cartridge. The vacuum draws the sample through the tubing and through the cartridge. The water is not collected but allowed to run to waste through the water pump.
6. Once all of the water sample has passed through it, the cartridge is washed with 10 ml of the 10 per cent methanol followed by 10 ml of the 20 per cent methanol and then finally washed with 10 ml of the 30 per cent methanol. The eluate from the three washes is discarded.
7. The cartridge is eluted with 3 ml of 0.1 per cent TFA in methanol. This is collected in a sample tube and dried on a hot block (45 °C) under a gentle stream of nitrogen gas. 8. Samples are resuspended in 0.1 ml methanol and placed in a microcentrifuge tube. A further 0.1 ml of methanol is used to rinse the sample tube and this is combined with the first aliquot.
9. This sample can now be analysed or it can be dried and stored in the freezer until required.
10. It is recommended that this procedure is carried out in duplicate for each water sample analysed.

   — Different systems can be used to pass the water sample through the SPE cartridge but a specially designed vacuum manifold apparatus is available for use with SPE cartridges. Most of these systems are designed so that a number of cartridges can be attached at one time (e.g. 10 or 20). Automated systems are also available, and peristaltic pumps have also been used successfully, but are not as practical.

— It is vitally important that all apparatus used is thoroughly cleaned to ensure no external contamination. Equipment used for trace enrichment should be dedicated to this procedure only, in order to reduce the opportunities for contamination.

Recently, an immunoaffinity purification method has been developed using an antimicrocystin-LR monoclonal antibody. This cleanup method was found to be remarkably effective in the removal of contaminants in the hepatic cytosol and enabled the analysis of microcystins and their metabolites, formed *in vivo* in mouse and rat livers, by HPLC and Frit-FAB LC/MS. Before the immunoaffinity purification, the spiked microcystins could not be accurately quantified due to the many impurities, whereas after the immunoaffinity purification, the impurities were effectively eliminated and the peaks of the microcystins were clearly detected. Although this immunoaffinity column is not commercially available, it has the potential to enable significant progress in a number of important areas of research; for example in metabolism studies and in elucidation of the fate of microcystins in the environment. Research into appropriate cleanup methods for other cyanotoxins has still to be carried out, although a similar approach to that which has been developed for microcystins may prove useful.

Toxicity Tests and Bioassays

There have been many biological detection methods developed for cyanotoxins that use the bioactivity of the toxins, such as potent hepatotoxicity, neurotoxicity, cytotoxicity, enzymatic activity and immunological interactions. However, for many years, the mouse bioassay alone has been used to determine bloom toxicity. Although this bioassay provides a measure of the total toxicity within a few hours, it is generally not very sensitive or specific. Considerable research efforts have been made to find suitable alternative methods to the mouse bioassay as a routine monitoring assay for cyanotoxins and many novel and sensitive methods have become available in recent years. However, no single method is currently available to replace the mouse for the detection of all cyanotoxins using a single assay, and further validation and comparison of methods is needed before general recommendations on their application can be given.

Different bioassays are described below. Users are advised to test their suitability for locally-prevalent cyanotoxins with other methods, such as chemical toxin analysis. Sensitivity and selectivity are important criteria for the selection of methods.

Mouse bioassay

Male Swiss Albino mice are the most used animals for toxicity testing for cyanotoxins.

Toxicity is tested by intraperitoneal injection (i.p.) of 0.1-1.0 ml of a lysate of cyanobacteria prepared either by sonication or by freeze-thawing of a cell suspension which has been sterilised by membrane ultra-filtration. Samples can be suspended in water or physiological saline solution which is preferred if the volume to be injected is 0.5 ml or greater.

Mice should be observed for 24 h and then killed by an approved method. The observation period must be extended to seven days where cylindrospermopsin is suspected and animals should be injected with sterile samples. This toxin demonstrates protracted symptoms which result from progressive organ failure, specifically liver and kidneys. At the end of the observation period post-mortem examination of tissue injury is performed. The observed symptoms and the results of the post-mortem are used to determine which cyanotoxin is present. However, where more than one type of cyanotoxin is present, the more rapid-acting toxin may mask other symptoms.

Toxicity is expressed as LD_{50} mg cell dry weight per kg mouse body weight. Values are classified as follows:

- > 1,000 non toxic
- 500-1,000 low toxicity
- 100-500 medium toxicity
- < 100 high toxicity

Note that in some countries non-toxic limits may be at least 2,000 mg cell dry weight per kg mouse body weight.

Invertebrate bioassays

A number of invertebrates have been investigated for use in routine bioassays for cyanotoxins. Of these, the brine shrimp (*Artemia salina*) has been the most popular, because no culture maintenance or specialist equipment is required. Brine shrimps have been exploited for many years for the detection of toxic secondary metabolites and are now commercially available as standardised test kits. Brine shrimp eggs are readily available from biological supply companies and can be stored for several years at -20 °C without loss of viability.

Brine shrimp bioassay

Apparatus

- Conical flasks, glass, 250 ml
- Microtitre plates, 96-well

- Incubator, 25 °C
- Desk lamp
- Pasteur pipette and bulb
- Pipette, 0.1 ml
- Universal bottle, 25 ml
- Dissecting microscope with low power objective

Reagents

- Brine shrimp eggs, stored in freezer
- Brine shrimp medium (BSM) stock solution, composed of:

  Sodium chloride (NaCl) 300 g

  Calcium chloride dihydrate ($CaCl.2H_2O$) 3 g

  Magnesium chloride hexahydrate ($MgCl_2.6H_2O$) 15 g

  Magnesium sulphate heptahydrate ($MgSO_4.7H_2O$) 5 g

  Potassium chloride (KCl) 8 g

  Glycine 60 g

  Disodium glycerophosphate 30 g
- Methanol or formalin

The stock solution chemicals are dissolved in 1.25 litres of distilled water, dissolving each chemical separately in the order shown. It is important to add the disodium glycerophosphate last to prevent an insoluble precipitate occurring. The stock solution should be stored in a brown glass bottle in the refrigerator (~4 °C).

Procedure

1. Mix 20 ml of BSM stock solution with 140 ml distilled water in a 250 ml conical flask.
2. Add 100 mg of brine shrimp eggs and incubate at 25 °C until they hatch (usually 36-48 hours). Once a regular hatching time has been established try to keep it constant.
3. Separate the hatched larvae from unhatched eggs and egg cases by transferring them into a shallow dish, e.g. Petri dish or slightly deeper.
4. Position a desk lamp to one side of the dish and allow the larvae to concentrate towards the light.

5. Using a Pasteur pipette carefully, using slow steady suction on the pipette bulb, collect the hatched larvae.
6. Place larvae in a universal bottle and repeat the collection process until most of the hatched larvae have been collected.
7. If a substantial number of unhatched eggs have been collected in error, the separation can be repeated by placing the collected larvae in a clean dish and illuminating.
8. Using a mechanical pipette with plastic tips pipette 0.1 ml of the larvae suspension into a microtitre well. Make sure the larvae suspension is regularly mixed because they tend to cluster rapidly in the bottle. Mixing is easier if the bottle is only half full.
9. Examine the well containing larvae under the microscope to determine the approximate number of larvae in 0.1 ml. Ideally this should be between 15 and 25 larvae. Dilute the sample with fresh media if the larvae are too concentrated. Repeat the pipetting a few times to ensure reasonably reproducible numbers of larvae are being deposited in each well and that they are alive and look healthy.
10. Fill as many wells with larvae suspension as will be required to carry out the bioassay.
11. To determine the toxicity of cyanobacterial cells, the sample must first be extracted. This can be done as described above from filter discs, but if the final extract is in methanol it must be diluted in BSM so that the final concentration of methanol does not exceed 5 per cent in the test well.
12. A dilution series of each sample should be prepared and 0.1 ml of the test solution added to the larvae in the wells in triplicate. The microtitre plate is then incubated for 18 hours at 25 °C after which the percentage mortality is calculated.
13. First, with the help of a microscope, count the number of dead or immobilised larvae in each well, then add a few drops of formalin or methanol and wait until all the larvae are dead. Now count the total number of larvae in each well and calculate the percentage mortality using a mean of the three replicates.
14. By plotting the concentration of cyanobacteria against the percentage mortality the LC_{50} value for each sample can be determined.

Care must be taken to exclude particulate material from the assay and therefore extracts must be filtered or centrifuged. High concentrations of cell extract cause false positives, probably due to oxygen depletion caused by the activity of bacteria, therefore sample cleanup is necessary to detect low concentrations of toxin successfully. Sample cleanup, e.g. the use of SPE, can also enhance the specificity of the assay because it selectively concentrates microcystins and nodularins, reducing interference by other compounds.

When 21 hepatotoxic bloom samples were assessed by this assay the results compared very favourably with both mouse bioassay and HPLC. This assay has not been fully evaluated for toxins other than microcystins, although there does appear to be a correlation between anatoxin-a content and toxicity.

Daphnia bioassays can successfully detect microcystins, although standardised culturing is extremely labour intensive. A commercially prepared test kit, similar to the brine shrimp assay in that it uses the resting stage of an aquatic invertebrate, *Thamnocephalus platyurus* has been found to be sensitive to a number of cyanotoxins. These kits are relatively expensive and have a limited shelflife (6 months), although the standardised format of this bioassay leads to highly reproducible results and low inter-laboratory variability. The use of mosquito adults and larvae have both been investigated as potential bioassays. Adults were injected and larvae immersed in aqueous extracts. Both methods were relatively sensitive but have not been widely adopted due to the difficulties of handling this organism. The other insect which has been shown to detect microcystins successfully in bloom samples is the fruit fly (*Drosophila melanogaster*). These organisms are easy to maintain in the laboratory, with no special equipment required. Toxin is administered orally by adding filter discs spotted with sample plus sucrose to tubes containing prestarved (24 h) flies. The flies were not, however, sensitive to neurotoxic *Aphanizomenon*.

Two invertebrate assays have been investigated for the detection of saxitoxins. Firstly, adult house flies injected with purified toxins and natural samples (shellfish extracts) gave results which compared well with toxicity determined by mouse. However, the flies are difficult to handle and require microinjection (1.5 μl) which is difficult to administer. A locust bioassay has recently been found to detect saxitoxins successfully in a range of samples, namely cyanobacteria and shellfish. Locusts are easy to handle because they can be readily immobilised by holding their rear legs. Samples are administered by injection (10 μl) and results are obtained within 90 minutes. The LD_{50} for pure saxitoxin was 8 $\mu g\ g^{-1}$ but the bioassay was not sensitive to microcystin-LR or anatoxin-a.

Bacterial bioassays

Bacterial bioassays have been investigated to determine if they can provide simple routine methods for cyanotoxin detection. The one that has received the most attention is the Microtox bioluminescence assay which indicates toxicity by a reduction in the light emitted by the test bacterium (*Photobacterium phosphoreum*). Initial investigations suggested that this system may be suitable for the rapid detection of microcystins in bloom samples although more detailed analysis revealed that the assay responded to unknown components of cyanobacterial extracts other than microcystins. Several studies have now been published that clearly indicate there is no correlation between response in the Microtox assay and cellular content of the known cyanotoxins. A second bacterial bioassay which used the inhibition of pigment (prodigiosin) formation in *Serratia marcescens* as an indication of toxicity has been proposed by Dierstein *et al.* This bioassay was thought to be useful for saxitoxins and microcystins. However, like the Microtox system, little correlation was found between actual content of known cyanotoxins and inhibition of pigment formation.

Biochemical assays

The protein phosphatase inhibition assay is a sensitive screening method for microcystins and nodularins which uses the biochemical activity of these toxins. One version is based on the quantitation of 32P-phosphate released from a radiolabelled substrate by the activity of the protein phosphatase enzyme. It is sensitive to sub-nanogram levels of microcystin and is a rapid assay allowing the analysis of many samples in a few hours. The method has been applied to detect microcystin class compounds in the marine environment, to extracts of liver tissue taken from Atlantic salmon afflicted with netpen liver disease and to hydrophobic microcystins from freshwater cyanobacteria. The method has also been successfully used for quantitation of microcystins in drinking water before and after water treatment. The detected amounts in raw and treated waters were estimated to be 0.12-0.87 and 0.09-0.18 $\mu g\ l^{-1}$, respectively. Although this method has been widely used in research, there is a reluctance in adopting it for the routine monitoring of microcystins because of the requirement to use radioactivity which necessitates specialised laboratory equipment and regulations.

An and Carmichael have used a colorimetric protein phosphatase inhibition assay which avoids the complications of using radioactive materials. Isobe *et al.* reported a firefly bioluminescence system for the detection of protein phosphatase 2A inhibitors, in which luciferin phosphate is hydrolysed to luciferin and inorganic phosphate by protein phosphatase 2A. The use of the protein phosphatase inhibition assay is extremely helpful to confirm biological activity, and hence toxicity, of microcystins in environmental samples. The non-radioactive bioassay may therefore be Used increasingly for the routine screening of water samples, as shown recently by Ward *et al.*

The biochemical activity of anatoxin-a(S) can be exploited in an enzyme-based assay to detect the inhibition of acetylcholinesterase (AChE), thereby providing an indication of the presence of this toxin. This is a sensitive method and is the only alternative to the mouse bioassay currently available for this toxin. The assay is not selective because it will also detect other toxicants, such as organophosphorus-based pesticides.

Immunological detection

The Enzyme-Linked Immuno Sorbent Assay (ELISA) technique is currently the most promising method for rapid sample screening for microcystins because of its sensitivity, specificity and ease of operation. Monoclonal antibodies raised against microcystin-LA were initially developed by Kfir *et al.* and offered a simple approach to a general immunoassay for microcystins. An ELISA technique was subsequently developed by Chu *et al.* as a practical method. This assay is based on polyclonal antisera raised in rabbits against bovine serum albumin conjugated to microcystin-LR. The antisera showed good cross-reactivity with microcystins-LR, -RR, -YR and nodularin, but less with -LY and -LA. The sensitivity of the assay showed approximately 50 per cent binding at a toxin concentration of 1 ng ml^{-1} which is appropriate for normal water quality testing. In fact, this method has been successfully employed for quantitation of cyanobacterial hepatotoxins in domestic water supplies and biomass extracts with detection limits of 0.2 μg l^{-1} and 0.25 μg g^{-1} for water and biomass samples, respectively.

Recently, Nagata *et al.* produced six monoclonal antibodies against microcystin-LR. Among them, M8H5 antibody showed cross-reactivity with microcystin-RR, microcystin-YR (44 per cent), microcystin-LA (26 per cent), [D-Asp^3]microcystin-LR (51 per cent), [Dha^7] microcystin-LR (48 per cent), glutathione conjugate of microcystin-LR (47 per cent), monomethyl ester of microcystin-LR (30 per cent), nodularin (46 per cent) and 6(Z)-ADDA microcystin-LR (< 4 per cent). Although the epitope of this antibody is not clear, the importance of the Adda moiety for antibody binding has been indicated. It should be noted that this antibody also reacts to the non-toxic monomethyl ester of microcystin-LR giving a false positive from the toxicological point of view. Using this monoclonal antibody, a more sensitive competitive ELISA method has been developed by Ueno *et al.* with detection limits of 0.05 μg l^{-1} for water samples. This method has been successfully applied to detect microcystins in drinking water in China.

Commercially, a polyclonal ELISA kit is available for microcystins. The antibodies are fixed to the walls of the wells of a microtitre plate. The first step involves binding of the calibrators a negative control and the samples to the antibodies in the wells. This is followed by addition of a microcystin-enzyme conjugate which binds to the remaining antibodies. After thorough rinsing, the concentration of bound enzyme is measured colourimetrically in an ELISA plate reader. The microcystin concentration is inversely proportional to the colour intensity.

Mammalian cells

Bioassays using mammalian cells have received attention as suitable replacements for mouse toxicity tests. The well documented fact that microcystins have caused acute liver damage has prompted studies using hepatocytes (liver cells). Freshly isolated rat hepatocytes were first investigated by Aune and Berg who reported good correlation between toxicity measured by leakage of the enzyme lactate dehydrogenase (LDH) from hepatocytes and results from mouse bioassay. More recently this method has received renewed interest due to legislative restrictions in the use of the mouse LD_{50}. Isolated rat hepatocytes have been incubated with pure toxin or bloom extracts for 4 h and 20 h and then the viability has been assessed using the MTT ((3,4,5-dimethylthiazol-2yl)-2,5-diphenyltetrazolium bromide) test. Toxicity was found to be timedependent with the LC_{50} for microcystin-LR being 0.20 μg ml^{-1} at 4 h reduced to 0.05 μg ml^{-1} after 20 h. Different LC_{50} values were found for microcystin variants, most notably microcystin-RR was found to be at least an order of magnitude less toxic in this assay. This is consistent with *in vivo* toxicity data.

An *in vitro* fibroblast cytotoxicity assay for microcystins has been investigated by Codd *et al.* as a replacement for the mouse bioassay. The assay, using V79 hamster fibroblast cells, responded to microcystins with the results correlating reasonably well with those of the mouse bioassay. However, the assay demonstrated a number of false positives and, of more concern, false negatives were also recorded.Another cell-based assay proposed in the past has used blood cells, with agglutination of the cells being reported as an indicator of microcystin level. Although this bioassay did appear promising, it has since been found to be a poor indicator for microcystins.

Two *in vitro* cell bioassays have been found to be successful in detecting saxitoxins and, like the immunoassay, they were developed primarily for monitoring toxins in shellfish. First, a neuroreceptor binding assay was developed that uses radiolabelled saxitoxin and works on the basis of competitive displacement. The initial protocol has subsequently been refined and the data obtained correlate well with the mouse bioassay. A neuroblastoma cell line technique for sodium channel blocking activity has also been developed. This assay, which is currently undergoing evaluation for its suitability as a replacement for the mouse in shellfish monitoring, is now available as a commercial test kit.

Analytical Methods for Cyanotoxins

Analytical methods use the physicochemical properties of cyanotoxins such as molecular weight, chromophores and reactivities due to the functional groups in the molecules. Physicochemical methods used for cyanotoxin detection are summarised in Table 13.2, which also indicates that the initial capital expenditure to establish most of these methods is high.

Another approach for further confirmation and identification of microcystins has been proposed using a LC-linked protein phosphatase assay. Essentially, this method makes use of HPLC to separate and identify tentatively the cyanotoxins present, and then monitors the bioactivity of the detected peaks using the protein phosphatase inhibition assay. This provides excellent confirmation of protein phosphatase inhibiting cyanotoxins, especially in complex samples.

Detection methods for microcystins and nodularins

The majority of analytical methods have been developed primarily for microcystins although because both nodularins and microcystins show similar physicochemical properties, nodularins can easily be analysed by the same methods. The most commonly-used analytical system for this class of cyanotoxins is HPLC. Combined with UV detection, HPLC has been used extensively for the detection of microcystins, but because this method relies on retention time for identification, microcystin standards are required. Detection by UV can be made more specific by using a photodiode array (PDA) UV detector but it has very limited ability to identify individual microcystins because almost all microcystins show a similar UV spectrum. Recent advances in detector hardware can now provide high resolution spectra that detect very slight variations in chemical composition and can be used in conjunction with advanced spectral matching software. These developments may assist in the identification of microcystins by spectral match data in conjunction with retention times. However, a fundamental problem still exists in the availability of standards. With over 60 microcystins known, it is currently impossible to create a definitive spectral library, which is a limiting factor when using this method to identify unknown microcystins.

Typical HPLC analysis uses a reverse-phase C18 silica column with separation achieved over a gradient of water and acetonitrile, both containing 0.05 per cent triflouroacetic acid (TFA). The gradient has to cover a sufficient range of polarities (e.g. 30-70 per cent acetonitrile) to allow the analysis of all microcystins which are known to vary considerably in their polarities. Data is gathered at 238 nm and where PDA is used spectral information is collected between 200 and 300 nm. Use of HPLC-PDA can allow tentative identification of microcystins and this method was found to perform very well when over 20 samples were assessed by HPLC and compared with mouse bioassay data. No false negatives were observed and only one false positive was reported, the latter being attributed to a relatively low level of microcystin which failed to cause death in the mouse bioassay. Other solvent systems have also been successfully used including methanol/water and ammonium acetate/acetonitrile.

When further confirmation and identification of microcystins is required, more advanced methodology must be used. Liquid chromatography/mass spectrometry (LC/MS) is a very promising method because it enables the simultaneous separation and identification of microcystins in a mixture. The toxic fraction contains mainly two microcystins as shown by the mass chromatogram monitored at the characteristic ion m/z 135 derived from Adda, which has proved to be useful for the discrimination of microcystins from other types of compounds. The two peaks were readily identified as microcystins-RR and -LR according to the mass spectra and mass chromatograms at their $[M+H]^+$. An advanced Frit-FAB LC/MS method using a microbore column (0.3 mm internal diameter) enabled the identification of nanogram levels of microcystins in water and biological samples. This increase in sensitivity was achieved by splitting the mobile phase between the pump and the injector so that the total effluent could be introduced into the mass spectrometer.

A physicochemical screening method that is based on the detection of 2-methyl-3-methoxy-4-phenylbutyric acid (MMPB) as an oxidation product of microcystins has been reported. The MMPB was initially prepared by Lemieux oxidation, followed by analysis by gas chromatography (GC) with a flame ionisation detector or HPLC with fluorescence detection. However, it required tedious procedures such as extraction, cleanup, oxidation and post-treatment in order to eliminate the reagents used, and derivatisation for GC and HPLC analysis. An improved method using ozonolysis made it possible to reduce significantly the formation times of MMPB because the previously required extraction, cleanup and other procedures could be entirely eliminated. The resulting intact MMPB was directly analysed by thermospray (TSP) interface LC/MS and EI-GC/MS using selected ion monitoring. This new procedure, from the ozonolysis of the microcystins to analysis of MMPB at picomol levels, took only 30 minutes to perform. The quantification of bloom samples achieved by this method were consistent with those obtained by HPLC analysis, showing that the method provided a means of screening for microcystins, as well as for their accurate quantification. Additionally, the most remarkable feature of this method is the applicability to complex sample matrixes, including solid material such as animal tissue, without the requirement for any complicated processing.

The structural determination of microcystins and nodularins has been carried out by nuclear magnetic resonance (NMR) spectroscopy and recent advances in 2D NMR techniques have proved to be essential for the structural determination of known and unknown microcystins. Fast atom bombardment MS and liquid secondary ion (LSI) MS give a protonated molecule $[M+H]^+$, providing information about molecular weight with further structural information obtained by Tandem FAB MS (FAB-MS/MS) as used in recent studies. However, NMR and MS (except LC/MS) usually require relatively large amounts of sample (milligram quantities) and completely purified microcystins, therefore they are not used in routine monitoring.

For identification of microcystins with very small sample volumes (less than 1 mm^3 freeze-dried material). Matrix Assisted Laser Desorption Ionisation Time of Flight (MALDI-TOF) has recently been developed. This method provides the molecular mass of all of the peptides in a sample and thus gives strong indications of the microcystin variants present. Post Source Decay (PDS) spectra may be obtained, which are characteristic for different microcystins. A library is currently being established. For rapid qualitative assessment of microcystins and other cyanobacterial peptides, this method is highly promising. Quantitative assessment is not yet possible. Different methods provide different and often complementary information, therefore combined use of suitable methods is recommended according to the purpose and type of data required. This is necessary because none of the methods currently available provides all the information which may be required. Furthermore, individual laboratories must identify the techniques that are both suitable for their analytical requirements and use their own expertise and available technology for a recent review of chromatographic methods for microcystins. Initial screening of samples can check rapidly for the presence of microcystins in a small amount of sample using sensitive and simple methods.

For studies requiring enhanced precision, accuracy and sensitivity in detection of individual toxins, a multistage procedure is required. This may comprise initial screening of samples to check for the presence of microcystins in a small amount of sample using sensitive and simple methods, such as bioassays and MMPB method. The use of screening helps to reduce the number of samples which require full analytical investigation and thus reduces the laboratory commitment. Furthermore, rapid results from an initial screen may aid prompt regulatory responses regarding the suitability of water for human use. If a sample is positive in a screening test, it will be necessary to follow through with identification and quantitative analysis. Prior to this, cleanup and sample concentration is often very important for many samples because relatively low levels of microcystins, around 1-2 $\mu g\ l^{-1}$, are usually present in water samples. A tandem cleanup system using ODS silica gel and silica gel cartridges, facilitates the accurate analysis of trace amounts of microcystins in water. Finally, quantitation of microcystins is usually essential, and although the screening methods can provide an indication of total microcystin concentrations in samples, a separation and spectroscopic method such as HPLC with UV detection should be applied at this final step. Availability of authentic standards would strengthen the capability of HPLC with UV detection but otherwise LC/MS should be the method of choice despite the requirement of a more specialised technique.

Anatoxin-a

Three methods for analysis of anatoxin-a have been reported to date: GC/MS, GC with electron capture detection (ECD) and HPLC. One GC/MS method has been published as a confirmation of anatoxin-a in material taken from a toxic bloom that resulted in the deaths of 16 cows. In the procedure, following liquid-liquid extraction with *N*-acetylation, the resulting acetylated toxin was analysed by a capillary GC/MS. Although confirmation was based on interpretation of the mass spectrum, the detection limit was unclear. Another GC/MS method has also been established for *N*acetylated anatoxin-a and it was successfully employed for preliminary characterisation of neurotoxic cyanobacteria from Finland. The GC/ECD method provided a higher sensitivity than other analytical methods, it used an internal standard for accurate quantification, and it could be applied to analysis of field samples. Although GC/ECD requires a considerably more complicated cleanup operation and derivatisation prior to analysis, it has been used successfully for stability studies on anatoxin-a.

Derivatisation with N-pentafluorobenzylbromide was shown to achieve a sensitivity of 2.5 pg. It is possible to use HPLC with UV detection to analyse intact anatoxin-a which has a strong absorption at 227 nm. After extraction of a cyanobacterial suspension with chloroform, followed by re-extraction with 0.01N hydrochloric acid, the resulting extract is separated under the following conditions: column, ODS silica gel; mobile phase, methanol-0.01M perchloric acid (7:3); detection, UV (227 nm). Harada *et al.* reported an alternative approach using a reversed phase HPLC method with a methanol-10.01M ammonium chloride (1:9) solvent system. Mass spectrometry is a very effective method for identification of anatoxin-a and its derivatives. Conventional electron ionisation (EI) and chemical ionisation (CI) can yield molecular ions and protonated molecules, respectively. Ross *et al.* evaluated secondary ion mass spectrometry (SIMS), GC/MS, desorption CI and TSP-MS for detection of the toxin and its derivatives. The use of HPLC coupled with mass spectrometry using thermospray interface (TSP-LC/MS) has also been investigated. The latter method made possible a sensitive, specific and reproducible analysis of anatoxin-a and its non-toxic oxidation product when used in combination with a cleanup method including a solid phase extraction with a reversed phase carboxylic acid cartridge. Using this method, trace amounts of anatoxin-a were detected in three strains and two bloom samples collected in Japan.

Anatoxin-a(S)

Matsunaga *et al.* purified anatoxin-a(s) and determined its structure, although no analytical method has been developed. Traditional HPLC plus UV detection is not suitable because this cyanotoxin lacks a strong chromophore. Liquid chromatography/mass spectrometry may prove useful, but this analytical technique has not been evaluated yet for anatoxin-a(S).

Cylindrospermopsin

The first analytical method reported for this toxin consisted of a combination of a cleanup step using HP-20 and ODS silica gel cartridges followed by HPLC with photodiode array detection. This method was applied to a bloom sample collected in Japan which was thought to contain Cylindrospermopsin. However, the authors suggested that the mobile phase used in this study was not satisfactory because Cylindrospermopsin showed poor retention power on the ODS column and slight tailing, and their method requires further development to allow precise analysis. A recently published method describes the extraction of cells in 5 per cent aqueous acetic acid and analysis using Spherisorb ODS-2 with a 10 minute linear gradient from 0 to 5 per cent methanol followed by a further 10 minutes at 5 per cent methanol. This gave good retention, separation and peak shape. Cylindrospermopsin was found to have a characteristic UV absorbance spectrum between 200 and 300 nm, with a maximum absorbance at 262 nm.

Saxitoxins

The methods mentioned here have been developed primarily for the analysis of saxitoxins in the marine environment, particularly in shellfish. However, they have been found to be equally suitable for cyanobacterial samples. The most commonly used analytical method for the saxitoxins is HPLC with on-line post-column oxidation and fluorescence detection. Although this method has been found to be the most satisfactory to-date, it requires three different mobile phase systems to allow analysis of all the saxitoxins. Furthermore, there is limited availability of analytical standards for all saxitoxin variants, which are necessary for peak confirmation. Mass spectrometry with FAB, TSP, electrospray and ion-spray ionisation has been reported for the paralytic shellfish poison (PSP) toxins. The application of an LC/MS method for saxitoxins has been reported, although it appears that it will be difficult to achieve analysis of all PSP toxins within one LC/MS run. The combination of capillary electrophoresis (CE) with ion-spray ionisation has proved to be successful for the saxitoxins. Analysis using CE was found to give poor detection limits due to the very small injection volumes (< 10 nl), but it may be possible to overcome this problem with an on-column sample preconcentration system. Furthermore, developments in CE technology are anticipated in the near future which will increase flow cell volume, hence enhancing detection limits.

REFERENCES

Chen, D.Z.X., Boland, M.P., Smillie, M.A., Klix, H., Ptak, C., Andersen, R.J. and Holmes, C.F.B. 1993 "Identification of protein phosphatase inhibitors of the microcystin class in the marine environment". *Toxicon,* 31, 1407-1414.

Davio, S.R. and Fontelo, P.A. 1984 "A competitive displacement assay to detect saxitoxin and tetrodotoxin". *Analyt. Biochem.,* 141, 199-204.

Edwards, C., Lawton, L.A., Beattie, K.A., Codd, G.A., Pleasance, S. and Dear, G.J. 1993 "Analysis of microcystins from cyanobacteria by liquid chromatography with mass spectrometry using atmospheric-pressure ionization". *Rapid Commun. Mass Spectrom.,* 7, 714-721.

Harada, K.-I. 1996 Chemistry and detection of microcystins. In: M.F. Watanabe, K.-I.

Isobe, M., Sugiyama, Y., Ito, T., Ohtani, I.I., Toya, Y., Nishigohri, Y. and Takai, A. 1995 "New analysis method for protein phosphatase type 2A inhibitors using the firefly bioluminescence system". *Biosci., Biotechnol. Biochem.,* 59, 2235-2238.

Kiviranta, J., Sivonen, K. and Niemelä, S.I. 1991 "Detection and toxicity of cyanobacteria" by *Artemia salina* bioassay. *Environ. Toxicol. Water Qual.,* 6, 423-426.

Lawton, L.A., Campbell, D.L., Beattie, K.A. and Codd, G.A. 1990. "Use of a rapid bioluminescence assay for detecting cyanobacterial microcystin toxicity". *Lett. Appl. Microbiol.,* 11, 205-207.

9

Cyanobacteria and Human Health

Epidemiological evidence results from studies of human populations that have shown symptoms of poisoning or injury attributed to the presence of cyanotoxins in drinking water or other sources of water. This type of evidence depends upon good case definition, good characterisation of exposure and a reporting system that enables these data to be compared. Such evidence, discussed later, includes the fatal intoxication by microcystin of 50 dialysis patients in Brazil, and the hospitalisation of 140 children supplied with water from a dam containing *Cylindrospermopsis raciborskii* in Australia. Reports of gastro-enteritis after the appearance of cyanobacterial blooms in drinking water sources come from North and South America, Africa and Europe. Most cases of human injury attributed to cyanobacterial toxins have been studied retrospectively, and complete epidemiological data, especially regarding exposure (number of organisms, type and concentration of cyanotoxins), are rarely available. Nevertheless, epidemiological evidence is of special importance in directly demonstrating the link between toxin exposures and human health outcomes, which otherwise cannot be derived directly from animal experiments.

A large number of animal poisonings have been recorded, some of which have been reported in detail, while others are anecdotal. These are important in the overall body of knowledge on cyanobacterial toxicity, because they demonstrate effects under natural conditions. However, cyanotoxin exposure is rarely quantified and cannot be directly extrapolated to human populations.

Although animal toxicity tests are not performed under conditions of natural exposure and are undertaken with different species, they are of particular importance because they are conducted under controlled laboratory conditions. They provide plausibility for the role of cyanotoxins in human and animal poisonings, and provide information on their comparative toxicity. Oral toxicity testing is especially relevant, and has been carried out on large animals as well as on rodents.

The overall case for the human health relevance of exposure to cyanotoxins can only be assessed properly by combining all available evidence and understanding its strength and its weaknesses.Of all the cyanobacterial toxins currently known, the cyclic peptides represent the greatest concern to human health because of the potential risk of long-term exposure to comparatively low concentrations of the toxins in drinking water supplies. The cyclic peptide microcystins and nodularins are specific liver poisons in mammals. Following acute exposure to high doses, they cause death from liver haemorrhage or from liver failure, and they may promote the growth of liver and other tumours following chronic exposures to low doses. The alkaloid neurotoxins have shown only acute effects in mammals. Risk assessment will, therefore, be limited to acute exposure. However, there are a number of complicating factors. Firstly, while there is a great deal of published information on the mammalian and human toxicity and toxicology of saxitoxins/PSP toxins, the animal data for anatoxins is rather sparse, and completely lacking for human exposure. Secondly, some of the alkaloid toxins, PSP toxins in particular, are known to accumulate to high concentrations in marine and freshwater biota. In many areas of the world, the local human population will be consuming fish, shellfish and animals that have grown in, or nearby, water bodies contaminated with toxic cyanobacteria. Thus there may be several sources of oral exposure for neurotoxins in addition to drinking water or recreational water. The potential for transfer of PSP toxins, and possibly other neurotoxins, to humans via these routes may be significant in freshwaters.

The toxicity of pure cylindrospermopsin has recently been studied following the identification of the toxin in 1992. These studies have helped to characterise the histopathology of cylindrospermopsin after intraperitoneal injection. It is not known whether cylindrospermopsin will elicit histopathological damage following chronic exposure at low concentrations. Lipopolysaccharides (LPS) from cyanobacteria can elicit both allergic and toxic responses in humans, although little is known about their acute or chronic effects. The lipid part of the molecule is believed to cause the deleterious response in humans, while the sugar moiety is important in determining LPS solubility and micelle properties, and hence is important in modulating toxicity. A lack of knowledge regarding the occurrence and toxicity of cyanobacterial LPS, combined with the diversity of LPS structures within the cyanobacteria, should be considered by health officials and water managers, particularly when gastrointestinal and respiratory symptoms are reported from exposed humans.

HUMAN AND ANIMAL POISONINGS

Water supplies have been associated with gastrointestinal illness throughout human history with cholera, dysentery, and typhoid responsible for much human misery and death. The epidemiological evidence for human illness due to cyanobacterial toxins therefore has to be viewed against a background of alternative causes, with bacterial, viral or protozoal infections being the first causes to be investigated.

Short-term Effects in Humans

The recorded cases of gastrointestinal and hepatic illness that can be reliably attributed to cyanobacterial toxins in water supplies have all been coincident with either the breakdown of a natural cyanobacterial bloom or with the artificial lysis of a bloom by application of copper sulphate. Both mechanisms lead to cyanotoxin release from decomposing cells. Whereas treatment procedures might have removed cyanotoxins bound in intact cells, they were not effective in removing the dissolved cyanotoxins in these cases. The earliest reported cases of gastro-enteritis from cyanobacteria were in the population of a series of towns along the Ohio River in 1931. In these cases low rainfall had caused the water of a side branch of the river to develop a cyanobacterial bloom which was then washed into the main river. As this water moved downstream a series of outbreaks of illness were reported, which could not be attributed to infectious agents. In Harare, Zimbabwe, children living in an area of the city supplied from a particular water reservoir, developed gastro-enteritis each year at the time when a natural bloom of *Microcystis* was decaying in the reservoir. Other children in the city with different water supplies were not affected and no infectious agent was identified. The most lethal outbreak attributed to cyanobacterial toxins in drinking water occurred in Brazil, when a newly flooded dam developed an immense cyanobacterial bloom. Eighty-eight deaths, mostly children, were reported to have occurred.

Examples of illness following the use of copper sulphate to destroy a cyanobacterial bloom in a water storage reservoir have been described in the USA and in Australia. In each of these instances the cyanobacterial genera involved were identified, either at the time or subsequently. In one incident, 62 per cent of the population connected to a filtered, chlorinated water supply developed symptoms of gastro-enteritis within a period of five days. No pathogens were found, and it was concluded that abloom of cyanobacteria in an open storage reservoir which had over 100,000 cells per ml of *Schizothrix calcola, Plectonema, Phormidium* and *Lyngbya* was responsible. The reservoir had just been treated with copper sulphate.

In Armidale, Australia, the water supply reservoir had been monitored for blooms of toxic *Microcystis* for several years, when a particularly dense bloom occurred. Within three weeks the water supply authority treated the reservoir with 1 ppm (part per million) of copper sulphate, which killed the bloom. An epidemiological study of the local population indicated liver damage occurring simultaneously with the termination of the bloom.

A more severe outbreak of cyanobacterial toxicity in a human population occurred on an island off the north-eastern coast of Australia. Due to complaints of bad taste and odour in the water supply, which were attributed to a cyanobacterial bloom, the authorities treated the reservoir with copper sulphate. Within a week numerous children developed severe hepato-enteritis, and a total of 140 children and 10 adults required hospital treatment. *Cylindrospermopsis raciborskii* was identified as the cyanobacterium responsible for this episode.

Within human populations, for a variety of reasons, there will be individuals who are at a much greater risk of injury from cyanotoxins than the population as a whole. Children are the most obvious example, because they drink a higher volume of water in proportion to their body weight than adults. Individuals who already have injury to organs susceptible to cyanobacterial toxins are also at increased risk, such as people with hepatitis, liver cirrhosis, toxic liver injury from other sources, or kidney damage. Kidney dialysis patients, if exposed to microcystins in the water used for dialysis, are especially vulnerable because treatment exposes them intravenously to large volumes of water. In a disastrous incident in Caruaru, Brazil, 117 patients developed cholestatic liver disease and at least 47 deaths were attributed to dialysis with water containing cyanobacterial toxins. Examination of the carbon filter from the dialysis unit demonstrated microcystin-LR, as did the blood and liver tissue of deceased patients.

Chronic Effects in Humans

While acute toxicity is the most obvious problem in cyanobacterial poisoning, a long-term risk may also be present. Short exposures to toxins may result in long-term injury, and chronic low-level exposure may cause adverse health effects. Animal experiments have shown chronic liver injury from continuing oral exposure to microcystins. In particular the possibility of carcinogenesis and tumour growth promotion need careful evaluation, because both have been shown in animal experimentation.

The incidence of human hepatocellular carcinoma in China is one of the highest in the world, and studies have explored whether cyanobacterial toxins are part of a complex of risk factors which increase the incidence of this disease. The distribution of hepatocellular carcinoma varies geographically. In south-east China, rates of less than 15 incidents per 100,000 people are seen in some districts, compared with over 60 incidents per 100,000 people in adjacent localities. Two proven risk factors are infection with hepatitis B virus and intake of aflatoxin B1 from food items, such as corn. The third significant element of association was the source of drinking water. On a village basis, lower cancer mortality rates were seen when the water was drawn from deep wells, compared with much higher rates when the water came from ponds and ditches. Cyanobacteria are abundant in surface waters in south east China where the incidence of hepatocellular carcinoma is highest, and it has been proposed that microcystins in the drinking water are responsible for the higher incidences of cancer among drinkers of pond and ditch water. Very low levels of cyanotoxins were found in one study with limited sampling of some drinking water sources in China, but a more representative study would be needed to investigate this association. At present, all three risk factors are being reduced concurrently in China and liver cancer rates appear to be decreasing.

Injury from Recreational Exposure

There have been repeated descriptions of adverse health consequences for swimmers exposed to cyanobacterial blooms. Even minor contact with cyanobacteria in bathing water can lead to skin irritation and increased likelihood of gastrointestinal symptoms, Some severe skin reactions have been reported, particularly from contact with the marine cyanobacterium, *Lyngbya majuscula,* which causes deep blistering when trapped under the bathing suit of swimmers. Individual sensitivity to cyanobacteria in bathing waters varies greatly, because there can be both allergic reactions and direct responses to toxins. The cyanobacterial pigments can cause severe allergic reactions in sensitive individuals. Cyanobacteria have features in common with general airborne allergens and surveys have shown allergic responses to cyanobacteria in patients with naso-bronchial allergy. More extensive discussion on this aspect can be found in Ressom *et al.*

Animal Poisonings

Numerous cases of animal poisoning substantiate the concern of health hazards for humans exposed to cyanobacteria. The post mortem symptoms of toxicity can be characteristic of the poison involved. Animal deaths from cyanobacterial toxicity have been reported from North and South America, Europe, Australia and Africa. The major injury reported is hepatotoxicosis, i.e. liver poisoning. The cyanobacteria responsible have been *Microcystis aeruginosa, Nodularia spumigena* and *Oscillatoria (Planktothrix) agardhii.* In the most recent cases, post mortem examination has shown evidence of cyanobacterial ingestion as well as characteristic tissue injury in the liver.

PSPs paralytic shellfish poisons

The other main cause of livestock and pet deaths due to cyanobacterial toxins has been from acute neurotoxicity leading to respiratory failure, with no post mortem indications of organ injury. In one case the neurotoxin, anatoxin-a, was isolated from the stomach contents of a dog. Henriksen *et al.,* demonstrated lethality in wild ducks due to anatoxin-a(S). The cyanobacteria associated with deaths from neurotoxicity are Anabaena flos-aquae, Anabaena circinalis, Aphanizomenon flos-aquae and *Oscillatoria* spp. The toxins anatoxin-a, anatoxin-a(S) and saxitoxins have been implicated in different cases.

While the reported deaths have usually occurred shortly after the animals have ingested cyanobacterial scums, lasting injury with progressive mortality has also been seen in animal poisoning by *Microcystis aeruginosa.* The characteristic symptoms are those of liver failure with secondary photo-sensitisation, i.e. severe sunburn-like reactions.

Toxicological Studies

In order to set safe levels of toxicants or contaminants in food or drinking water, it is first necessary to determine the dose level in humans that is considered to be without adverse effects when taken daily over a lifetime; this is known as the Tolerable Daily Intake (TDI). Ideally, this value is derived from human studies, but often such studies are inadequate or non-existent. Alternatively, when there is an adequate experimental database, it can be derived from animal studies. One of the major shortcomings of animal studies is that differences in sensitivity between animals and humans vary, so that safety factors need to be incorporated to deal with this uncertainty. A further important extrapolation problem associated with animal studies relates to routes of exposure. Many toxins are more toxic when given by the intravenous or i.p. route of administration than by the oral route. This difference is evident when LD_{50} values are examined for various routes of exposure. Studies using the i.p. or i.v. route of administration require much less toxin and can be used to indicate relative acute toxicity and may provide information on the mechanism of toxicity. Such studies are used in bioassays, but they have little use in predicting toxicity after oral administration. Another problem with many natural toxins, including cyanotoxins, is obtaining sufficient pure toxin to conduct controlled experiments. Several studies, to be described, have used algal extracts rather than pure toxin. An advantage of this is that the extract more closely mimics the real environmental situation.

A disadvantage is that unless the extract is fully characterised for all the toxins that may be present and their concentrations, the assay may have limited predictive value for scenarios involving other conditions. Several steps are involved in determining the TDI from appropriate human or animal studies. The highest dose associated with the absence of adverse health effects is selected or, when no NOAEL is available, the lowest dose associated with adverse effects is selected. In order to extrapolate from animal data to the human situation, the next step involves the application of a number of uncertainty factors. In the absence of reliable data on inter- and intraspecies differences in sensitivity to a chemical substance, standard default uncertainty factors of 10 are generally used to give a composite uncertainty factor of 100. Additional uncertainty factors may be used to deal with inadequacies in the database and the severity of effects noted. When extrapolating from human data, it is also necessary to use some of these factors. With human data there are, in addition, often problems in determining exposure accurately and thus the dose causing adverse effects.

Microcystins

In the various reported incidents of poisoning in humans and livestock caused by cyanobacteria or their toxins, *Microcystis* is the most frequently cited organism. As a consequence, extensive experimental studies have been carried out on this cyanobacterium and its toxins. Although there is qualitative evidence of cause and effect, indicating human susceptibility, these data have so far been inadequate to establish a dose response relationship that can be used for assessing human risk in a variety of exposure scenarios. The animal studies described below have been conducted in an attempt to address these issues.

Metabolic disposition

There have been no pharmacokinetic studies with orally administered microcystins. After i.v. or i.p. injection of sublethal doses of variously radiolabelled microcystins in mice and rats, about 70 per cent of the toxin was rapidly localised in the liver. Plasma half-lives of microcystin-LR, after i.v. administration, were 0.8 and 6.9 minutes for the first and second phases of elimination, but the concentration of radioactive label in the liver did not change throughout the later part of the six-day study period.

In this study, about 9 per cent of the dose was excreted early via the urinary route, with the remainder being excreted slowly (~1 per cent per day) via the faecal route. Based on the protective effect of microsomal enzyme inducers, it is evident that the liver plays a large role in the detoxification of microcystins. Time-dependent appearance and disappearance of additional chromatograph peaks, thought to represent detoxification products, were seen in urine, faeces and liver cytosol fractions. Three metabolic products have since been identified, a glutathione conjugate, a cysteine conjugate and a conjugate with the oxidised ADDA diene.

Microcystin-LR does not readily cross cell membranes, and hence does not enter most tissues. After oral uptake it is transported across the ileum into the bloodstream through a bile acid type transporter present in hepatocytes and cells lining the small intestine and is concentrated in the liver as a result of active uptake by hepatocytes. It is covalently bound to a 40 kdalton protein (protein phosphatase 2A and possibly protein phosphatase 1) in the hepatocyte cytosol. Some other microcystin congeners are more hydrophobic than microcystin-LR and may cross cell membranes by other mechanisms, including diffusion.

Acute and subacute toxicity

Microcystin-LR is highly toxic. The LD_{50} by the i.p. route ranges from 25 to 150 μg kg^{-1} body weight (bw) in mice. The oral LD_{50} is 5,000 μg kg^{-1} bw in one strain of mice, 10,900 μg kg^{-1} bw in a different strain of mice, and higher in rats. This indicates that, even by the oral route, microcystin-LR displays acute toxicity in rodents. There is no evidence of hydrolysis of microcystins by peptidases in the stomach and it is apparent that a significant amount of microcystin-LR passes the intestinal barrier and is absorbed. Similarly, the oral LD_{50} of *Microcystis* extracts in mice was 50- to 170-fold higher than the i.p. LD_{50} of the same extracts.

The i.p. LD_{50}s of several of the commonly occurring microcystins (microcystin-LA, -YR, and -YM) are similar to that of microcystin-LR, but the i.p. LD_{50} for microcystin-RR is about tenfold higher. However, because of differences in lipophilicity and polarity between the different microcystins, it cannot be presumed that the i.p. LD_{50} will predict toxicity after oral administration.

The microcystins are primarily hepatotoxins. After acute exposure by i.v. or i.p. injection of microcystin, severe liver damage is characterised by a disruption of liver cell structure, a loss of sinusoidal structure, increases in liver weight due to intrahepatic haemorrhage, haemodynamic shock, heart failure and death. Other organs affected are the kidneys and lungs and the intestines.

In a recent study, a single oral dose of microcystin-LR at 500 μg kg^{-1} bw was given to 5 and 32 week old mice, and liver pathology was examined 2-19 hours later. In 62 per cent of the older mice, there was clear microscopic evidence of liver injury. Furthermore, microcystin-LR and a metabolite were detected in the livers. On the contrary, none of the young mice developed liver injury. In untreated control mice, an examination of gastric mucosa and small intestine revealed age-related disruption of surface cell structure. The authors suggested that this disruption may markedly influence gastro-intestinal absorption of microcystin-LR and hence explain the observed age-dependent liver toxicity in exposed mice.

Intranasal installation of microcystin-LR in mice resulted in extensive necrosis of the epithelium of the nasal mucosa of both the olfactory and respiratory zones, progressing to destruction of large areas of tissue down to levels of deep blood vessels. The LD_{50} by this route of administration was the same as the i.p. LD_{50}, and dose-dependent liver lesions were observed. The same authors also demonstrated cumulative liver damage after repeated dosing. While a single dose of 31.3 μg kg^{-1} bw did not result in an increase in liver weight, repeated daily administration over a period of seven days caused a 75 per cent increase in liver weight, which was almost as high as the effect of a single intranasal dose of 500 μg kg^{-1} bw.

Repeated oral administration

Pure microcystin-LR. For assessing possible chronic human health effects, studies involving repeated oral administration of pure microcystins at various dose levels are most desirable. In a 13-week gavage study in mice, conducted under good laboratory practice (GLP), pure microcystin-LR was administered orally to groups of 15 male and female mice at 0, 40, 200, or 1,000 μg kg^{-1} bw per day for 13 weeks. The NOAEL was 40 μg kg^{-1} bw per day. At the next highest dose level there were slight liver tissue changes in some male and female mice. At the highest dose level, all male and most female mice showed liver changes, including chronic inflammation, focal degeneration of hepatocytes and haemosiderin deposits. In male mice at the two highest dose levels, serum transaminases were significantly elevated, serum gamma glutamyl transferase was significantly reduced, and there were small but significant reductions in total serum protein and serum albumin. In female mice, at the highest dose level, only the changes in transaminases were observed. *Microcystis extract.* In an oral dosing study, *Microcystis aeruginosa* extract was supplied to mice at five concentrations in their drinking water for up to one year. At the higher concentrations there was increased death, increased bronchopneumonia (which was endemic in the colony), and chronic liver injury. There was no evidence of tumourigenic changes in the liver, despite the liver injury caused by chronic oral exposure to microcystins. No clear NOAEL was established.

In another study in pigs, *Microcystis aeruginosa* extract was given to groups of five pigs in their drinking water for 44 days at dose levels calculated from potency estimates using the mouse i.p. bioassay to be equivalent to 280, 800 or 1,310 μg kg^{-1} bw per day of microcystins. The extract contained at least seven microcystin variants, with microcystin-YR tentatively identified by high pressure liquid chromatography (HPLC) as the major constituent. A LOAEL of 280 μg kg^{-1} bw per day of toxins was identified, with general liver injury observed at the two higher dose levels. At the lowest dose level, one pig was affected. The authors determined the potency of their extract by the mouse i.p. LD_{50} bioassay, by HPLC analysis and by the *in vitro* phosphatase inhibition assay. Summation of the peak areas from the HPLC identification of microcystin variants, standardised against microcystin-LR, indicated that the LOAEL equated with 100 μg microcystin-LR equivalents per kg bw per day.

Developmental effects

In an investigation on the effects of microcystin-LR on embryonic and foetal development of the mouse, groups of 26 time-mated female mice were dosed once daily by gavage with aqueous solutions of pure microcystin-LR from days 6 to 15 of pregnancy, at dose levels of 0, 200, 600, or 2,000 μg kg^{-1} bw per day. Maternal clinical signs, body weights and food consumption were recorded. Only treatment at 2,000 μg kg^{-1} bw per day was associated with maternal toxicity and mortality. Seven of the 26 females died and two were sacrificed because of distress during the dosing period. On day 18 of pregnancy the remaining females were killed. At the highest dose level, there was a retardation of foetal weight and skeletal ossification, but no foetal deaths. Apart from this there was no effect of microcystin-LR on sex ratio, implantation, postimplantation loss, or on external or visceral or skeletal abnormalities. The NOAEL for any aspect of developmental toxicity was 600 μg kg^{-1} bw per day. In a scoping study, eight 20-week old male and female mice which had received an extract of *Microcystis aeruginosa* in their drinking water since weaning were mated; extract exposure continued throughout pregnancy. Examination of litters showed no effects on weight, number or sex ratio of offspring, compared with control litters. However, there was some evidence of hippocampal injury and reduced brain size in 7 out of 73 of the five-day old young from parents who had received the *Microcystis* extract, and in none of the controls.

Carcinogenicity

Microcystin administered alone. In a recent study, microcystin-LR, administered i.p. to mice, induced neoplastic liver nodules. Animals were given the toxin at 20 μg kg^{-1} bw, 100 times over 28 weeks. At autopsy, nodules up to 5 mm in diameter were observed in the livers of all exposed mice. Some mice were kept a further two months after cessation of dosing, and autopsy showed that liver nodules persisted. In the same study, mice orally administered microcystin-LR, at a dose level of 80 μg kg^{-1} bw, 100 times over the same time period showed no evidence of liver injury or nodule formation. Previous work had shown that microcystin-LR given 20 times i.p. to mice, at 25 μg kg^{-1} bw, over 10 weeks did not induce liver nodules. *Microcystin interaction with tumour initiators.* It is generally understood that some chemicals can initiate the cancer process while other classes of chemicals are able to promote the appearance of cancer after initiation has occurred. Microcystins have been tested for tumour promoting activity.

In vivo animal experiments. There has been some evidence of tumour promotion in animal studies. In a modified, two-stage carcinogenesis mouse skin bioassay, a single dose of 7,12-dimethyl benzanthracene was applied to the dorsal skin of groups of 20, three-month old, female mice. After one week, groups of treated and control mice received either water alone, water with *Microcystis* extract, or croton oil applied to the skin. At 52 days after DMBA exposure, there was a 1.6-fold increase in the number and a sevenfold increase in the mean weight of skin papillomas per mouse in animals given the cyanobacterial extract compared with those given water alone. It was concluded by the authors that oral consumption of *Microcystis* extract from drinking water may act directly or indirectly as a tumour promoter.

The mechanism of action is not clear because microcystin-LR, at least, has difficulty penetrating epidermal cells. In this study, there was considerable liver damage, which could affect the interpretation of these findings. In a short-term two-stage carcinogenicity bioassay, groups of 10-19 seven-week old male F344 rats were initiated by i.p. injection with 200 mg kg^{-1} bw diethylnitrosamine (DEN), followed by partial hepatectomy. In two separate experiments, twice-weekly i.p. doses of 1 and 10 μg kg^{-1} microcystin-LR and of 10,25 and 50 μg kg^{-1} microcystin-LR were then administered during eight weeks and this resulted in a dosedependent increase in the number and area of GST-Ppositive liver foci. The i.p. NOAEL in this study was 1 μg kg^{-1}. Microcystin-LR given without DEN initiation showed no induction of liver foci. Macroscopic nodules, histologically diagnosed as neoplastic nodules, were seen in this study when DEN was followed by microcystin-LR at 10 μg kg^{-1} and 50 μg kg^{-1} partial hepatectomy.

In another tumour initiation and promotion assay aimed at evaluating possible tumour promoting effects in the upper small intestine, two doses of 40 mg kg^{-1} bw of the initiator N-methyl-N-nitrosourea (MNU) were orally administered to C57 black mice, one week apart, followed by drinking water containing various levels of *Microcystis* extracts, estimated to be equivalent to 0, 1.2 or 4.2 mg kg^{-1} bw per day of microcystins for up to 22 weeks. Time to 50 per cent survival was about 17 weeks in controls and 15 weeks in extract-exposed mice. No primary liver tumours were seen in any group and there was no evidence of microcystin-induced promotion of lymphoid or duodenal tumours. *In vitro* studies on interaction of microcystins and tumour initiators are discussed in the section on *in vitro* studies below.

Genotoxicity

No mutagenic response has been observed for purified toxins derived from *Microcystis* in the Ames *Salmonella* assay with or without S9 activation. The *Bacillus subtilis* multigene sporulation test was also negative with regard to mutagenicity using both the 168 and hcr-9 strains. An earlier study with purified *Microcystis* extract also elicited a negative response in the Ames test. The purified toxins from a *Microcystis* species tested against human lymphocytes suggested that the toxins may be clastogenic as indicated by dose-related increases in chromosomal breakage.

In vitro studies

Extracts from *Microcystis aeruginosa* blooms containing mainly microcystins from a lake in China were tested in a two-stage celltransformation assay using Syrian hamster embryo (SHE) cells. In this assay, the microcystin extract had no initiating activity when followed by the tumour promoter TPA. With methylcholanthrene as the initiator (0.5 μg ml^{-1}), followed by bloom extract, a dose-related increase in transformation frequency was observed. Primary hepatocyte cultures in the presence of picomolar and nanomolar concentrations of microcystin-LR showed selective cell toxicity and selective cell proliferation depending on the ploidy of the cells.

Mechanism of action

Microcystin-LR was found to be a potent inhibitor of eukaryotic protein serine/threonine phosphatases 1 and 2A both *in vitro* and *in vivo*, and this effect has become the basis of one of the bioassays to detect its presence. Substances that inhibit these protein phosphatases are considered to be non-phorbol ester (TPA)-type tumour promoters. Other substances that act in a similar way to microcystins are okadaic acid, nodularin, tautomycin and calyculin. The protein phosphatases serve an important regulatory role to maintain homeostasis in the cell. Protein phosphatase inhibition results in a shift in the balance towards higher phosphorylation of target proteins, such as tumour suppresser proteins. This is a major post-translational modification which can result in excessive signalling and may lead towards cell proliferation, cell transformation and tumour promotion. In liver cells, intermediate filaments of the cytoskeleton are hyperphosphorylated leading to cellular disruption. In monolayer cell cultures this leads to detachment from adjacent cells, and involves actin filament contraction. The inhibition of protein phosphatase 2A by microcystin-LR can be effectively reversed in the presence of polyclonal antibodies against microcystin-LR; such antibodies can also protect *in vivo* against microcystin-LR toxicity as shown with i.p. co-administration studies in mice. The implications of protein phosphatase inhibition in humans, due to low level chronic exposure to microcystins, are not known.

Additional study requirements

Further short-term studies are needed to understand better the genotoxic and tumour promoting potential of microcystins. Such studies would also be useful as a preliminary to the design of appropriate chronic or other *in vivo* studies to assess their carcinogenic potential. Regarding the possible role of microcystins in tumour promotion, further studies are especially needed to establish a dose-effect relationship for nodule induction with microcystin alone using various routes of administration.

Derivation of a tolerable daily intake (TDI)

Most of the relevant data on microcystin toxicity have come from either reported human injury related to consumption of drinking water containing cyanobacteria or from limited work with experimental animals. Although the cyanotoxins have not been reviewed by the International Agency for Research on Cancer (IARC), their standard evaluation procedures lead to the conclusion that, at present, the human evidence for microcystin carcinogenicity is inadequate and the animal evidence is limited. In such instances, the current practice for deriving a TDI is to use a LOAEL or NOAEL divided by appropriate safety or uncertainty factors as described in the Addendum to the *Guidelines for Drinking Water Quality*.

A 13-week mouse oral study with pure microcystin-LR has been considered the most suitable for the derivation of a guideline value for microcystin-LR. In a study by Fawell *et al.* a NOAEL of 40 μg kg^{-1} bw per day was determined, based on liver histopathology and serum enzyme level changes. By applying a total uncertainty factor of 1,000, a provisional TDI of 0.04 μg kg^{-1} bw per day has been determined for microcystin-LR. This TDI is supported by the results of a 44-day pig study, in which pigs were given *Microcystis* extract in their drinking water, resulting in a LOAEL of 100 μg microcystin-LR equivalents per kg bw per day. To this LOAEL an overall uncertainty factor of 1,500 was applied, arrived at by using 10 for intra-species variability, 3 rather than 10 for inter-species variability, 5 for extrapolating from a LOAEL to a NOAEL and 10 for the less-than-lifetime exposure. This resulted in a provisional TDI of 0.067 μg kg^{-1} bw per day. The lower of these two values, 0.04 μg kg^{-1} bw per day, has been used in deriving a provisional guideline value.

Nodularin

Compared with *Microcystis* and microcystins, there have been fewer reported incidents of human and livestock disease involving *Nodularia* spp. and nodularin.

Experimental animal data

The toxicity and liver pathology induced by nodularin is similar to that caused by microcystins, with hepatocyte necrosis and haemorrhagic diathesis. Nodularin inhibits protein phosphatases 1 and 2A with the same potency as microcystin-LR. In a two-stage liver carcinogenesis experiment in male F344 rats initiated with DEN and without partial hepatectomy, repeated i.p. administration of 10 μg nodularin per kg bw induced GST-P-positive foci more effectively than microcystin-LR. Nodularin alone also induced some foci. In addition, nodularin was capable of activating several proto-oncogenes of the *fos* and *jun* family, which are considered to play a role in tumour promotion.

Derivation of a tolerable daily intake

There are no studies available that use oral administration of nodularin, thus a TDI cannot be set. Nevertheless, several experimental studies cited above indicate that nodularin has similar toxicity to microcystin-LR. It may be appropriate, therefore, to consider nodularin in an analogous fashion to microcystins in terms of human health risk assessment.

Anatoxin-a

The potent neurotoxin, anatoxin-a, from *Anabaena flos-aquae* has frequently been involved in animal and wildfowl poisoning.

Experimental studies

Anatoxin-a is a nicotinic (cholinergic) agonist that binds to neuronal nicotinic acetylcholine receptors. It has been suggested that the activation of presynaptic nicotinic acetylcholine receptors by anatoxin-a results in an influx of Na^+, producing sufficient local depolarisation to open voltage sensitive Ca^{++} and Na^+ channels. The latter may then amplify the response, activating further Ca^{++} channels. As a result of this depolarisation there is a block of further electrical transmission, and at sufficiently high doses this can lead to paralysis, asphyxiation and death. Anatoxin-a is more potent than nicotine or acetylcholine in evoking type 1A or type 2 current responses in rat hippocampal neurones, and it is more potent than nicotine in its ability to evoke the secretion of endogenous catecholamines from bovine adrenal chromaffin cells through their neuronal-type nicotinic receptors. Similar to nicotine, anatoxina was more potent than noradrenaline in releasing dopamine from striatal nerve terminals from rat superfused hippocampal synaptosomes. *In vivo* studies in the rat showed that the toxin stimulates the sympathetic system through the release of catecholamines from nerve endings.

Acute toxicity

In mice, the i.p. LD_{10} (lowest dose causing death) of anatoxin-a was 250 μg kg^{-1} bw and the i.p. LD_{50} of anatoxin-a is 375 μg kg^{-1} bw. The i.v. LD_{50} is less than 100 μg kg^{-1} bw. The oral LD_{50} for anatoxin-a is greater than 5,000 μg kg^{-1} bw, and the intranasal LD_{50} in mice is 2,000 μg kg^{-1} bw.

Subacute toxicity

In two studies, anatoxin-a was administered to rats orally in the drinking water for 54 days at 0.51 or 5.1 ppm or by i.p. injection at 16 μg per rat per day for 21 days. No toxicity was observed. In a dose range-finding study for a four-week oral toxicity experiment, groups of 2 male and 2 female mice were administered anatoxin-a hydro-chloride by gavage at dose levels of 1,500, 7,500 or 15,000 μg kg^{-1} per day for five days. All mice at the highest dose, and one female at the intermediate dose, died within three minutes of their first dose. Males at the intermediate dose were hyperactive after their third dose. All other animals survived, and no treatment-related signs of clinical toxicity, or changes in body weight were observed. No treatment-related changes were observed at necropsy.

In a 4-week study, groups of 10 male and female mice were administered anatoxin-a hydrochloride by gavage at dose levels of 0, 120, 600 or 3,000 μg kg^{-1} bw. In the course of the experiment, one male receiving 600 μg kg^{-1} bw per day and one female receiving 3,000 μg kg^{-1} per day died. No signs of clinical toxicity or histopathological abnormalities were observed in these two animals, and no cause of death could be identified. It was concluded by the authors that the possibility that these deaths were treatment-related could not be excluded, although this was considered to be unlikely. There were no other treatment-related findings. Although the NOAEL for anatoxin-a hydro-chloride in this study was determined to be 120 μg kg^{-1} bw per day, equivalent to 100 μg kg^{-1} bw per day of anatoxin-a, the toxicological database was considered to be inadequate to use these results for setting a formal TDI.

Reproductive effects

Doses of 200 or 125 μg kg^{-1} bw anatoxin-a were given i.p. to hamsters one or three times per day, respectively, at days 12 to 14 of pregnancy (after organo-genesis), and the dams were sacrificed at day 15. The treatment given three times per day caused foetal malformation (hydrocephaly) in all foetuses in one of 10 litters, and stunted growth in almost all litters; treatment given once per day resulted in stunted growth. No maternal toxicity was observed.

Groups of 10 and 12 time-mated female mice were given anatoxin-a hydrochloride by gavage at 0 or 3,000 μg kg^{-1} bw per day respectively, on days 6 to 15 of pregnancy. The anatoxin-a dose was considered the maximum dose that could be tolerated. No treatment-related effects were observed in the dams or offspring, although there was a slight decrease in foetal weight compared with controls. No data on the genotoxic potential of anatoxin-a are available.

Derivation of a tolerable daily intake

Anatoxin-a is a potent acute neurotoxin. The available data indicate that significant concern with regard to chronic toxicity is unlikely, but the data base is insufficient for derivation of a TDI.

Homoanatoxin-a

Homoanatoxin-a is a potent neuromuscular blocking agent with an i.p. LD_{50} in mice of 250 μg kg^{-1} bw. Toxicosis in the lethal dose range leads to severe body paralysis, convulsions and death by respiratory arrest in 7-12 minutes. Experiments with rat phrenic nerve hemidiaphragm preparations demonstrated that the physiological effects of homoanatoxin-a are related to those observed for *d*-tubocurarine. Recent studies have shown that homoanatoxin-a enhances the influx of Ca^{++} ions in the cholinergic nerve terminals. No TDI can be derived.

Anatoxin-a(S)

Anatoxin-a(S) is an organophosphate produced by the cyanobacteria *Anabaena flosaquae* and *A. lemmermannii*. This toxin blocks acetylcholinesterase activity in a manner analogous to organophosphate insecticides. The i.p. LD_{50} in mice is 20 μg kg^{-1} bw with muscle weakness, respiratory distress (dyspnea) and convulsions (effect on seizure threshold) preceding death. There are no oral toxicity studies for this toxin.

Derivation of a tolerable daily intake

A lack of adequate experimental data or human data precludes the derivation of a TDI.

Saxitoxins

Effects in humans

Saxitoxin and some of its analogues are produced by *Anabaena circinalis* in Australian freshwaters and *Aphanizomenon flos-aquae* in the USA. The saxitoxin group has been the cause of paralytic shellfish poisoning (PSP) in people. Several species of dinoflagellates produce PSP toxins that accumulate in molluscs which filter-feed on these organisms.

People who have consumed shellfish containing high levels of PSP toxins may suffer from this acute illness. The signs and symptoms of PSP in humans may range from a slight tingling and numbness about the lips to complete paralysis and death from respiratory failure. More than 1,000 cases of PSP have been reported since the early 1900s in North and Central America and there have been 109 deaths. In a recent epidemic in Guatemala, the mortality rate in children was 50 per cent, while for adults it was 5 per cent. No PSP-like illnesses have been reported in humans from the consumption of drinking water containing saxitoxins, although there have been fatalities in sheep after ingesting a PSP toxin-containing bloom of *A. circinalis*.

Experimental studies

Mode of action. Of the various PSP toxins, only saxitoxin has been studied in detail for pharmacological effects, partly because the other toxins are usually not available in sufficient quantities for such studies. Nearly all the systemic actions of saxitoxin can be explained by its pharmacological effect on nerve axon membranes. This involves a wide spread blockage of sodium ion channels of the excitable membranes of nerves, thereby affecting impulse generation in peripheral nerves and skeletal muscles. This results in generalised nerve dysfunction as measured by electromyography. In mammals, these effects lead to paralysis, respiratory depression and respiratory failure. Direct cardiac effects are usually minimal. *Metabolism.* No data are available on PSP toxin absorption, distribution, metabolism and excretion.

Acute toxicity

The principle of the standardised mouse bioassay developed by Sommer and Meyer is measurement of time to death after i.p. injection. In that study, the authors suggested that signs characteristic of PSP, such as dyspnea, could be observed after the i.p. administration of toxin. Hypotensive effects have been observed to accompany the respiratory depression, implicating both central and peripheral nervous system actions.

Acute toxicity studies have been conducted in several species with extracts containing PSP toxins obtained from the Alaskan butter clam. Using this preparation, Wiberg and Stephenson determined the LD_{50} values for male mice, using three routes of administration. In addition, the i.p. LD_{50} was determined in female mice. The PSP toxin extract was much less toxic when administered by the oral route than by the i.v. or i.p. routes. Increasing the pH of the injection medium (> 3.8) or the addition of sodium ions affected i.p. toxicity, while the addition of sodium ions did not influence oral or i.v. toxicity. Similar comparative LD_{50}s for different routes of administration were obtained in rats; in addition new born rats were about tenfold more susceptible than adults after oral administration. For a number of animal species tested, the oral LD_{50} for PSP toxin ranged from 128 μg kg^{-1} bw PSP toxin in guinea pigs to 420 μg kg^{-1} bw in mice.

Prior exposure to non-lethal doses of PSP toxin seems to lower the susceptibility of rats to lethal doses of PSP toxin. In a study using Sprague-Dawley rats, the oral LD_{50} value for the purified PSP extract was determined. One group of rats was given a non-lethal dose of PSP toxin 14 days before the test. The LD_{50} for the pre-treated rats was about 50 per cent higher than that for untreated rats. This finding corroborates the fact noted by Prakash *et al.* that fishermen who habitually eat shellfish containing low levels of PSP toxins appear to be less susceptible to developing PSP.

There is a wide range in i.p. toxicity of the various PSP toxins as tested following the Association of Official Analytical Chemists (AOAC) mouse bioassay. Potencies of these toxins are usually expressed in mouse units (MU) per μmol of toxin. Saxitoxin is one of the most toxic of the PSP toxins and the C toxins are the least toxic. Potencies may also be expressed as saxitoxin equivalents. Inter-conversions during storage, cooking or digestion may modify the i.p. and oral toxicity. No subchronic or chronic animal studies with PSP toxins are available.

SAFE LEVELS AND SAFE PRACTICES

Throughout their lifespan, humans are exposed to many chemical substances, both beneficial and harmful. It is not possible to eliminate exposure to all toxins in our environment. Of the harmful chemicals, some are anthropogenic and others occur naturally. Although cyanotoxins occur naturally, their presence and abundance are, to a large extent, influenced and increased by human action. The extent to which cyanotoxins pose a human health risk depends on human exposure to these toxins. Reducing human exposure to cyanotoxins may be achieved either through preventing the occurrence of hazardous cell densities of toxic cyanobacteria, or by placing barriers that would reduce human exposure to the cyanobacterial hazard.

Establishing and controlling safe practices and procedures to reduce or prevent microbiological problems are common in the food industry, where the process is known as Hazard Analysis of Critical Control Points (HACCP). There is also a long history of similar activities in relation to drinking water supply, where the process is referred to as "sanitary inspection". This type of approach is also starting to be applied in dealing with some natural toxins, such as fungal toxins (mycotoxins). Safe practice guidelines can assist in reducing exposure, even for substances for which a guideline value cannot be set, or where implementation of a guideline value is very difficult.

Major routes of exposure to cyanotoxins include oral and dermal routes through drinking water and recreational water use. However, the very limited available information suggests that inhalation in aerosols may be an equally important route given that the toxicity of microcystin following intranasal instillation approaches that for intraperitoneal injection. In some countries, cyanobacterial dietary supplements may constitute a major route of oral intake for a small sub-population, if the cyanotoxin levels in the supplements are not controlled. A specific route affecting a susceptible sub-population may occur in dialysis clinics.

The extent or duration of exposure throughout the year is shorter if water resources are populated by cyanobacteria at specific seasons. In temperate climates, water bodies dominated by the genus *Microcystis* usually exhibit a bloom season of 3-5 months, whereas in milder climates, such as in Australia, some of South America and South Africa, this genus may bloom for 6-10 months. Other taxa, such as *Planktothrix agardhii*, may show perennial mass development even in temperate climates. This is the case in some lowland regions of north-western Europe, especially in years with mild winters without ice cover.

TOLERABLE EXPOSURES

Tolerable daily intake

The repair mechanisms of the body are continuously active and ensure that cells and tissues are replaced as they are damaged by normal "wear and tear", as well as by external factors such as toxic chemicals. The tolerable daily intake (TDI) is the amount of a potentially harmful substance that can be consumed daily over a lifetime with negligible risk of adverse health effects.

WHO Guideline Values for Drinking Water Quality

The World Health Organisation (WHO) *Guidelines for Drinking-water Quality* represent a scientific consensus, based on very broad international participation, of the health risks presented by microbes and chemicals in drinking water. This scientific consensus is used to derive "guideline values" which are associated with guidance on monitoring and management aspects. The guideline values themselves are based upon a number of assumptions that may be amended locally or nationally according to specific circumstances.

A guideline value for lifetime consumption of a chemical contaminant of drinking water is usually calculated by applying the derived TDI to a typical daily water intake in litres *(L)* by an individual of a given body weight *(bw)*. The proportion *(P)* of total daily intake of the contaminant which is ingested from the drinking water needs to be considered, because some intake may come from food or by inhalation from air. The guideline value is then calculated as:

$$\text{Guideline value} = \frac{\text{TDI x bw x}}{\text{L}}$$

A provisional guideline value of 1.0 μg l^{-1} has been adopted by WHO for microcystin-LR. In order to derive this, an average adult body weight of 60 kg and an average water intake for adults of 2 litres per day was used, as is standard practice. The provisional TDI of 0.04 μg kg^{-1} bw per day was used and the proportion of the TDI allocated to drinking water (*P*) was assumed to be 0.8. The resulting concentration (0.96 μg l^{-1}) was rounded to 1.0 μg l^{-1}. In water containing cyanobacterial cells, this guideline value should be applied to the total cell-bound and extracellular microcystins.

There were insufficient data to derive a guideline value for cyanotoxins other than microcystin-LR. For saxitoxins (STX), the guideline of 80 μg STX equivalents per 100 g (this is the conventional way of expressing this value) shellfish, used in North America for closing shellfish growing areas for harvesting, may provide preliminary orientation.

Exceeding the provisional guideline value of 1 μg l^{-1} for microcystin-LR can be tolerated. This may occur if, for example, discontinuation of exposure is expected in the near future due to implementation of measures to eliminate cyanotoxins from drinking water or cyanobacteria from the water resource. In such instances of guideline exceedances, it may be appropriate that information is communicated to the public, and especially to particularly susceptible sub-populations.

Assessment of microcystins other than microcystin-LR in relation to the guideline value

There are more than 60 different analogues of microcystin and, in many regions, microcystin-LR is not the most commonly occurring microcystin. It may not even be amongst the microcystins detected. The expression and interpretation of quantitative results from analysis or assay for these toxins in relation to the WHO guideline value for microcystin-LR may, therefore, be problematic. The most widely used analytical technique for the detection and quantification of individual microcystin variants for which quantitative reference materials are available is high pressure liquid chromatography (HPLC). For toxin variants where reference materials are available, HPLC results can be truly quantitative. For HPLC peaks that identify microcystin variants for which no standards are available but the result has been derived from a comparison with the concentration of the standard for microcystin-LR, the estimates can be expressed as "concentration equivalents" (CE) of microcystin-LR. It is therefore possible, using HPLC, to derive an approximate concentration of total microcystins in a sample expressed in terms of microcystin-LR CE. However, some of the observed HPLC peaks may relate to toxicologically inactive or weakly toxic variants. In terms of "toxicity equivalents" (TE) of microcystin-LR the actual toxicity of an unknown sample reported as microcystin-LR equivalents (CE) is likely to be less than the same concentration of pure microcystin-LR, because microcystin-LR is one of the most potent microcystins, at least in acute terms. Thus, the microcystin CE approach would result in a "worst case" estimate of toxic microcystin concentration.

A toxin concentration measured from a water or bloom sample by a bioassay may give a toxin concentration value which can be expressed as microcystin-LR TE, provided the assay has been calibrated using microcystin-LR as the quantification standard. This measure of total microcystin concentration expressed as microcystin-LR TE will be closer in actual toxicity to a solution of the same concentration of pure microcystin-LR. The position of enzyme assays, such as *in vitro* protein phosphatase inhibition assays, in this context is currently unclear. Differences in toxicity of different microcystins include variation in their ability to enter intact cells and in their capacity to inhibit protein phosphatases.

In practice, it is important to report quantitative analytical results for samples containing several microcystins with the above qualifications in mind and to indicate the method and assumptions used for deriving the quantitative value. If it is necessary to calculate quantitative values for total microcystins in a sample, then the value should be qualified as either microcystin-LR CE or TE. This information can be used as a preliminary orientation of the hazard presented by the sample in relation to the guideline value.

Recreational Water Exposure

Three potential routes of exposure to cyanotoxins can be distinguished: direct contact of exposed parts of the body, including sensitive areas such as the ears, eyes, mouth and throat, and the areas covered by a bathing suit; accidental swallowing; and inhalation of water. Cases of illness from accidental swallowing and inhalation of *Microcystis* have been reported and provide direct evidence of harm to recreational water users from cyanobacterial blooms in the recreational water bodies.

Health effects observed in the prospective epidemiological study of Pilotto *et al.* occurred at low cyanobacterial cell densities. These related clearly to the cyanobacterial cell population, but not to the concentration of microcystins. Thus, this hazard appears to be due to additional, or other unidentified, cyanobacterial metabolites or compounds from associated bacteria, even at moderate levels of exposure.

Intake through oral ingestion or inhalation

Incorporation of toxins through swallowing, contact with nasal mucosa, or by inhalation are likely to be important routes of exposure to cyanotoxins during water-contact sports. Well-documented evidence from one animal experiment and one case of multiple human illness indicates that inhalation and resorption through nasal and pharyngeal mucous membranes may present a high risk in water sports involving intensive submersion of the head and inhalation of aerosols.

Experimental results indicate a hazard of cumulative liver damage by repeated microcystin intake, as can occur during a holiday with daily bathing at a recreational site with a high density of microcystincontaining cyanobacteria. Sub-acute liver injury is likely to go unnoticed, because signs of liver injury are only apparent after severe injury. In addition, the dose-response curve for liver injury from microcystins is relatively steep. There may be little evidence of acute liver damage when levels are close to those that could lead to severe acute toxicity, and thus exposure at such levels is likely to be continued by people if they are uninformed of the risk, thereby increasing the risk of cumulative liver damage.

Risks of ingestion are particularly high for children playing in shallow near-shore water where scums tend to accumulate. Because the hazard of microcystin uptake is directly related to the levels of toxins in the water and the volume of water ingested or inhaled, the range in these levels needs to be recognised in deriving guidelines for recreational water safety.

Direct contact

Allergic and toxic dermal reactions of varying severity are known from cyanobacteria as well as from freshwater algae, but have not been documented extensively. Bathing suits, and particularly diving suits, tend to aggravate such effects by accumulating cyanobacterial cells, thereby enhancing the disruption of cells and hence the liberation of cell contents onto the wearer's skin. Reports from the USA have recorded allergic reactions from recreational exposure, and the cyanobacterial pigment phycocyanin was shown to be responsible in one case. Severe dermatitis, resembling skin bums, has been reported from marine bathing in the presence of cyanobacteria dislodged from rocks, particularly after storms in tropical seas.

SAFE PRACTICES

The placing of barriers that reduce exposure to a cyanotoxin hazard is an important measure and involves identifying "critical control points" and implementing measures for their monitoring and control. In the case of cyanobacteria, critical control points might include, for example, noting the tendency of a water body to develop blooms, scums or mats. Monitoring schemes need to be developed that are capable of detecting proliferation of cyanobacteria and drinking water treatment technology needs to be in existence that is capable of preventing human exposure if cyanobacteria occur in source waters.

Drinking Water

A drinking water supply safe from cyanotoxins will either draw upon a resource which does not harbour cyanotoxins, or have treatment in place that is likely to remove cyanobacterial cells as well as removing cyanotoxins. However, in many circumstances a potential cyanotoxin hazard can be managed effectively without the necessity of advanced treatment processes, through water resource management techniques and removal of intact cells. Most of the reported incidents of human injury that have raised awareness of the importance of cyanotoxins in drinking water have involved the inappropriate treatment of water supplies, such as the use of copper sulphate in dealing with an established bloom of cyanobacteria.

A very effective approach to safe practices may involve changing the drinking water source. In a number of regions, surface waters are used for reasons of easy access and tradition, although groundwater of high quality is available. Exploring options of improving practices of drinking water abstraction with low technological input may lead to health benefits. In China, a high prevalence of endemic primary liver cancer was related to several factors: hepatitis B, aflatoxins in the diet, and drinking surface water polluted with cyanobacteria likely to contain microcystins. Changing the drinking water source from shallow, eutrophic ponds and ditches to groundwater was a major element of a package of measures which showed some success in improving health.

Recreational Waters

Recreational water use is likely to be a major route of exposure to cyanotoxins in some parts of the world. Whereas similar approaches to resource protection apply as for drinking water, there are very few further management options available once cyanobacteria proliferate or accumulate in a recreational water. Because adequate surveillance is sometimes difficult and management options, except precluding or discouraging use, may be scarce, a large share of the responsibility for safe practices lies with the users of a bathing site. The provision of adequate information to the public becomes, therefore, a major responsibility of public authorities. The growth of cyanobacteria in lakes and rivers used for recreational purposes has been well recognised as a public nuisance. Water blooms of cyanobacteria may be associated with unpleasant odours and the offensive appearance of lake shores, especially when scums of the organisms accumulate and decay. Areas with extensive cyanobacterial scums or accumulated detached mats on bathing beaches may be avoided by swimmers and other water users because of the obviously unpleasant environment, particularly if locally anaerobic water conditions or cyanobacterial toxins cause fish-kills, further emphasising the unattractiveness of water contact. In temperate climates, cyanobacterial dominance is most pronounced during the summer months, when the demand for recreational water is highest. In some regions, cyanobacteria have been abundant for more than a generation and visitors have accepted this water quality as "normal" for their region. Multiple anecdotal observations of children playing with scum material have been reported.

Health impairments from cyanobacteria in recreational waters must be differentiated between the chiefly irritative symptoms caused by unknown cyanobacterial substances, and the more severe hazard of exposure to high concentrations of known cyanotoxins, particularly microcystins. A single guideline therefore, is not appropriate. Rather, a series of guidelines associated with incremental severity and probability of adverse effects has been defined at three levels as described below.

1. *Relatively mild and/or low probabilities of adverse health effects* : For protection from health outcomes not due to cyanotoxin toxicity, but due to the irritative or allergenic effects of other cyanobacterial compounds, a guideline level of 20,000 cyanobacterial cells per ml can be derived from the prospective epidemiological study by Pilotto *et al.* Whereas the health outcomes reported in this study were related to cyanobacterial density and duration of exposure, they affected less than 30 per cent

of the individuals exposed. At this cyanobacterial density, 2-4 μg l^{-1} of microcystins may be expected if microcystinproducing cyanobacteria are dominant, with 10 μg l^{-1} being possible with highly toxic blooms. This level is close to the WHO provisional drinking water guideline value of 1 μg l^{-1} for microcystin-LR which is intended to be safe for life-long consumption. Thus, health outcomes due to microcystin are unlikely and providing information for visitors to bathing sites with this low-level risk is considered to be sufficient. Additionally, it is recommended that the authorities are informed in order to initiate further surveillance of the site.

2. *Moderate probability of adverse health effects:* At higher concentrations of cyanobacterial cells, the probability of irritative symptoms is elevated. Additionally, cyanotoxins may reach concentrations with potential health impact. To assess risk under these circumstances the data used for the drinking water provisional guideline value may be applied. Swimmers involuntarily swallow some water while bathing and the harm from ingestion of bathing water will be comparable with that from a drinking water supply with the same toxin content. A swimmer can expect to ingest up to 100-200 ml of water in one session, sail-board riders and water skiers would probably ingest more.

A density of 100,000 cyanobacterial cells per ml is a guideline for a moderate health alert in recreational waters. At this density, 20 μg l^{-1} of microcystins are likely, if the bloom consists of *Microcystis* and has an average toxin content per cell of 0.2 pg, or 0.4 μg microcystin per μg chlorophyll *a* but toxin levels may approximately double if *Planktothrix agardhii* is dominant. This toxin concentration is equivalent to 20 times the WHO provisional guideline value for microcystin-LR in drinking water, but would result in consumption of an amount close to the TDI for an adult of 60 kg consuming 100 ml of water while swimming. However, a child of 15 kg consuming 250 ml of water during extensive playing could be exposed to 10 times the TDI. The health risk will be increased if the person exposed is particularly susceptible. Therefore, cyanobacterial densities likely to cause microcystin concentrations of 20 μg l^{-1} should trigger further action. Non-scum-forming species of cyanobacteria, such as *Planktothrix agardhii,* have been observed to reach cell densities corresponding to 200 μg l^{-1} of chlorophyll *a* or even more in shallow water bodies. Transparency in such situations will be less than 0.5 m when measured with a Secchi disk. *Planktothrix agardhii* has been shown to contain a very high cell content of microcystin and therefore toxin concentrations of 200-400 μg l^{-1} can occur without scum formation.

An additional reason for increased alert at 100,000 cells per ml is the potential of some frequently occurring cyanobacterial species to form scums. These scums may increase local cell density and thus toxin concentration by a factor of 1,000 or more in a few hours, thus rapidly changing the risk from moderate to high for bathers and others involved in body-contact water sports.

Cyanobacterial scum formation presents a unique problem for routine monitoring carried out at the usual time intervals of one or two weeks, because such monitoring intervals are unlikely to detect hazardous maxima. Because of the potential for rapid scum formation at a cyanobacterial density of 100,000 cells per ml or 50 μg l^{-1} chlorophyll *a*, intensification of surveillance and protective measures are appropriate at these levels. Daily inspection for scum formation and measures to prevent exposure in areas prone to scum formation are the two main options.

Intervention is recommended to trigger effective public information campaigns educating people on avoidance of scum contact. Furthermore, in some cases, restriction of bathing may be judged to be appropriate. An intensified monitoring programme should be implemented, particularly looking for scum accumulations. Health authorities should be notified immediately.

HIGH RISK OF ADVERSE HEALTH EFFECTS

Abundant evidence exists for potentially severe health hazards associated with scums caused by toxic cyanobacteria. No human fatalities have been unequivocally associated with oral ingestion of scum, even though numerous animals have been killed by consuming water containing cyanobacterial scum material. This discrepancy can be explained by the fact that animals would drink higher volumes of scum-containing water, compared with the small amounts of scum accidentally ingested by humans during bathing.

Cyanobacterial scums can represent a thousand-fold to million-fold concentration of cyanobacterial cell populations. It has been calculated that a child playing in a *Microcystis* scum for a protracted period and ingesting a significant volume could receive a lethal exposure, although there are no reports that this has actually occurred. Based on evidence that the oral LD_{50} of microcystin-LR in mice is 5,000-11,600 μg kg^{-1} bw, for a child of 10 kg the ingestion of 2 mg of microcystin or less could be expected to cause liver injury, because concentrations of up to 24 mg l^{-1} of microcystins have been published from scum material. Substantially higher enrichment of scums is occasionally observed, and accidental ingestion of smaller volumes of these could cause serious harm. Anecdotal evidence indicates that children, and even adults, may be attracted to play in scums. The presence of scums caused by cyanobacteria is a readily detected indicator of a high risk of adverse health effects for those bathers who come into contact with the scum. Immediate action to control scum contact is recommended for such situations.

Swimmers may be in contact with benthic cyanobacteria after a storm breaks off clumps of filaments, or cyanobacterial mats naturally detach from the sediment and are accumulated on the shore. Some marine beaches have been reported to have widespread problems due to a benthic marine cyanobacterium, *Lyngbya majuscula*, growing on rocks in tropical seas and causing severe blistering when trapped under the bathing suits of people swimming following a storm. This response may be due to acute toxicity; *Lyngbya* can produce irritant toxins. Measures of cyanobacterial population cell density, will not detect these hazards. Instead, this type of hazard calls for critical and well-informed observation of bathing sites, coupled with a flexible response. It is difficult to define "safe" concentrations of cyanobacteria in recreational water in relation to allergenic effects or skin reactions, because individual sensitivities vary greatly. Aggravation of dermal reactions due to accumulation of cyanobacterial material and enhanced disruption of cells under bathing suits and wet suits may be a problem, even at all densities below the guidelines described above. Further information related to monitoring and management of recreational waters is available in Bartram and Rees.

Other Exposure Routes

Renal dialysis

Renal dialysis patients are at great risk when water used for dialysis contains contaminants such as cyanotoxins. For these patients large volumes of water are used and the route of exposure is similar to the i.v. route, which allows for a much greater uptake of toxin than following oral ingestion.

The WHO *Guidelines for Drinking-water Quality* do not consider the especially high quality of water needed for dialysis treatment, intravenous therapy or other clinical uses. The treatment processes used at conventional surface water treatment plants are normally effective in removing cyanobacterial cells, but are not effective in removing or destroying dissolved cyanotoxins, especially from water supplies with a high organic content and cyanobacterial dominance. Consequently, clinics and hospitals with special water needs, such as for dialysis treatment or for transfusions, may need to provide additional water treatment to remove the cyanotoxins. Such treatment ranges from granular activated carbon filtration, followed by reverse osmosis, to more elaborate treatment including membrane filtration. The extent of treatment necessary depends on the quality of the municipal water supply. Continuous monitoring of performance and equipment is essential to ensure adequate quality of the water. On-site water treatment systems in clinics and hospitals require rigorous monitoring and regular maintenance, including back-flushing of filters and recharge of activated carbon, according to manufacturers' specifications. It is important that manufacturers' specifications should be assessed for their adequacy for maintaining performance under local conditions. Activated carbon, for example, may be exhausted for its ability to remove cyanotoxins long before it reaches saturation for removal of other organic compounds, and some manufacturers may be unaware of this.

Irrigation Water

The use of water from sources containing cyanobacterial blooms and toxins for spray irrigation of crops presents potential health hazards through several exposure routes, including uptake into the food chain. Workers or passers-by may inhale toxins with spray drift, and skin contact with cyanobacteria and dissolved toxins may also occur. Questions therefore arise about the health significance of spray irrigation with water containing cyanobacterial toxins. Animal experimentation has indicated that microcystin uptake through nasal mucosa may be considerable. When considered together with the skin irritations, respiratory distress and nasal mucosal irritations observed after recreational exposure, the indicators are that occupational exposure to spray irrigation water should be avoided if the water contains cyanobacterial toxins. Incidental exposure of humans and animals to such spray irrigation water, for example by downwind drift, should also be avoided.

There are several indications that terrestrial plants, including food crop plants, can take up microcystins. Mustard seedling development is inhibited if microcystin-LR is presented to the roots in aqueous solution. Microcystins have several perturbatory effects on plant physiology and metabolism, when sufficient levels of toxin enter the plant cells. Plant protein phosphatases show high susceptibility to inhibition by microcystin-LR *in vitro*. Plant sucrose metabolism is inhibited if microcystin-LR is administered in solution by injection into the transpiration stream. Inhibition of whole leaf photosynthesis by French Bean plants occurred after topical exposure of the leaves to an aqueous solution of microcystin-LR during greenhouse studies. The degree of whole leaf photosynthesis inhibition increased with subsequent brief exposures to the toxin in solution at 48-hour intervals, eventually becoming irreversible. These results were observed at dissolved microcystin-LR concentrations which can be found in untreated waters containing cyanobacterial blooms (20 $\mu g\ l^{-1}$ of toxin); leaf necrosis occurred at higher exposure levels. The relevance of these findings for field situations is currently unclear.

In addition to the possibility of internal accumulation of microcystins, irrigation may lead to accumulation of toxins on the external surfaces of edible plant material. The toxins are deposited when the water dries on the plant surface between irrigation periods or when the water becomes trapped in the centres of, for example, salad plants. Further research is needed into the uptake and fate of microcystins and other cyanobacterial toxins by food plants and the persistence of the toxins on plant surfaces.

Cyanobacteria Sold as Dietary Supplements

In some countries cyanobacteria are sold as dietary supplements, with the number of users of these products estimated to be well over a million in North America alone. Large-scale production of cyanobacteria and microalgae started some 50 years ago. Much of the early research work dealt with the basic photosynthetic properties of microalgae, their possible therapeutic, antibiotic and toxicological properties and their potential as an agricultural commodity. The microalgae biomass industry now provides significant biomass for pigments and speciality chemicals used primarily in the food industry. The bulk of this microalgal biomass comes from two filamentous genera of cyanobacteria: *Spirulina* including *S. platensis* and *S. maxima* and *Aphanizomenon flos-aquae.* While *Spirulina* is grown in artificial outdoor ponds, mainly in southern California, Hawaii, Thailand, Taiwan and Japan, *Aphanizomenon* is at present harvested from a natural lake. Production of food-grade *Spirulina* exceeds 1 × 10^6 kg a^{-1}. *Aphanizomenon* production is also substantial. As a consequence of the consumption of these products many quality control issues arise. One such issue concerns the possible production of cyanotoxins by cyanobacterial genera used for dietary supplements. In particular, *Aphanizomenon flos-aquae* has been shown to be capable of producing saxitoxins and the neurotoxin anatoxin-a. While no saxitoxins have been detected in *Aphanizomenon flos-aquae* marketed as a dietary supplement, it is appropriate to monitor all supplements in order to ensure safety. In natural lakes, mixtures of species often occur in cyanobacterial blooms. In particular *Microcystis* and *Anabaena,* which usually contain microcystins, can both occur in association with *Aphanizomenon.* The microcystins may then become part of the biomass harvested for human consumption. Failure to monitor and regulate these toxins in cyanobacterial biomass used as part of a human diet could lead to an increased risk for the consumer.

Cyanobacteria and Vibrio cholerae

Islam described detection of *Vibrio cholerae* inside the mucilaginous sheath of *Anabaena variabilis* for up to 15 months after artificial exposure and also noted that *V. cholerae* 01 did not lose toxigenicity during the association. Field studies have also detected *V. cholerae* in the mucilaginous sheath of *Anabaena* sp. from a pond in Dhaka when it could not be detected in association with microalgae collected from the same environment (*Euglena* sp. and *Phacus* sp.). The association between *V. cholerae* and cyanobacteria remains poorly understood, but it has been noted that vibrios may produce mucinase and it has been suggested that exchange of oxygen and carbon dioxide may permit a symbiotic relationship. Other workers have noted that motile bacteria can easily discriminate heterocysts from vegetative cells and attach to the heterocyst vegetative cell junction, following which both host and epiphytes start growing. The bacteria rarely penetrate cyanobacterial cell walls. Evidence is accumulating that association with mucilaginous cyanobacteria may be an important factor in inter-epidemic survival of *V. cholerae.* The implications of this for the control of cholera in humans remain unclear and it should be noted that the evidence relates to mucilaginous cyanobacteria, and *Anabaena* in particular. No studies have suggested or apparently investigated a relationship with cyanobacterial toxigenicity.

Tastes and Odours

Cyanobacteria have, for a long time, been recognised as a nuisance in the drinking water industry because of the ability of several taxa to produce earthy and musty smelling compounds, notably geosmin and 2-methyl isoborneol (2-MIB), for which the odour detection thresholds of less than 10 ng l^{-1} are remarkably low among sensitive individuals.

The cyanobacterial genera that are known to produce geosmin are *Anabaena,* Aphanizomenon, Lyngbya, Microcystis, Oscillatoria, Phormidium, Schizothrix and *Symploca*. All these (except *Symploca)* are also known to include toxinforming species and strains. Because of this, the possibility of using odour compounds as an early warning for the development of toxin-producing cyanobacteria blooms has been considered. However, there is no evidence of a correlation between toxin production and the production of taste- and odour-producing compounds that would provide a warning of toxicity.

Evidence from the literature on the capability of various cyanobacterial species to produce both toxins and taste and odour compounds has been summarised by Kenefick *et al.* and does not indicate that cyanobacterial species which produce toxins invariably also produce taste and odour. Nevertheless, some characteristic tastes and odours may indicate the presence of cyanobacteria, and toxic cyanobacteria frequently occur without noticeable tastes or odours. In Alberta, Canada, 89 bloom samples from 10 lakes were analysed for the presence of microcystin-LR, and the taste and odour compounds geosmin, 2-MIB and β-cyclocitral. The latter compound, which is only mildly odorous compared with geosmin and 2-MIB, was reported to be produced in large quantity by *Microcystis* spp. The results showed that all but three of the bloom samples had detectable levels of microcystin while none had detectable levels of 2-MIB. Several samples had detectable levels of geosmin, but there was no clear relationship between the presence of geosmin and the presence of microcystin-LR. In the case of β-cyclocitral, there was a significant correlation, at the 1 per cent level.

However, this relationship is of no practical use for providing early warning of the presence of microcystin-LR, because the relationship is not consistent and β-cyclocitral is not odorous enough to act as a sensitive surrogate for microcystin-LR. Although there have been some rare occasions when cyanobacterial isolates have been found to produce geosmin, microcystin and anatoxin-a simultaneously, as for *Anabaena lemmermannii* from Lake Hallevann in Norway, such reports are exceptions. The biochemical pathways to the biosynthesis of microcystin, anatoxin and saxitoxin are becoming understood. These show no connection between toxin production and the production of the alcohols geosmin and 2-MIB. It is therefore very unlikely that the production of the taste and odour compounds are biochemically connected to the production of the cyanotoxins.

Just as the presence of earthy or musty odours in water indicates the presence of cyanobacteria and/or actinomycetes, taste and odour problems can be used as a warning of the need for further investigation in the event that the occurrence of cyanobacteria could result in the presence of cyanotoxins. It is, however, important to recognise that the converse does not apply: lack of taste and odour by no means implies the absence of cyanobacteria.

REFERENCES

Astrachan, N.B., Archer, B.G. and Hilbelink, D.R. 1980. "Evaluation of the sub-acute toxicity and teratogenicity of anatoxin-a". *Toxicon,* 18, 684-688.

Bagu, J.R., Sykes, B., Craig, M., Holmes, C.F.B. 1997. "A molecular basis for different interactions of marine toxins with protein phosphatase 1". *J. Biol. Chem.,* 272, 5087-5097.

Carmichael, W. 1997. "The cyanotoxins". *Adv. Botan. Res.,* 27, 211-256.

Cohen, S.G. and Reif, C.B. 1953. "Cutaneous sensitization to blue-green algae". *J. Allergy.* 24, 452-457.

Dillenberg, H.O. and Dehnel, M.K. 1960. "Toxic water bloom in Saskatchewan 1959". *Can. Med. Assoc. J.,* 83, 1151-1154.

Eriksson. J.E, Meriluoto, J. and Lindholm, T. 1986 "Can cyanobacterial toxins accumulate in aquatic food chains?" In: *Proceedings of the 4th International Symposium of Microbiol Ecology,* Ljubljana (Yugoslavia), 658-658.

Falconer, I.R. 1991. "Tumor promotion and liver injury caused by oral consumption of cyanobacteria". *Environ. Toxicol. Water Qual.,* 6(2), 177-184.

10

Management of Cyanobacterial Hazards

Risk management is a relatively recent discipline, in which developments are still occurring very rapidly. Various model schemes for risk management have been portrayed, most of which have some common elements. These include the need for an information base upon which to make decisions; the need to make decisions based on often inadequate information; the need to compare and "value" different and often very diverse outcomes with one another; and the need for broad participation in the process in all its stages.

The implementation of risk management will vary greatly according to the political, social and economic context in which it takes place. Whilst it is often seen as a highly rational process, it should be recognised that the scientific basis for many of its elements is actually often extremely weak and procedures for valuation of health effects are generally poorly developed. More importantly, no scientific assessment will support effective risk management if it fails to address the perceptions and priorities of the society concerned. Public acceptance of cyanobacterial-related turbidity and discoloration in recreational waters in some countries illustrates this point. Thus, a commitment to communication and participation is an essential element of any effective risk management plan. The movement towards a more comprehensive approach to risk management is inhibited by four key impediments, sometimes referred to as "institutional failures":

- *Data limitations* are of two kinds: a lack of historical record on which to base current change and to recognise trends, and the inadequacy of scientific understanding on which to base judgements about cause and consequence. Such data limitations are common in the management of cyanobacterial and eutrophication problems. A common question that faces water managers is whether a cyanobacterial problem is getting worse, due for example to cultural eutrophication, or whether the problem has always been present but has remained unnoticed or, perhaps, unreported. Following widespread dissemination of information concerning the potential toxicity of cyanobacteria, blooms may start to be reported more frequently by, or to, relevant authorities.

— *Poor frameworks for analysis* impede comprehensiveness in assessment and decisionmaking necessary to make informed judgements. Cyanobacteria are good examples. Until recently, a general lack of knowledge and poor communication by scientists were a serious impediment to sound management action - a constraint which is now being overcome in some countries.

— *Inadequate regulatory principles* which lay too much discretion at the door of the safety official and which lead to discrepancies in levels of safety provided for different groups in society. In some countries, cyanobacteria-infested drinking water sources are the only ones available to the poor or geographically isolated.

— *Insufficient consultation* procedures restrict participation of different interests that have a legitimate role to play in the determination of risk and its management. Lack of communication between government and industry sectors is especially detrimental in managing cyanotoxin issues. For example, health impact assessment generally requires health sector participation, resource management is generally under the control of environmental and/or local authorities, and abatements such as water treatment may be undertaken by local authorities or the private sector.

COMMUNICATION AND PARTICIPATION

In many cultures, certain characteristics lead to some risks being perceived as high priorities. These include:

— Risks which are "new" or unfamiliar to the population.

— Risks which are perceived to have been caused by a third party.

— Risks for which the consequences are poorly characterised and understood.

— Risks with potentially widespread or catastrophic and irreversible consequences.

— Risks outside the control of the individual.

— Risks in which the population or individual has had little possibility of influencing official responses.

Risks with these characteristics are perceived as high priorities when compared with other risks which are perhaps of equal or greater importance to health but which are voluntarily undertaken and familiar. The example of the social acceptance of smoking across much of the globe, despite widespread knowledge of the adverse health effects associated with it, is a good example of this.

A brief review of the public perception of cyanobacteria as a health hazard indicates that they will often be perceived as a high priority, perhaps higher than rational analysis of the available data would otherwise indicate. Newspaper headlines have been printed proclaiming *"Water More Toxic Than Cyanide"* after cyanobacteria were found in a particular water body. While cyanotoxins are, on a per-unit-mass basis, more toxic than cyanide, this does not mean that drinking water containing a low concentration of cyanotoxins is anywhere near as dangerous as drinking a lethal solution of cyanide, as the newspaper headline implied. Being a "new" issue, with which the general population is not familiar, the blame may be assigned to municipalities, public service utilities, or farmers. Individual choice does not enable risk avoidance and entire communities may be affected through their drinking water supply. Experience indicates that, in some countries, public concern regarding cyanobacteria has often been greater than the corresponding concern from the professional community.

In other countries exactly the opposite perception exists - the public has become accustomed to "green" or "smelly" water and may disbelieve scientific warnings of risk. Improved understanding of risk assessment and management, and an improving understanding of effective intervention, has demonstrated that differences in public and professional perceptions are not an error by either party that needs to be corrected (for example through attempts at altering public perception or discrediting scientific attempts to assess objectively and to describe risks), but are legitimate viewpoints to be taken into account through open channels of communication and especially through enabling public participation in risk management. It should be recalled that the factors which lead to a low public tolerance of risk often include the lack of familiarity and the inability to exert influence. Effective communication and provision of mechanisms for participation will, therefore, often address these directly.

HAZARD IDENTIFICATION AND HAZARD CHARACTERISATION

In the context of risk management, a hazard is usually understood to mean the property of a substance to cause harm. Many substances are hazardous but will not necessarily lead to harm unless circumstances lead to human exposure. Even after exposure, an adverse health outcome is not necessarily certain, but rather a probability. A hazard is therefore defined as an intrinsic property of a biological, chemical or physical agent to cause adverse health effects under specific conditions. This definition implies some certainty that under similar conditions the agent will cause similar adverse health effects.

The term "risk" refers to a probability that exposure to a hazard will lead to a specific health outcome and is usually expressed as a frequency in a given time. Thus, for example, the WHO *Guidelines for Drinking-water Quality* define guideline values as concentrations of specific chemicals estimated to lead to a negligible additional risk for the consuming population. Where such a point of reference is adopted, it is generally referred to as the "acceptable risk", although the term "tolerable risk" is preferred by some people. Hazard identification involves the identification of known or potential adverse health effects associated with a particular agent, based on studies conducted under specific conditions, such as the species tested and the experimental conditions. Epidemiological studies and animal toxicity studies are ranked as providing the greatest predictive information, and this is followed by *in vitro* assays and qualitative structure activity relationships (QSAR) predictions.

Hazard characterisation is the extrapolation phase of risk assessment aimed at making a predictive characterisation of the hazard to humans based on animal studies under low exposure conditions. The endpoint of hazard characterisation is the estimation of a "safe dose" such as a tolerable daily intake (TDI) or equivalent. In general, TDIs are only determined when there is likely to be a threshold in the relationship between dose and effect, based in part on theoretical knowledge of the mechanism of action. For genotoxic carcinogens it has been accepted that there is no threshold dose below which effects, such as initiation of the carcinogenic process, will not occur. When such chemicals cannot be completely avoided, mathematical models have been used to estimate, through extrapolation, the possibility of adverse effects at low doses. The dose corresponding to a risk level of 1 in 1×10^5 or 1 in 1×10^6 has been considered as posing a negligible or tolerable risk.

With cyanotoxins, risk assessment is made more difficult by the paucity of scientificallysound toxicological and epidemiological studies. The available animal data are limited, particularly in the case of chronic or long-term effects of cyanotoxins. The lack of data is reflected in the fact that a WHO guideline has been agreed only for one group of cyanotoxins and only as a provisional guideline value for the toxin most commonly studied. Uncertainties about the environmental fate of cyanotoxins add to the difficulty.

EXPOSURE ASSESSMENT AND RISK CHARACTERISATION

Exposure assessment can relate to exposure to cyanobacterial toxins during an outbreak of disease, or it can be an assessment of the likely exposure of people through consuming drinking water or swimming in lakes known to contain certain levels of cyanotoxins or cyanobacteria.Risk characterisation is the qualitative and/or quantitative estimation, including the attendant uncertainties, of the severity and probable occurrence or absence of known or potential adverse health effects in an exposed population. This estimation is based on hazard identification, hazard characterisation and exposure assessment. If it is calculated as the "probable risk", the estimated risk takes on more meaning than it deserves, because of uncertainties in the process. Alternatively, risk characterisation can be taken as establishing levels of daily exposure over a lifetime at which the risk is "negligible".

POLICY DEVELOPMENT

The processes of hazard identification, hazard characterisation, exposure assessment and risk characterisation may be readily viewed as rational, scientific activities. In contrast, policy development takes account of both rational assessment and societal values. It therefore requires the valuation of a specific health outcome. Most definitions of risk assessment therefore combine a frequency estimation with some valuation of the seriousness of the consequences. Whilst several approaches have been proposed for the rational comparison of the "value" of different adverse health outcomes it should be recalled that such valuation is principally driven by public perception and societal concerns, and rational analysis may be of very limited relevance. Experience with the enforcement of recreational water quality standards leading to restrictions on bathing has shown that public reaction may vary between the extremes of proclaiming that "there is nothing wrong with the water, we have been swimming in it for years without any illness, the authorities do not know what they are talking about" to "this water has killed a pet dog and must be doing me harm too - something must be done about it immediately".

Knowledge of the characteristics of a hazard, the local occurrence of the hazardous conditions and an assessment of the seriousness of the outcomes of exposure, provides the basis for development of policy. However, other factors should be considered. These may include: the seriousness of other hazards and associated health outcomes that might compete for limited resources; the cost and effectiveness of remedial and preventative actions; and the availability of technical solutions and of experience in their application in the country or region. For effective policy development the above mentioned factors must be brought together.

POLICY IMPLEMENTATION

One of the most frequent failures of policy development is the formulation of legislation in the absence of consideration of its sound and sustainable implementation. In the field of water supply, there has been an increasing recognition by governments of the general need to ensure availability of water supplies, rather than costly treatment only for favoured localities. A number of actions are available to governments with which to support policy implementation, the most obvious being regulatory enforcement. This implies a capacity to monitor the implementation of the regulations and a will to enforce compliance when the regulations are not met. Other mechanisms for implementation include voluntary codes, conflict resolution, economic instruments and public information and participation. Most frequently, multiple actions will be used and will interact with one another to encourage safer behaviours and practices.

EVALUATION OF MANAGEMENT PLANS AND ACTIONS

The types of actions described above should, ideally, be combined in appropriate preventative, remedial and contingency plans according to local circumstances and should constitute part of a declared policy for control of adverse human health impacts from cyanobacteria. Thus, they would be integrated into water sector policy more generally, established at national level and operated at local levels supported by legislative frameworks, trained staff and effective institutions with proactive strategies for awareness raising and information dissemination. These plans must be periodically reviewed. In particular, after an incident, it is useful to reflect on the parts of the contingency plan that worked well and those parts that did not function effectively. With long-term, preventative or remedial management actions it is crucial to put in place a system of monitoring that will enable the efficacy of the actions to be evaluated. There may be long delays before any benefit of a management action is perceived in lakes and reservoirs - in some cases 5-10 years or longer.

It is important to consider the tenets and principles of Adaptive Environmental Management (AEM), particularly with long-term management actions. Decisions and actions are often made on the best available advice, recognising that with any complex ecological problem, such as with a toxic cyanobacterial bloom, the information available to guide a specific management action will always be limited and inadequate. It is important, therefore, to monitor the outcome of the management action, and then to modify or revise the action depending upon the response, or as new scientific information and techniques become available.

SITUATION ASSESSMENT

Assessing the risk posed by toxic cyanobacteria, or the potential for development of cyanobacterial blooms, and linking this to effective measures for the protection of public health within available resources, is complex. Situation assessment may be proactive, for example to determine whether contingency planning is required or to inform long-term action, such as pollution control to minimise bloom formation; or it may be reactive, for example to assist in interpretation of specific local events or conditions to inform emergency or incident response.

An important factor in situation assessment is understanding the adequacy of available information with which to make the assessment. In many, if not all, cases epidemiological evidence of cyanobacteria-related health effects would not be available because of poor or non-standardised, or poorly differentiated, diagnoses; lack of awareness of cyanotoxins as potential causes of symptoms; and inadequate reporting systems, research programmes, or information analysis. In most situations a limited range of information is available to assist in identifying whether a problem or potential problem exists.

Drinking Water Supply Information

The monitoring of water bodies and supply systems for cyanobacteria and cyanotoxins is not yet common practice in most countries in the world. There are a number of critical control points in the potable water supply system where testing for cyanotoxins and intact cyanobacterial cells should be carried out if significant cyanobacterial populations occur in the source water. These may include the water storage reservoir or river; the treatment plant raw water intake; key points in the treatment process; and in the final treated drinking water, depending on local circumstances.

One requirement is to be aware of which members of the community receive drinking water from which water supply. This information is usually readily available in cities and large towns but this may not be the case in rural areas. In addition, the existence of water treatment systems and their effectiveness in cyanotoxin removal should be ascertained. For small community supplies there may be little or no water treatment, and this must be taken into account by health authorities when assessing any potential risk situation.

Visual monitoring for scums around bathing areas considered to be at-risk is crucial for protecting human health and is quite easy to perform. Operators of recreational sites can be involved in assessing changes of situation. Furthermore, participation of the public in monitoring for cyanobacteria, and particularly for scums, is a very effective approach. This involves education on appearance and toxicity of scums, as well as on recognition of cyanobacteria at high cell densities when they do not form scums. An example of a message to convey this may be:

> "If you walk into the water up to your knees, carefully, without stirring up sediment, and cannot see your feet because of a greenish discoloration, don't swim and inform the local authority using the following telephone number".

In many countries and regions, the current situation is that public awareness is negligible and knowledge of the risk posed by bathing in cyanobacterial scums is lacking. Eye witness accounts in Australia, Germany and Japan report swimmers deliberately covering themselves in cyanobacterial scum for the sake of an "exciting" photograph, and one well-documented poisoning case in the UK resulted from army cadets undertaking kayak "Eskimo rolls" in cyanobacterial scum. If public education not only addresses personal health risk but also stimulates a sense of responsibility in reporting scums, valuable support for risk assessment at recreational sites can be gained.

Environmental Information

Environmental information, including physical, chemical and biological data can aid in the prediction and assessment of the likelihood, or presence, of cyanobacterial mass developments, including blooms, scums and mats. It can also help predict and assess types of cyanobacterial toxins and their location.

Historical records and local knowledge

Consultation of historical records, if available, can indicate whether a water body has been prone to cyanobacterial bloom development. Useful information is sometimes available from the local community, including descriptions of the water body and examples of human health incidents, livestock mortalities and fish-kills associated with blooms and scums. However, a lack of historical and local evidence of blooms cannot be taken as assurance that cyanobacterial blooms have not occurred, or will not occur, because data may be lacking and recognition of cyanobacterial blooms and associated problems may have been inadequate and because increases in cultural eutrophication may be ongoing.

Physical data

In the case of the many species of planktonic cyanobacteria whose growth is favoured in warm, thermally-stratified environments, the onset of favourable growth conditions is indicated by a rise in surface water temperature above about 18 °C and the establishment of persistent thermal stratification. For example, the Queensland Department of Natural Resources in Australia undertook a two-year survey of thermal stratification in its major water supply reservoirs to aid in its assessment of reservoirs potentially at risk of developing cyanobacterial blooms. The study enabled reservoirs to be categorised as seasonally strongly stratified, weakly stratified or well mixed, with the strongly stratified bodies being considered most at risk. In addition, the seasonal timing and persistence of stratification was used as a trigger to increase monitoring effort.

Hydraulic mixing and transport processes

The ratio between the depth of the mixed layer and the depth to which sufficient light for photosynthesis penetrates, strongly influences cyanobacterial mass development and the extent to which the populations may be dominated by particular cyanobacterial types. Data on flushing rates in lakes as well as river flow rates are useful because planktonic cyanobacteria do not usually attain high population densities in highly flushed environments with retention times of less than 5-10 days, or in the open channels of flowing rivers. If river flows are reduced due to drought and/or excessive abstraction of water, cyanobacterial bloom development can be anticipated provided nutrient concentrations and light penetration are adequate.

Chemical data

The mass development of cyanobacteria is dependent on the nutrient concentrations in a water body. The relationship between mean chlorophyll *a* concentrations and annual mean phosphorus concentrations provides a valuable basis for assessing the likelihood of planktonic biomass development. Inputs and concentrations of nitrate and ammonia should also be considered because these can influence growth rates, maximum biomass and phytoplankton species composition. The ability of several toxigenic cyanobacterial genera to fix dissolved atmospheric nitrogen under aerobic conditions (e.g. *Anabaena*,

Aphanizomenon, Nodularia), but not others (e.g. *Microcystis, Oscillatoria,* emphasises the need to take physical, chemical and biological factors into account when attempting to predict the likelihood of cyanobacterial mass development.

Biological data

Long-term and within-year monitoring records are useful in contributing to the assessment of the likelihood, onset and persistence of cyanobacterial mass developments. Such long-term data sets are not widely available, and often their value may not be apparent to managers who may see long-term monitoring as difficult to justify. This may be the case particularly in water bodies that have no history of cyanobacterial problems. Health authorities responsible for the quality of recreational waters and drinking water resources in many countries may not be sufficiently informed of data available in environmental authorities or local research institutions. Establishing such contacts is strongly recommended for assessing potential cyanobacterial risks. In addition to monitoring for total phytoplankton biomass and cyanobacterial genera or species distribution and numbers, information on other biota in the reservoir, lake or river can be useful. This could include the types and abundance of phytoplankton grazers and of zooplanktivorous fish.

Health Information

Information on the health of the population is collected in nearly all countries for the purpose of providing assistance in the prevention and control of disease. Often included in this information are records of outbreaks of gastro-enteritis and, where possible, their causes. Most sources of gastro-enteritis are infectious organisms, although in most cases of gastro-enteritis a specific cause is not identified. When a substantial outbreak of gastro-enteritis occurs it will generally be investigated in order to determine the source and the causative organisms. Faecal, food and water samples may be screened for a variety of possible pathogens, and only if no pathogens have been identified will the possibility of toxicity in the water or food be investigated. As awareness of cyanobacterial toxicity increases, the likelihood of these toxins being considered as a possible cause for clinical illness increases.

Routine monitoring for the presence of cyanobacterial cells or cyanotoxins in drinking water is undertaken in only a few countries at present, and then only by some water supply companies or authorities. In those countries where there is an awareness of the problem, monitoring of reservoirs after the onset of a cyanobacterial bloom has been detected is more common. As a result of the absence of routine monitoring, reports of gastro-enteritis outbreaks that have been later attributed to cyanobacterial poisoning have been made in the absence of cell counts or toxin measurement at the time of the event.

It is unlikely that an outbreak of illness will be related to cyanobacteria in the drinking or bathing water unless a specific local investigation is conducted. A link between data gathered by health authorities and cyanobacterial data obtained from water monitoring will be required if acute gastro-enteritis caused by cyanotoxins is to be understood and avoided. This may be established by the reporting of monitoring data collected by water supply agencies to health authorities. In outbreaks of gastro-enteritis in which no pathogen has been detected, it may be useful to look at the geographical distribution of cases to see whether the drinking water distribution system is the likely source. Investigation of the presence of cyanotoxins can follow, or proactive investigation for a subsequent bloom can be introduced. Until cyanobacterial monitoring and cyanotoxin analysis are more widely established, it will remain difficult to correlate clinical findings with the toxic effects of cyanobacteria.

Other Data

In addition to the drinking water, recreational water environment and health information that may assist in developing a situation assessment, additional information may be gleaned from veterinary records of animal deaths and *post mortem* examinations. While in many cases it may not be possible to attribute unequivocally animal deaths, or even poor water quality, to toxic cyanobacteria, consistent relationships between these observations and particular water bodies at certain times of the year may be indicative of water bodies with potential cyanobacterial problems.

Management Actions, the Alert Levels Framework

An Alert Levels Framework is a monitoring and management action sequence that water treatment plant operators and managers can use to provide a graduated response to the onset and progress of a cyanobacterial bloom. Circumstances and operational alternatives will vary depending upon the source of the water supply and the analytical and water treatment facilities available.

Vigilance Level

The Vigilance Level encompasses the possible early stages of bloom development, when cyanobacteria are first detected in unconcentrated raw water samples. The indicative value for the Vigilance Level is the detection of one colony, or five filaments, of a cyanobacterium in a 1 ml water sample, although this threshold may be adapted according to local knowledge and prior monitoring history. Taste and odours may become detectable in the supply as the population develops above the Vigilance Level, but their absence does not indicate absence of toxic cyanobacteria. For example, recognition of the highly odorous earthy/muddy smelling compound geosmin, produced by *Anabaena*, may occur at concentrations less than 1,000 cells per ml. In contrast,

Microcystis produces weakly odorous compounds that will be detected only at cell concentrations 100-1,000 times higher than this, and are therefore unlikely to be noticed in raw water.

The presence of cyanobacteria in low numbers constitutes an early warning for potential bloom formation; therefore, when the Vigilance Level is exceeded, it is generally appropriate to increase the sampling frequency to at least once a week, so that potentially rapid changes in cyanobacterial biomass can be monitored.

Alert Level 1

The Alert level 1 threshold is derived from the WHO guideline for microcystin-LR and the highest recorded microcystin content for cyanobacterial cells. In other words, threshold is the cyanobacterial biomass level, assuming the species is a potential producer of microcystin, where raw water microcystin concentration could exceed the WHO guideline. Consequently, Alert Level 1 conditions require an assessment to be made (preferably based on an analysis of total toxin concentration in the raw water by a quantitative technique such as high pressure liquid chromatography (HPLC) or enzyme linked immuno sorbent assay (ELISA), as to whether the concentration of potentially toxic cyanobacteria in the raw water supply can be reduced or whether the water treatment system(s) available are effective in reducing toxin concentrations to acceptable levels.

Alert Level 1 conditions require consultation with health authorities for ongoing assessment of the status of the bloom and of the suitability of treated water for human consumption. This consultation should be initiated as early as possible and continue after the results of toxin analysis on drinking water become available. Clearly, as the biomass of potentially toxic cyanobacteria increases in the raw water, so does the risk of adverse human health effects in the absence of effective water treatment systems. Therefore, on-going monitoring for cyanobacterial biomass and toxin concentrations is essential. The monitoring programme, which should be at least weekly in frequency, may be extended throughout the source water body to establish the spatial variability of the cyanobacterial population and of toxin concentration. It may also be appropriate at this time to issue advisory notices to the public through the media or other means. Government departments and authorities with possible interests or legal responsibilities should also be contacted, as should organisations or facilities that treat or care for special "at risk" members of the public.

For toxic cyanobacteria other than those that produce microcystin it will be necessary for local health authorities to undertake a detailed risk assessment based on the inherent hazard of the toxin detected, its concentration in treated drinking water, and the exposure expected. If Alert Level 1 biomass levels are maintained, but toxins or toxicity are not detected in cyanobacterial or raw water samples, regular monitoring should continue to ensure that toxic strains or species do not develop over ensuing weeks or months.

Alert Level 2

The threshold for Alert Level 2 describes an established and toxic bloom with high biomass and possibly also localised scums. The sampling programme will have indicated that the bloom is widespread with no indication of a cyanobacterial population decline in the short term. Conditions in Alert level 2 are indicative of a significant increase in the risk of adverse human health effects from the supply of water that is untreated, or treated by an ineffective system, even for short-term exposure. The need for effective water treatment systems and on-going assessment of the performance of the system thus becomes of heightened importance. Whereas hydrophysical measures to reduce cyanobacterial growth may still be attempted in this phase, application of algicides can enhance the problem by causing high concentrations of dissolved toxins as a consequence of cell lysis. Whereas filtration systems may have removed cell-bound toxins, dissolved toxin is more likely to break through treatment systems.

If effective activated carbon or other advanced treatment is not available, Alert Level 2 conditions should result in the activation of a contingency water supply plan which is appropriate for the operator and the user or community. This may involve switching to an alternative supply for human consumption, the implementation of contingent treatment systems, or in some circumstances the delivery of safe drinking water to consumers by tanker or in bottles. More extensive media releases, and even direct contact with consumers via letterbox delivery of leaflets with appropriate advice to householders, may be necessary. Where advice is provided to the public because of a cyanobacterial hazard to human health, it will generally be advisable to emphasise that the water would still be suitable for purposes such as washing, laundry, toilet flushing, etc. Withdrawal of a piped drinking water supply because of a cyanobacterial toxin hazard is usually not justifiable because the adverse health effects resulting from disruption of supply are likely to outweigh substantially the cyanobacterial toxin risk itself.

Monitoring of the bloom should continue in order to determine when the bloom starts to decline so that normal supply can be resumed. The sequence at Level 2 may follow through to deactivation of alert level conditions with media releases, and advice to government departments and health authorities to confirm this. It is possible that the collapse of a bloom, or a management action such as flushing, could lead to a rapid decline from Level 2 back to Level 1 or beyond. Likewise the sequence might escalate rapidly, bypassing Level 1 to Level 2, if adequate monitoring and early warning information is not available. Cyanobacterial populations in natural water bodies, especially in hot climates, may double in size in less than two days. Therefore, monitoring frequency needs to take such potentially rapid population growth rates into account.

PLANNING AND RESPONSE

The intensity and scope of management actions to control human health hazards associated with cyanobacteria may vary widely. In the simplest case, a one-off study may indicate little or no cause for concern or it may lead to an information dissemination and public awareness and education campaign. An initial situation assessment that indicates significant risk calls for the establishment of a monitoring programme, formulation of contingency plans, changes in pollution control and water supply management, and a more substantial public information and participation strategy. Which type of response is appropriate in a given situation will depend on a number of complex and interconnected factors, of which rational assessment of human health risk is one part. Other important aspects include, for example, technical and institutional capacities, economic considerations and social values. The principal management actions that may be taken fall into three main groups:

- *Water resource and catchment management.* Most long-term actions are intended to minimise nutrient inputs to water bodies, both from external and internal sources, as well as altering the hydro-logic conditions in such a way as to prevent or reduce conditions favourable for cyanobacterial bloom formation. Other actions may include manipulation of the aquatic food web to increase grazing pressure on cyanobacteria.
- Remedial measures in drinking water supply. Some of these actions may be seen to be applicable in short time frames and may therefore be deployed in response to a specific situation; other are also medium- or long-term actions, such as installing safeguards in drinking water supply that will assist in controlling risks while such safeguards remain installed and operated. Other actions are associated with contingency plans to be activated in case of need.
- *Public information and awareness-raising.* A well-informed public, aware of the potential and real risks of toxic cyanobacteria, is important for many reasons. These reasons include an improved surveillance capacity through public participation; for example swimmers and other users can contact local authorities if they see cyanobacterial scums, and householders can report unusual odours in their drinking water supply. Several Australian States have community-based "WaterWatch" or "Stream-Watch" monitoring programmes that are undertaken by high school students and community groups. Experience has shown that more aware communities, which are adequately informed by appropriate authorities, will be less likely to react inappropriately in the event of a bloom.

Few remedial measures specific to recreational water sites are available in addition to those for minimising bloom formation, increasing public awareness, and informing the public. In some instances, fencing-off or the placing of a boom to prevent accumulation of scums may be considered for densely used and highly impacted areas, in addition to public information and awareness-raising.

Proactive management, that is taking action before significant cyanobacterial proliferation has occurred, is generally preferable to reactive management, that is placing controls in place once significant cyanobacterial proliferation has begun. Few countries presently operate monitoring programmes capable of identifying potentially hazardous conditions and early warning systems have not as yet been developed, even though previous monitoring data may indicate annual "high risk" months or periods. In practice, therefore, management of cyanobacterial hazards typically occurs in three ways, preventative management plans, contingency plans and planning in emergency situations.

IMPLEMENTATION OF MANAGEMENT PLANS

Control of hazards to human health from cyanobacteria is part of a wider picture of water quality and quantity management, environmental protection, policy formulation and general development concerns. Implementation of programmes to address cyanobacteria-related health hazards therefore takes place within the larger framework of other local, national, and international plans and activities relating to water.

Vested Interests

Every water body has associated with it a set of persons and organisations with a vested interest and an involvement. Vested interests can be associated with the level of water; with the contents of the water; with the quality of the water; with the quantity of water; with its motility; with the preservation of the water body itself; or with some combination of these.

Few rivers, lakes, estuaries and seas can support all vested interests equally and so compromises amongst conflicting activities are inevitable. Diverting too much water for irrigation lowers water levels for transportation and may increase the salinity and mineral content of the water, reducing the numbers offish and causing stagnation. It may also increase the likelihood of algal and cyanobacterial bloom formation in river reaches downstream of irrigation reaches, because many cyanobacteria species prefer slowflowing river conditions for bloom development. Allowing unrestricted transportation of dangerous cargoes may damage the ecosystem and wildlife of the surrounding area. Limiting the construction of dams to preserve a riverine environment may mean insufficient energy production and water inputs for industry, domestic needs and mining. Activities particularly likely to increase cyanobacterial hazards in water resources include:

- Land-use, land-use changes and abstraction practices, which may have an impact through diffuse and point sources of nutrients and through changing run-off regimes.
- Pollutant discharges and especially nutrient-rich municipal wastewaters and sewage discharges, which are likely to increase eutrophication.

- Flow control and river flow reduction through, for example, damming, which may increase opportunities for cyanobacterial proliferation by increasing hydraulic retention times and the propensity for water column stratification.
- Abstraction, whether constant, seasonal or intermediate, which may tend to decrease flow rates and also increase opportunities for cyanobacterial proliferation by increasing retention times.

As with any other activity in relation to water resources, taking action in response to a cyanobacteria risk requires the recognition and balancing of vested interests. For example, increasing stream flows to decrease hydraulic residence time will have beneficial effects on drinking water and animal health, but may have detrimental effects on recreational use and agricultural interests.

Responding to an environmentally-related health issue, such as a toxic cyanobacterial blooms, implicates and involves many with vested interests who can be defined broadly as representing the public, the media, the government and the private or corporate sector. The largest and most vociferous vested interest group will often be the general public, who may demand that the quality of their drinking or recreational water be returned to a level that they deem acceptable, as soon as possible. Sometimes the water quality standard that the public expects may seem unrealistic. Farmers may also represent a vociferous lobby who may insist that the water necessary for farms and livestock be again made safe so that productive activity may resume. However, they may also be a strong lobby against reduction of diffuse pollution input by arguing that changes in agricultural practices, e.g. fertiliser application, would inevitably reduce crop yields. Water suppliers often have a strong vested interest in a better resource quality because that increases the safety of their product and reduces treatment costs, but agencies may tend to avoid publicity on this issue in order to avoid a negative image and adverse public comment. Thus, the greatest driving force for action will nearly always come from the user sector.

The media can play a key role in acting as vehicles for public awareness, thereby triggering public outcry. They can also be vital for advising the public of potential health risks, for informing the public whether appropriate actions have been taken and of the efficacy of these actions.

Notwithstanding governments' primary public safety role, where they exist, private companies involved in water supply and wastewater treatment processes will also have a key role in any response. Collaboration with such companies will be necessary whether planning is of an emergency or long-term strategic nature. However, private companies may view the monitoring data they generate as proprietary or commercial and their active participation needs to be encouraged, whereas governments may view information they hold as being in the public domain.

Other important sectors with a vested interest in many countries are tourism and recreation. Hotels, holiday resorts, water skiing and fishing-based tourism operations, for example, can be affected profoundly by adverse publicity and recreational restrictions brought about by toxic cyanobacterial blooms. Involvement of these sectors will be extremely important if local acceptance of management plans is to be obtained. The fact that private companies operate on a profit motive does not necessarily place them at odds with governments' public safety goals, because safe water is also a prerequisite for most, if not all, kinds of tourism and recreation activities, and is seen increasingly as a priority amongst tourists and other recreational water users. Where a cyanobacterial bloom occurs, private companies, governments, the media, the public and others with a vested interest will express and seek support for their particular interests, some of which may conflict with one another. As a result, multi-sectoral cooperation is essential.

Multi-sectoral Involvement, Agencies and their Roles

Governments have ultimate responsibility for safeguarding public health and their role will be to minimise the damage and to assess the circumstances that led to the cyanobacterial problem (s) in the first place. Although governments usually take the leading role in co-ordinating policy amongst vested interests, they may later turn over leadership to a multi-sectoral committee or group to continue developing longer-term plans or strategies.

Human health issues affect many different aspects of environmental policy and management and therefore programme co-ordination is both desirable and necessary for effective action in environment and health. Yet in many countries, environmental health programmes are assigned to different specialised agencies, making co-ordination difficult to achieve. Chapter 18 of Agenda 21, the plan of action arising from the United Nations Conference on Environment and Development outlines the problem as follows:

> "The fragmentation of responsibilities for water resources development among sectoral agencies is proving, however, to be an even greater impediment [to effective action]... than has been anticipated. Effective implementation and co-ordination mechanisms are *required*".

In addition to single-issue government or quasi-governmental agencies in a particular country, there may be national or regional co-ordinating groups which have as their mandate to bring together, or to attempt to bring together, the various sectors, user groups and agencies involved in water management. There may also be water users associations or co-operative societies with interests in water. Any one or several of these agencies and co-ordinating groups may be involved when a cyanobacteria problem arises.

The co-ordination necessary for effective action on environment and health issues is difficult to achieve, not only because it takes place amongst those with vested interests, agencies and groups representing divergent agendas, but also because the co-ordination itself is complex. That is, it does not simply consist of linking different institutions towards a common goal. It also requires combining inputs from different areas of expertise; facilitating collaboration between policy-orientated institutions and regulatory institutions; and fostering co-operation among sectors that positively and negatively influence environmental health quality, including industry, agriculture, local government and transport. Experience in the UK and Australia has shown that an effective strategy to achieve this complex co-ordination is the formation of interdepartmental task forces, with representatives from all relevant government agencies and other key groups with a vested interest. Such task forces have been crucial to the success of emergency and long-term management of toxic cyanobacterial blooms.

Policy Tools

A number of tools may be deployed by governments in the pursuit of water management and health protection policy. Although sometimes discussed in isolation, these tools are generally inter-related, and policy implementation is most effective when a range of tools is employed. The principal interventions and tools relevant to the control of cyanobacterial hazards include:

— *Consensus through conflict resolution.* Conflict resolution involves negotiation, through formal and informal processes, guided or unguided by an expert, which may be binding or non-binding, depending on the agreement of the parties before entering into the negotiation. Each party's interests are aired, then jointly evaluated, ranked and then balanced. Ideally, this process precludes, or at least places in abeyance, any subsequent conflict, because it is through consensus that a plan of action is developed and agreed upon.

— *Guidelines and recommendations.* An enhanced role for experts informs the development of guidelines and recommendations. Such guidelines and recommendations suggest methods of complying with desired policies. They are generally not binding, but rather serve an educational and advisory purpose, in that those studying them learn what actions to take. To be effective, each guideline or recommendation should be explained and justified.

— *Non-binding agreements, including industry codes.* Non-binding agreements, such as industry codes, are similar to guidelines and recommendations, except that they may be more formal. Although not binding, such codes have a high success rate, especially where industry members are involved in their formulation. Such involvement is desirable because industry members know their capacities and, furthermore, once they have signed up to a code there is "peer pressure" by other members to adhere to it. In addition, there might have been reluctance to follow codes developed by third parties, such as environmental and advocacy groups. Such concerns are assuaged where the codes have been developed following wide consultation. As with consensus building through negotiation, non-binding agreements involve all parties in creating acceptable compromises, thus ensuring greater success.

— *Economic measures.* A set of measures, both supportive and penalising, can be implemented to encourage the reduction of water pollution. Generally, such economic incentives and disincentives are contained in legislation, although they may also appear in non-binding agreements such as industry codes. Supportive measures can be direct subsidies for water protection technologies or practices, or various forms of tax reductions. Penalties involve fees for discharge, which can be calculated according to the load of individual hazardous substances, with each substance tagged with a sum per unit of loading.

— *National legislation.* Because of its formality and inflexibility, law may not necessarily be the best avenue for implementing policy. Nevertheless, it should be recalled that in situations where there are competing interests, legislation offers certain benefits. It provides stability in a complex and changing environment and those with vested interests know where to go to seek the rules that apply to them. In addition, at least at the national level, legislation provides a coercive effect upon those that have neglected to take action despite having agreed to do so.

All legislation must be underpinned with a firm policy background so as to ensure that it will be effective once enacted. As one expert explains:

> "Simply to promulgate laws and designate responsible officers will not suffice. It is first necessary to build an effective political base of opinion and power and then to design adequate systems, assign authority, provide resources and translate plans into action; finally the development of the systems must be monitored, so that they can be adapted as required. In most countries sectoral bureaucracies have to be brought under a considerable degree of control and given clear, consistent and persistent policy direction, if co-ordination schemes are not to flounder."

In most countries, before new legislation is introduced by a sector ministry, policy approval must be sought and obtained from the Cabinet or Council of Ministers. This ensures that the process of formulation of legislation is not an academic exercise, undertaken with no thought for the realistic chances for enactment and implementation. The existence of laws which are either unachievable or unenforceable means that financial or human resources have been diverted from other, perhaps more important, priorities. Moreover, the existence of such legislation reflects negatively on the entire legal system.

— *International and interstate plans and agreements.* Actions taken to further one interest may have effects on other vested interests. Where a body of water crosses international or state lines, such actions have international and, at times, diplomatic implications. Few rivers and lakes remain within national or state boundaries, adding a diplomatic or more complex legislative dimension to many water resources management issues. There is a clear trend towards development of international plans and agreements which organise water management around particular water bodies.

In addition to plans and strategies, many water bodies are governed by bi- or multilateral agreements which embody the customary rules and policy priorities of all the countries with an interest in the river, lake or sea. Such agreements may address issues such as reducing pollution, exchanging information on the level and volume of water, carrying out flood management, early warning systems, monitoring changes in the path of watercourses, sharing costs and managing dispute resolution mechanisms. Few of these are of direct relevance to the control of cyanobacterial hazards beyond policies relevant to the control of eutrophication in general. Nevertheless, some international agreements specifically include provisions for notification of downstream countries of pollution events in order to enable them to take corrective action. Including bloom formation as an issue requiring notification of downstream countries might assist in control of cyanobacterial hazards.

At the international level there is no "supranational" enforcement mechanism for nations that decline to follow even the agreements they have signed. Nonetheless, international agreements do offer certain benefits, namely that they are formal, written, and embody two or more countries' policy compromises. They are also usually vetted and approved by a majority of the national legislature before being signed.

Legislation, Regulations, and Standards

Making generalised observations about legislative frameworks is difficult, not least because of the diverse forms they may take but also because of the varied environments in which they operate. Nonetheless, it is possible to identify certain principal characteristics of the legislative framework within which actions related to the control of cyanobacterial health risks may be effected.

Forms of legislation

The term "legislation" or "laws" refers broadly to all legal texts which are promulgated by the legislative arm of the state exercising its legislative powers under the Constitution, or by the executive arm of the state exercising either its own executive powers under the Constitution or powers delegated to it from the legislative branch. A basic law is usually introduced by a sector ministry and enacted by Parliament or other equivalent national lawmaking body.

The term "subordinate legislation" refers to all legal instruments promulgated by the executive branch exercising delegated legislative power. Subsidiary regulations to a particular law enacted by the national legislature are elaborated and issued by the Minister with responsibility for the subject matter. There may also be schedules or annexes to the regulations, consisting of even more detailed subject matter. Under the principle of supremacy, laws adopted by the national legislature nearly always take precedence over subsidiary regulations as well as over legal instruments adopted by the legislative branches of the country's political subdivisions. In some countries, however, a variant of this provides that legislation promulgated by the national legislature has primacy for certain subject matters, whereas legislation by the local legislatures takes precedence for others. In other constitutional systems, such as the USA or Australia, any powers not expressly granted to the federal government in the Constitution are the responsibility of the states. In addition to laws and regulations, there may be other administrative legal texts such as rulings, circulars, guidelines, standards, administrative notes and decisions. Such legal instruments, although non-binding, assist the governments in achieving the aims of the umbrella legislation while at the same time providing guidance for private groups as to how government authorities will go about enforcing and implementing the law.

Water laws and standards

Enactment of a basic water law is important in countries where there are laws covering different subject matters and enacted at different times containing conflicting provisions relating to water. This might occur, for example, amongst a country's environment protection law, electricity law and public health regulations. However, during the process of updating or replacing legislation on water, it should be kept in mind that the trend in many countries is towards considering water issues in conjunction with other issues, such as the conservation and use of land, rather than viewing them in isolation.

In order to be effective in addressing cyanobacterial hazards, the law should refer to the jurisdictions, responsibilities, and authority of specified competent agencies in relation to water and to their relationships to one another. Thus it becomes the instrument which brings together the diverse parts of the water medium from which problems arise, as well as the diverse groups associated with their control. These groups include water users, those constructing and administering water supply systems, those regulating different aspects of water and having responsibilities in related domains, and agencies involved in data and information management and sector planning. It is important for the water law to reflect the policy priorities and political realities of the particular country. For example, in the case of drinking water supply, legislation in the established market economies is orientated principally towards the regulation of established water supplies administered by recognisable entities where extensive water supply infrastructure is in place. This may have little relevance to countries where the goal is principally to support the best use of available resources in the provision of incrementally improved service standards to the population as a whole. Such differences argue strongly for developing the policy for the particular country in light of its particular legislative and policy framework, rather than relying on some sort of "model" legislation which would not reflect the realities at hand.

Water laws generally follow the basic principle of national legislation, i.e., that a basic law is indeed basic, whereas any details, which may have to be changed in response to scientific advancements or other exigencies, are contained in the regulations or other subsidiary legislation. The rationale is that regulations are more easily changed because they are elaborated by the responsible Ministry and can be easily revoked, amended or replaced, rather than having to go through the Parliamentary process. Among the details which will normally form part of the subsidiary legislation of a water law are standards of various kinds. For example, there may be standards on minimum treatment requirements linked to source type and quality.

In some countries, service quality standards are contained in a "sanitary code" or "code of good practice". As mentioned above in the discussion of industry codes, such standards are most likely to receive support in their implementation if consultation with the affected entities has occurred. Other standards consist of quality targets, or what are known as water quality standards. Comprehensive guidance on the setting of drinking water quality standards is available in the *Guidelines for Drinking-water Quality*. The *Guidelines for the Safe Use of Wastewater and Excreta in Agriculture and Aquaculture* are also available and further guidelines for safe recreational-water environments are in preparation.

Awareness Raising, Communication and Public Participation

The understanding that cyanobacterial blooms and toxins present hazards to human and animal health is a prerequisite for anticipating, avoiding or reducing their adverse effects.

Raising awareness on the causes and effects of cyanobacterial blooms serves two goals: protection from the health hazards presented by the toxins, and long-term reduction of toxic bloom development through public participation. Early recognition of blooms and scums facilitates better management of the blooms and associated problems and helps to reduce their impact on the community as a whole. Case histories and risk assessments point to the need for greater awareness of cyanobacterial bloom and toxin hazards amongst not only water users, but also professional groups. Awareness raising can serve several functions:

- To alert professionals and water users to the presence of hazardous accumulations of cyanobacteria and toxins and to the risks involved in drinking and using water containing blooms and toxins.
- To inform health-care professionals of the circumstances and exposure routes that lead to intoxications.
- To inform water users of the causes of cyanobacterial bloom development and of ways in which they can assist in reducing bloom formation.

— To inform environmental regulators, water user and water supply organisations about cyanobacterial toxin hazards, and to help identify and implement appropriate policies on water access, use, consumption and treatment.

Professional Group Awareness

Professional groups are often no more cognisant of the circumstances leading to cyanobacteria-associated health problems than are the general public. The kinds of professional groups that require up-to-date information on cyanobacterial blooms and toxins in order to manage effectively the associated problems include water treatment and supply authorities and companies, medical and veterinary practitioners, public and environmental health authorities, and national/state environmental agencies. Initiatives to increase the awareness of these groups and organisations have centred around conferences, workshops, correspondence in widelyread professional journals, review-type papers, and the large-scale publication of handbooks and reports. Such educational materials have been developed for the water supply, treatment and environmental sectors; the human health-care sector, and the veterinary sector. In countries intending to address cyanobacterial risks, these communication routes should be established and developed further.

Public Awareness

In some parts of the world, such as Australia, the UK and Scandinavia, cyanobacterial awareness programmes have operated for several years. However, in other countries with water resources and supplies containing toxic cyanobacterial blooms, awareness and the availability of information are more limited. These deficiencies can be combated in several ways. When cyanobacterial populations in water bodies exceed threshold levels, media warnings should generally be issued to the public. Such warnings and notifications may be issued through the newspapers, radio and television, and may require media targeting and press conferences. The content of any announcement varies according to individual bloom situations and the purposes for which the affected water body is used. For example, if a cyanobacterial bloom occurs in a municipal water supply source, media releases may be appropriate at lower cell populations than if the waters are used only for recreational purposes.

The information supplied in the warnings should be prompt, concise and should include details of expected changes in the quality of supply, such as tastes, odours, discolouration and the actions being taken to alleviate the problems. Where recreational activities, livestock watering, and further water uses other than human supply, are involved, media warnings may also contain a brief description of the bloom and how it can be recognised. Other information which may be supplied in the warnings and notices, if appropriate to the bloom situation, includes:

- Possible health problems including gastrointestinal upsets, skin rashes, and eye irritations.
- Warnings not to use the water for swimming, bathing, or showering.
- Advice on finding alternate sources of water for animals and preventing pets from swimming in affected areas.
- Notification that boiling the water will not destroy the toxins.

The community often has several concerns when a cyanobacterial bloom occurs in its drinking water supply or recreational amenity. Commonly-asked questions concern the use of water in evaporative air coolers; its use for drinking, laundry, dish washing, washing fruits and vegetables and cooking; whether fish, shellfish and crustaceans harvested from cyanobacterial bloom-containing water are suitable for human consumption; and where to obtain alternative water supplies. Information or fact sheets prepared in advance can be circulated upon request. It may be appropriate to display warning signs and notices at major points of public access to affected water bodies. For eutrophic waters with occasional bloom problems, provision of basic information to visitors is particularly likely to contribute to protecting them from health hazards, because the rapid appearance and disappearance of scums cannot always be detected in time for the responsible authorities to implement specific measures.

In such situations, health protection must, to some extent, rely on the users' own informed judgement. Warning signs should be prominent and simple, with symbols or minimum text to indicate no swimming, paddling, drinking, livestock watering, pets in the water, or other activities deemed by the responsible agencies to present unacceptable exposure risks. The signs may be removed when the blooms decrease to acceptable levels, or they may be kept in place throughout the year in areas with year-round problems. If the latter option is taken, the gradual decrease in visual impact of warning signs and notices needs to be borne in mind.

Brochures and fact sheets are other tools which are increasingly used to convey to the general public and specialist water user groups information on the appearance and hazards presented by cyanobacterial blooms and scums. Examples are available from several countries which have experienced bloom-related problems, such as the UK, Australia, Denmark, Finland, South Africa, and the USA.

Brochures and fact sheets intended for the public should be simple and free of scientific jargon, and should include brief descriptions with the following kinds of information:

- The nature and occurrence of cyanobacteria.
- The health hazards presented by cyanobacteria and their toxins.
- The recognition of cyanobacterial blooms and scums.

— The propensity of cyanobacteria to produce toxins.

— Precautions and steps to be taken to avoid potential health problems.

— Whom to contact if the presence of a cyanobacterial bloom or scum is suspected.

Special fact sheets for farmers, regarding protection of farm water supplies for animals, and also for water sports organisations, have also been used. Professional associations and user-group networks, such as national water sports bodies, provide a useful means of spreading information. Posters and displays in public places and association premises are additional tools for conveying information to the public on cyanobacteria and for increasing community awareness. Videos are another useful method of educating the public about cyanobacterial problems. They can be used to present the causes and consequences of cyanobacterial blooms, as well as to provide good visual examples of waters containing blooms and scums. They may also outline the strategies proposed and used by authorities to manage the problems, and to inform the public about what, as individuals, they can do to help. For example, to inform the public and to enable them to initiate action at a community level, Australia's Murray-Darling Basin Commission initiated and supported the development of a "study circle" kit for adults entitled "Blooming Blue-Green Algae". The kit contains a video, an audiotape, posters and printed information on cyanobacteria which enables people to discuss and find out more about cyanobacteria.

Education on the causes of blooms can begin through programmes run in schools. Students may be encouraged to participate in simple water quality investigations, such as in the Australian "Streamwatch" programme. Easy-to-use kits are employed to enable the early detection of impending cyanobacterial blooms. The effectiveness of the tools discussed above is only as good as the efforts made to publicise and to distribute them. The need for contact points may be especially important in remote locations where monitoring waters by government agencies or water body owners is not possible and thus the assistance of the community is vital.

PREVENTATIVE MEASURES

Cyanobacterial bloom formation can be avoided by measures which address their growth requirements, i.e. plant nutrients and light. The basis for excessive growth of cyanobacteria and other phytoplankton organisms is enrichment of aquatic ecosystems with plant nutrients. This process is termed eutrophication. The key nutrient in many cases is phosphate. In some systems, not all of the phosphate available is actually used for phytoplankton growth because other resources limit the maximum possible biomass. These may be light intensity or availability of nitrogen. Furthermore, other biota can affect the growth of cyanobacteria and phytoplankton organisms: submerged aquatic plants may compete for nutrients, and grazing by zooplankton may reduce the stock of many phytoplankton organisms and also of some cyanobacteria.

The key management action for abatement of Cyanobacterial blooms is to address the source of the problem by control and reduction of external nutrient loading to the water body, and thus of the concentrations within it. Measures addressing light availability directly or targeting the community structure of the biocoenosis have been successful chiefly in less eutrophic situations. For highly eutrophic water bodies under restoration by reduction of nutrient loading, such measures may accelerate and enhance success.

Control and reduction of nutrient loading usually focuses on phosphorus, but measures addressing phosphorus may be designed to reduce nitrogen input simultaneously. Targets for nutrient concentrations can be achieved by following basic principles of good catchment management with respect to agriculture and sewage treatment.

A decision tree for application of different management approaches in order to control phytoplankton growth is given in Reynolds. It leads from restriction of all sources of phosphorus enrichment to enhancing flushing or sediment removal, biomanipulation and artificial destratification. Reynolds points out that decision points for application of measures are still being quantified by current research.

Carrying Capacity

The concept of the carrying capacity of the resources in a given ecosystem to sustain a population has proved very helpful in planning measures to control the size of that population; Applied to cyanobacteria, this means asking questions such as:

- How much biomass can be sustained on the basis of the amount of nitrogen available?
- How much biomass can be sustained on the basis of the amount of phosphorus available?
- How much biomass can be sustained with the amount of light that penetrates into the water?

At any one point in time, it is likely that one of these three resources will limit the possible amount of biomass at a lower biomass level than the others. However, the limiting resource may change seasonally, for example at higher latitudes it changes in relation to the angular height of the sun and day length, and in tropical climates it frequently changes in relation to turbidity changes caused by seasonality of the flow regime. During winter or in turbid situations, light is usually the limiting factor, whereas the available nitrogen and phosphorus could have allowed a higher level of biomass. As light intensity increases in spring or as water becomes clearer, phytoplankton organisms begin to multiply and incorporate available nitrogen and phosphorus into their biomass, often up to the point where either of these resources is depleted in the environment and further growth is not possible. If nutrient concentrations are excessively high, phytoplankton may reach a density that causes such a high level of turbidity that light availability limits any further growth, and in these situations populations will be light- rather than nutrient-limited. For planning and management, it is important to be able to estimate which of the key resources is likely to control phytoplankton biomass in any given system. In other words, the questions are:

— Which resource determines the carrying capacity for phytoplankton?

— How high is the carrying capacity?

For a first step in answering this, it is not important to differentiate between cyanobacteria and other phytoplankton, because the maximum amount of phytoplankton possible can be equal to the maximum amount of cyanobacteria possible, once cyanobacteria have become dominant. One approach at estimating the carrying capacity in relation to nitrogen and phosphorus is to look at the relative amounts of these nutrients in phytoplankton biomass, known as Redfield Ratio by mass. These ratios are:

C	H	O	N	P
42	8.5	57	7	1

Among these components, hydrogen and oxygen are never limiting in aquatic environments. Carbon is available as carbon dioxide (CO_2) and is consumed by photosynthesis. Consumption of dissolved CO_2 enhances diffusion of atmospheric CO_2 into the water, a process which takes time. Carbon limitation has been extensively investigated and has been summarised by Reynolds. He concluded that CO_2 limitation can occasionally have an impact, particularly in soft-water lakes with low bicarbonate alkalinity, but that these situations are generally brief and do not substantially limit the maximal amount of biomass possible.

Nitrogen

Nitrogen may enter water bodies as leachate from soils, as run-off from animal feedlots, and from untreated or biologically-treated sewage, unless treatment includes nitrification and denitrification. Phytoplankton can take up inorganic dissolved nitrogen in the form of nitrate, nitrite and ammonia. In some arid continental regions, nitrogen is found to be the chief factor limiting phytoplankton growth. The relevance of nitrogen to limitation of cyanobacterial biomass is under debate, because a number of cyanobacterial taxa can compensate for its lack by fixing atmospheric nitrogen at rates of up to 175 kg $ha^{-1}a^{-1}$. Thus, lack of dissolved inorganic nitrogen may actually support the dominance of such species as *Anabaena* and *Aphanizomenon*. However, these taxa also occur under conditions of surplus inorganic nitrogen. More importantly, nitrogen fixation is a process requiring high amounts of light energy and will not be effective in very turbid waters. Thus, in a given water body the maximum amount of biomass that can grow, in addition to the biomass already present, can be estimated from the Redfield Ratio on the basis of the concentrations of dissolved inorganic nitrogen.

Phosphorus

Phosphorus, like nitrogen, enters water bodies from untreated and from biologically treated sewage, and further treatment steps are required to eliminate it. Phosphorus is biologically available as phosphate, which binds to soil particles more effectively than nitrate. Thus, the main entry route into water bodies from land areas is as surface run-off and with erosion. Although biomass needs only about one seventh of the amount of phosphorus as it needs of nitrogen, phosphorus is the resource which most frequently limits phytoplankton growth in aquatic environments. Cyanobacteria and many other phytoplankton organisms have developed storage mechanisms for phosphate. These enable them to store enough phosphate for 3-4 cell divisions. As a consequence, one cell can multiply into 8-16 cells without requiring any further phosphate uptake, and biomass can increase by a factor of 10 or more even when dissolved phosphate is entirely depleted. For this reason, the amount of biomass that can grow in addition to the biomass present cannot be predicted from the concentrations of dissolved phosphate.

Phosphorus is naturally abundant only in very few aquatic ecosystems. Furthermore, its inputs to aquatic environments are often easier to control than nitrogen inputs. Methods for elimination of phosphorus from domestic sewage are well developed and currently more cost-effective than nitrification and denitrification. Measures to protect soils from erosion can also be very effective against loss of phosphorus, whereas control of nitrate leachate from over-fertilised soils may be more difficult. Nitrogen limitation may be to some extent compensated by fixation of atmospheric nitrogen by cyanobacteria, whereas there is no comparable compensation mechanism for phosphorus.

Available Light Energy

Light energy is a critical resource indirectly affected by nutrient concentrations. Light arriving at a water surface is partly reflected, and the remainder is very quickly absorbed by the water itself as well as by the dissolved substances and by the suspended particles in the water. An upper limit of phytoplankton cell density is reached when the cells shade each other to such an extent that further growth is no longer possible because the individual cells do not receive enough light. This level can be estimated following Lambert-Beer's basic law of exponential extinction with increasing thickness of the water layer. This law can be expressed as:

$$I_z = I'_0 e^{-z\varepsilon}$$

where I_Z is the intensity at depth Z

I'_o is the surface intensity

ε is the vertical extinction coefficient, which in turn is the sum of extinction by the water itself and the substances dissolved in it such as humic acids ε_w, the algae suspended in the water ε_a, and other particles suspended in the water ε_p.

The average amount of light I^* available to a phytoplankton organism entrained in vertical mixing of the entire water body or within the upper, warm water layer is the square root of the intensity at the surface I_o and at the bottom of the mixed layer or of the water body I_m. This relationship illustrates the decisive influence of depth on light availability and enables estimation of the carrying capacity for phytoplankton biomass. For otherwise clear water with the sum of ε_w and ε_p being only 0.2, Reynolds uses chlorophyll as a measure of phytoplankton biomass and demonstrates that at 1 m depth and a daily insulation of 10^{-3} mol photons m^{-1} s^{-1} a maximum of 670 μg l^{-1} of chlorophyll may be sustained, whereas at a mixed depth of 10 m, only 49 μg l^{-1} are possible, and if mixing occurs down to 30 m, carrying capacity declines to only 3 μg l^{-1} chlorophyll. At these phytoplankton biomass levels, turbidity has also increased, and not enough light can penetrate to enable further growth. Nutrient availability often influences light limitation. If nutrients are limiting, phytoplankton cannot grow to density levels that reach the light-determined carrying capacity. If nutrient concentrations are excessive, phytoplankton will reach the biomass limit determined by light. Further increase of nutrient concentrations will then have no further effect on phytoplankton biomass. This is often the case in hypertrophic water bodies. Turbid situations where the light-determined carrying capacity has been reached are often dominated by cyanobacteria, because at low light intensity these have a higher growth rate than many other phytoplankton organisms.

Target values for total phosphorus within water bodies

In determining target values of phosphorus within water bodies to control cyanobacterial blooms, two questions are important:

- What phytoplankton biomass density can be expected at a given concentration of total phosphorus?
- At what threshold concentration of total phosphorus does phytoplankton density create a turbidity level high enough to reach the light-determined carrying capacity, and thus switch a water body from total phosphorus control of biomass to control by light limitation?

Finding answers to these questions requires a clear definition of total phosphorus. In the past, the soluble phosphate fractions have frequently been addressed when dealing with eutrophication issues. This has some predictive value if it can be measured in seasons where very little phytoplankton is present to consume dissolved phosphate, and if inputs are fairly constant throughout the year. However, the carrying capacity for phytoplankton biomass is more reliably analysed in terms of the total amount of phosphate, i.e. the sum of phosphate bound in biomass and phosphate dissolved in the water, known as total phosphorus.

For predicting phytoplankton density from total phosphorus concentrations several models have been developed. The most comprehensive statistical model was established through an international cooperative study organised by the Organisation for Economic Co-operation and Development (OECD). The concentration of chlorophyll *a* was used as an easy-to-analyse measure for phytoplankton density. Data for annual mean values of chlorophyll *a* and for maxima of chlorophyll *a* were compiled from a wide variety of phosphorus-limited lakes and related to annual mean concentrations of total phosphorus. The resulting regressions were almost linear and highly significant. The result roughly means that per microgram of total phosphorus, an annual mean phytoplankton biomass corresponding to 0.25 μg of chlorophyll *a*, and a maximum of up to 1 μg of chlorophyll *a*, may be expected.

Regressions can be applied as management tools to predict the average and the maximum phytoplankton biomass range likely at a given concentration of total phosphorus. However, this approach has been criticised because these models integrate the behaviour of a number of lakes rather than the response of any one lake to changes in phosphorus concentrations. It must be emphasised that regression B is useful to estimate the maximum phytoplankton biomass at a given phosphorus concentration. However, the estimate given by the double logarithmic regression is only rough, the scatter of points within the 95 per cent confidence limits covers a factor of 10. Maximum ratios of chlorophyll to total phosphorus of 2 are still within this limit. This scatter reflects the effects of other environmental factors controlling phytoplankton biomass, particularly depth and mixing conditions, and losses due to grazing of algae and some cyanobacteria by zooplankton. The carrying capacity for phosphorus will not have been reached in all of the lakes among the wide variety used for regression B.

Predictability of maximum biomass levels is enhanced by combining this model of phosphorus-determined carrying capacity with a measure for light-determined carrying capacity by introducing the threshold concentrations where phosphorus limitation switches to light limitation. This requires knowledge of the depth of the water body and, if it is thermally stratified, knowledge of the depth of the warm upper mixed layer, also termed "mixing depth" Z_m. If data on light extinction are available as outlined above, average light intensity through the mixing depth I^* can be calculated. In absence of such data, light availability can be estimated by regarding Z_m in relation to the depth of light penetration. If mixing is deep in relation to light penetration, cell or colony densities cannot become very high, because the deeply entrained cells would be spending too much time in the dark. If mixing is shallow, cells are frequently moved near the surface, where enough light can penetrate, even through a dense suspension, to enable extensive proliferation.

The maximum biomass density which can be reached is higher in lakes with shallow mixing depths than in deeply mixed lakes, because the latter reach their light-determined carrying capacity at a much lower level of biomass.

When designing programmes to reduce cyanobacterial proliferation in highly eutrophic lakes, total phosphorus concentrations must be reduced below the threshold value for phosphate limitation of biomass in a given lake in order to have an effect; this threshold value will depend upon the depth of the lake. However, once phosphorus levels have been brought below the threshold value the OECD model can be applied. For example, if a shallow drinking water resource has total phosphorus levels of 600 μg l^{-1} and cyanobacterial densities corresponding to 200 μg l^{-1} chlorophyll *a*, and if restoration measures are applied that achieve total phosphorus levels of 200 μg l^{-1}, the annual mean chlorophyll *a* concentration is likely to decline to 70 μg l^{-1}. Although this is a step in the right direction, cyanobacterial biomass is still very high and problems of bloom formation are not yet resolved.

Experience collected during the past two decades with restoration of water ecosystems shows that phosphorus control for abatement of cyanobacterial blooms should target concentrations at least as low as 30-50 μg l^{-1} total phosphorus. In many water bodies, substantial reduction of cyanobacterial and algal population density can be expected at these concentrations when compared with higher concentrations. However, significantly lower total phosphorus concentrations may be required, particularly in deep lakes, in order to prevent blooms of some stratifying ecostrategists in the long term. Information on the prevalent ecostrategists within the cyanobacterial population will help to predict the success of management measures.

Target Values for Total Phosphorus Inputs to Water Bodies

The regression showed a close correlation between annual means of in-lake concentrations and the annual means of inlet concentrations in relation to the residence time of the water:

$$TP = 1.55[\text{Pinlet}/(1+\sqrt{\text{residence time}})]^{0.92};\ r = 0.93,\ n = 87$$

The "safe loadings" given in Table 1 were derived from this model. The model may serve for preliminary scaling of measures to reduce inputs of phosphorus. For prediction of the total phosphorus concentrations in a given water body, estimates of inputs from inlets, from surface run-off, from urban storm-water outfalls and from atmospheric precipitation are needed. Acquiring all of the necessary data may be difficult, and approaches to estimates are discussed in the context of lake and reservoir diagnosis by Cooke *et al.*

Table 1. Permissible and dangerous inputs for phosphorus and nitrogen for different depths and for a renewal time of 2 m^3 m^{-2} a^{-1}

Mean depth (m)	*Permissible inputs*		*Dangerous inputs*	
	P (g m^{-2} a^{-1})	*N (g m^{-2} a^{-1})*	*P (g m^{-2} a^{-1})*	*N (g m^{-2} a^{-1})*
<5	<0.07	<1.0	>0.13	>2.0
<10	<0.1	<1.5	>0.2	>3.0
<50	<0.25	<4.0	>0.5	>8.0
<100	<0.4	<6.0	>0.8	>12.0
<150	<0.5	<7.5	>1.0	>15.0
<200	<0.6	<9.0	>1.2	>18.0

Currently, very little information on relationships between inputs and in-lake concentrations is available from tropical and subtropical aquatic ecosystems. Future research must investigate whether relationships established for water bodies in temperate climates apply, or whether changes are necessary. Differences may be expected, especially with respect to sediment-water interactions and mineralisation rates, because these depend strongly upon temperature and upon thermal stratification. If inputs exceed critical values for a given system, increase of the concentrations of total phosphorus within that system are likely. In turn, cyanobacterial proliferation is likely, and management measures are then needed to reduce phosphorous inputs.

Sources and Reduction of External Nutrient Inputs

In most cases, eutrophication is enhanced by anthropogenic activities. The three major sources of external nutrient inputs are run-off and erosion from fertilised agricultural areas, erosion resulting from deforestation, and sewage. Exceptions may occur and are illustrated by the example of Lake Victoria. This large lake has an area of 68, 000 km^2. Tributaries supply about 15 per cent of its water and 85 per cent originates from precipitation. Burning of field stubble is widely practised and leads to substantial air pollution. Thus, 60 per cent of the phosphorus load is estimated to enter Lake Victoria through precipitation. Only 40 per cent originates from sewage and run-off into the tributaries. Replacement of the practice of burning stubble would substantially reduce this load.

Sustainable approaches aim at reducing nutrient loads at their source or as close to the source as possible. If this is not feasible, approaches to reducing inputs from a main tributary may be considered. Reduction of external inputs beneath the threshold expected to be effective is an important basis for the success of further in-lake restoration measures.

The first questions to ask in designing programmes for abatement of eutrophication by improving agricultural practices and/or by introducing or improving sewage treatment are:

— How high is the phosphorus input from wastewater to a given water body (excretion of 2-4 g P per person per day may be assumed)?

— How high is the input from agriculture and run-off from other surfaces?

— Down to what concentrations must phosphorus input be reduced in order to reduce concentrations in the receiving water body beneath a total P threshold likely to be effective in the given water body?

Answering these questions requires specific evaluation of the resources to be protected, of their catchment land use, and of the water and effluent drainage network. Ideally, nutrient inputs and the relative share of different nutrient sources should be estimated and catchment characteristics, such as soil type, run-off potential and vegetation cover should be considered. In a region with nutrient-rich, erodible soils and reduced vegetation cover or natural eutrophication, reductions in inputs will not be possible to the same extent as in a region with sandy soils, flat relief and dense tree cover. Thus, the same measures and similar investments into reducing inputs are likely to be more successful in reducing eutrophication in a potentially oligotrophic ecosystem than in a naturally eutrophic one.

In many cases, quantitative assessments of inputs will not readily be available because this requires detailed analysis of hydrological conditions, as well as nutrient concentrations and their variations over time in all of the main tributaries. Such investigations require time and resources. Whereas inputs from point sources like sewage effluents are relatively easy to assess, diffuse inputs from agriculture are often very difficult to quantify. Managers are likely to be confronted with water bodies for which almost no limnological data are available, and perhaps not even the depth contours of the water body are known. Planning and implementation of resource protection measures may be delayed for several years before reliable data become available. The dilemma for managers is whether to begin with measures to reduce obviously substantial inputs, without having the data basis to predict whether the measures taken will reduce nutrient concentrations below the threshold effective for controlling cyanobacteria, or whether to delay planning and decision-making until data become available.

In some countries, user friendly "decision support" software programs have been compiled for use by managers and community groups in identifying the main sources of nutrients from a catchment, and for identifying possible actions e.g. off-river disposal of wastewaters or (re) construction of riparian buffer strips to protect against inputs from erosion. There is an array of computer models that may be used to simulate the hydrodynamic and transport conditions in a catchment system. Trudgill gives examples of available models and profound discussions of various processes that may be included in the compilation. Since numerous parameters are required for calibration of these models, the complexity of the model should be tailored to the extent of the data base available. For applying these models, the study should include the entire watershed and not just the lake water body.

No general recommendation can be given to resolve the dilemma between the necessity of adequate planning data, and the need to implement obvious measures without delay. Although numerous restoration measures around the world have proved to be ineffective due to insufficient diagnosis and evaluation by scientists and managers, most of these were measures addressing ecological balances within the water body. In contrast, measures addressing reduction of external nutrient inputs are not likely to be applied mistakenly. In the worst case they may prove to be insufficient and require further action - either further reduction of inputs, or in-lake action. An effective alternative to quantitative assessment of loading, is a "common-sense" or qualitative approach. This begins with studying maps and geographical information to identify main tributaries, slopes critical for erosion, precipitation patterns and land use. Detailed and critical inspection of the catchment area may provide an excellent basis for recognising priority actions, some of which may be implemented at low cost. Such inspection is generally of underestimated value. Qualitative assessments should include identification of sewage outfalls.

While investments in internal or "in-lake" control measures can be wasted without sufficient pre-restoration evaluation, reduction of external nutrient inputs will be at least a first step in the right direction. Managers are encouraged to implement nutrient control measures, even if the data base is not sufficient to predict the quantitative impact on concentrations within the water body.

Domestic wastewater

Wastewater emissions have frequently been managed by a philosophy of using water as a medium for transporting wastes out of the locality without considering the impairment of the function of the water as a resource further downstream, or of the possible enrichment of lakes with pollutants and nutrients.

Many developing and least developed countries are extending their coverage of drinking water supplies in order to improve human health. Evaluation of experience shows that this usually leads to a several-fold per capita increase in water consumption and thus also in the amount of waste-water generated. Improvement of human health therefore also requires development of wastewater collection and treatment, as highlighted by the World Bank:

"For urban water supply, experience indicates that the collection and proper treatment of sewage must be an integral part of water supply projects. Bringing water into a city without taking sewage out exposes the population - and particularly the poor -to increased pollution."

Possible exposure to pathogens is a major reason why collection and treatment of wastewater is important. Another reason is that cyanobacteria proliferate in eutrophic lakes and rivers fertilised by wastewaters.

Replacement of phosphorus in laundry detergents can typically reduce phosphate loads in sewage by 50 per cent at best, the remaining 50 per cent being inevitable because it originates from human excreta. Wherever sewage outfalls are considered to be a significant input of phosphorus to a water resource, phosphate elimination, alternative treatment approaches, or sewage diversion is necessary. Criteria for adequate technology largely depend upon population structure and on geographic conditions.

If population density is low, and the flushing rates of water bodies or phosphorus absorption capacity of the soils is high, nutrient elimination from sewage may not be necessary - high nutrient concentrations in water resources in such situations are more likely to originate from agriculture. Treatment methods adequate for protection from infectious agents may also be sufficient in these circumstances. If domestic wastewater is used in agriculture, health risks should be avoided by following the WHO guidelines for the use of wastewater in agriculture and aquaculture. A special aspect of such situations is tourism. Low population areas affected by tourism may need special consideration because the population may increase temporarily several-fold and overload sewage treatment capacities. In temperate regions, the tourism season may coincide with the cyanobacterial growth season.

Sparsely populated regions with water bodies highly susceptible to eutrophication, e.g. because of long water retention times, will require more carefully designed sanitation systems to protect these resources. Supplementary treatment techniques may be necessary, particularly during the tourist season.

In many densely populated areas, municipal sewage is the source of at least half of the total phosphorus inputs to rivers and lakes. In large urban areas, treatment of wastewater collected in sewerage systems requires industrial-scale plants for the protection of surface waters. Untreated municipal sewage contains more than 10 mg l^{-1} of phosphorus.

Biological treatment oxidises organic matter, but does not substantially reduce phosphorus content. Where municipal sewage constitutes a significant source of phosphorus pollution, removal of phosphorus at treatment plants is necessary. Heavy seasonal tourism may also be a problem in such situations because it increases the demand on the capacity of treatment plants and sewerage and also causes substantial fluctuations in the sewage load. Sewerage and treatment for fluctuating amounts of sewage present specific technical difficulties. Lake Balaton is an example of such a situation, where the number of tourists during July and August is twice that of the local population.

In lowland regions, discharge of domestic and industrial wastewater may amount to 50 per cent and in dry seasons to almost 100 per cent of the total flow of the recipient river, e.g. River Thames in London and Havel River in Berlin. Such slow flowing rivers are suitable habitats for cyanobacterial growth but the river water may also be needed for production of drinking water. In such situations, phosphorus elimination by simultaneous chemical precipitation or "bio-P" is not adequate, because concentrations in the recipient water body will almost equal the outfall concentrations. Treatment objectives therefore are 0.03-0.05 mg l^{-1} P and this can be attained by adding a further treatment step for removal of phosphorus in small, slowly settling floes. The removal of pathogens is usually also a further important objective of filtration treatment in regions with such intensive water use. Various methods of filtration over sand or gravel and pumice beds are available, and new methods of membrane filtration are also being developed. The costs of such treatment technology may appear intimidating at first glance because of the necessary investments. However, even sophisticated procedures involving filtration need not cost more than US$ 0.15-0.30 per m_3 of treated water, and this is usually only a small fraction of the costs of drinking water. Effective resource protection in such densely populated areas will save the cost of drinking water treatment, for example by saving the necessity of activated carbon filtration.

Eutrophication due to sewage outfalls has been recognised as a widespread problem since the 1970s. Abatement was begun in that decade with several isolated projects, such as diversion of sewage around lakes with specific touristic value and the introduction of phosphorus precipitation in a few treatment plants. Comprehensive programmes began in the mid 1980s. In Europe, an international convention for the protection of the North Sea triggered introduction of phosphorus and nitrogen elimination in larger sewage treatment plants. Recently, the implementation of measures to eliminate nutrients in sewage has started to show substantial success:

- Denmark achieved a 79 per cent reduction of phosphorus inputs from sewage between 1985 and 1995 and further reduction is expected in 1997 when two plants in Copenhagen go into full operation.
- Switzerland achieved a 60 per cent decline in total phosphorus inputs from municipal wastewater.
- The USA and Canada together achieved a 50 per cent reduction in phosphorus concentrations in Lake Ontario and the west basin of Lake Erie, with significant impact upon "algal blooms" and accumulations of filamentous cyanobacteria on shorelines, by the construction and upgrading of sewage treatment plants along the Great Lakes. However, in order to maintain the quality now achieved despite the

expected increase of population density by the year 2011, and in order to meet the quality targets for Hamilton Harbour, tertiary sewage treatment is considered necessary.

Charlton emphasises the importance of reliable performance of sewage treatment plants: "Part of the difficulty in understanding sewage problems is the mistaken belief that sewage treatment plants, once built, will perform as planned, with no *operational problems*". Steady degradation of performance during critical summer months has led to considerable phosphorus inputs, the highest being 2.7 times the target. Internal assessment, optimisation of performance, and identification of staff with the job and the achievement of effluent target concentrations, are crucial for reducing effluent loads.

Agriculture and Erosion

Losses of phosphorus and nitrogen from deforested, agriculturally-used areas into surface waters are the other major factor enhancing eutrophication. As with phosphorus pollution from wastewater, this problem has increased exponentially in some parts of the world since the 1950s. The causes are structural changes in agriculture involving intensification by tillage of larger plots, extensive application of mineral fertilisers, and the establishment of large-scale animal husbandry. Simultaneously, other major parts of the world suffer substantial lack of phosphorus in topsoils. Zehnder points out a striking global imbalance between surplus phosphorus in most of the industrialised regions and a lack of phosphorus in most developing regions. On a global scale, phosphorus must be perceived as a limited resource. In regions with phosphorus deficiency, controlled fertilisation in combination with adequate protection from erosion, would help maintain fertility of tropical and subtropical soils, so that further deforestation would no longer be necessary in order to gain new farmland. Further, development of sewage treatment methods that allow reclamation of phosphorus for reuse as fertiliser should be a long-term target for sustainable handling of this resource.

Excessive use of fertilisers and manure has created eutrophication problems in lakes and rivers. In regions with intensive agriculture and slowly flowing rivers with little discharge, e.g. in much of north-western Europe, water bodies without cyanobacterial problems have become scarce. Heavy surplus fertilisation has been enhanced by the widespread trust in phosphorus retention in soils (soils as "savings banks" for phosphorus), but soil erosion and surface runoff have proved to be major pathways into surface waters, particularly through storms and intensive rainfall shortly after application of fertiliser or manure. The extent of these losses is site-specific and largely depends upon geographic and hydrological conditions: "It *has been suggested that up to 90 per* cent of the annual phosphorus losses occur from only 5 per cent of the land during only one or two storms, especially in areas where surface runoff and erosion are the *dominant routes for phosphorus losses*". The importance of leaching, first recognised as a pathway for nitrate inputs, is being recognised increasingly for phosphate in some types of soils, e.g. sandy, acidic soils with a high degree of saturation of their phosphorus adsorption capacity. For the Netherlands, Oenema and Roest estimate 300,000-400,000 ha of phosphorus-leaking sandy soils to be pollution "hot spots" requiring high priority in identification and remediation.

In tropical and subtropical regions, eutrophication of lakes and reservoirs due to inputs associated with erosion is greatly enhanced. Closing cycles by reuse of manure as a nutrient resource in agriculture, especially if combined with changing practices of land tillage and deforestation, can contribute to reduction of nutrient pollution. Such approaches are sustainable alternatives to considering animal slurry as waste, and reduce the costs of purchase of fertilisers. In this context, performance of large-scale livestock farming in industrial dimensions requires regulation just as for industry:

> "The output of waste from many cattle and pig units measures up to that of a large town and needs to be managed with at least the same care as that accorded to human waste. At one time the only environmental requirement placed on farmers was that they should observe good agricultural practice. This is no longer a satisfactory basis for environmental protection. Agriculture must be put on the same basis as other major industries with strict controls on the quality and quantity of effluent discharges."

Apart from closing nutrient cycles, sustainable biological methods of production have been developed in many pilot projects and are becoming increasingly popular. Model projects have demonstrated that productivity of biological or "organic" farming methods is not substantially lower than that of conventional farming, provided methods are adapted adequately to the given geographic conditions. The economic balance of "biological" or "organic" farms is frequently equally good because of reduced expenditure for agrochemicals and, in some cases, better prices for the product. Cooperation between water supply agencies and farmers has supported this development in Germany and has shown success in improving resource quality.

In drinking water catchments, it is particularly important that agriculture follows "best agricultural practices". This can be encouraged by designating protection zones around the drinking water source and regulating practices allowed or prohibited within these protection zones. Sophisticated models distinguish two to three degrees of protection, depending upon the relative impact of the respective part of the catchment upon the water quality. In addition to agriculture, other activities which impact water quality, such as forestry, fisheries and tourism, may be regulated in drinking water protection zones.

Countries with a traditional rural society may have options for reclaiming historical agricultural experience and combining it with modern approaches to sustainable "best agricultural practice" in order to attain high outputs of high quality products at low environmental and health impact. Such an integrated approach requires continuous development, evaluation of experience and training.

Countries with large-scale industrialised agriculture may have problems in implementing change. In contrast to the success achieved in phosphorus elimination by wastewater treatment, progress in abatement of agricultural phosphorus pollution has at best been modest. For Switzerland, Wehrli *et al.* estimate that, while phosphorus emissions from sewage have been reduced by 60 per cent during the past 15 years, losses from agricultural areas into water bodies have rather increased and the need for a new agricultural policy is just beginning to be widely perceived. Some of the reasons for this delay in awareness and action are:

— Losses of phosphorus from farmland are rarely perceived as economic losses.

— Effective measures will differ regionally or even from farm to farm, and management practices must be optimised locally rather than administered generally. This requires shifts of attitudes.

— Even where programmes and regulations exist, their implementation and control may be difficult.

A basic change in attitude is required. In some countries this appears to be developing slowly as a new generation of farmers with better training (including education on sustainable farm management and ecological impacts) takes over, and as consumer awareness for quality criteria and the ecological impact of products is growing. A wide array of measures can be used by government authorities to support such developments. Examples are training and advice to farmers, eco-audits on products, subsidies for setting land aside from use, subsidies or tax redemptions during periods of transition to organic farming methods, pollution taxes and legislation to enforce water protection.

Treatment of Drinking Water Reservoir Inlets

Where drinking water reservoirs with one major inflow have a large share of diffuse, nonpoint source inputs and a strong need for rapid remediation, reduction of nutrients in the inflow may be the most effective option. Pre-reservoirs with retention times of at least several days can reduce total phosphorus inputs by 50-65 per cent. Retention times should allow incorporation of phosphorus into algal biomass and sedimentation of that biomass, but should not be large enough for slow growing taxa, such as cyanobacteria, to establish dominance. Sediment dredging may be necessary at intervals of several years in order to counteract the re-release of phosphorus.

The Kis-Balaton reservoir in Hungary is an example of a special wetland and shallow reservoir system of 60 km^2 designed to retain phosphorus and reduce inputs to the tenfold larger Lake Balaton. Water is retained for one month in an intricate system between coffer dams and reed zones. Phosphorus retention has been successful, but flooding with stormwater from the Zala River has caused pulsed phosphorus inputs to the lake.

If the largest nutrient share originates from a single major inflow, phosphorus stripping facilities can be very effective in reducing inputs. Successful examples, with different degrees of technological sophistication, are the Wahnbach Reservoir, Lake Schlachtensee, Lake Tegel, and the Haltern Reservoir.

Internal Measures for Nutrient and Cyanobacterial Control

In planning restoration measures for lakes and reservoirs, it is important to realise that substantial time lags may occur between measures to reduce external inputs and the results achieved in the water body. Feedback mechanisms within the ecosystem require time to reach a new equilibrium. Hypertrophic aquatic ecosystems have specific positive feedback mechanisms which stabilise trophic state and cyanobacterial dominance and therefore resilience effects are not uncommon, even after substantial reduction of inputs below thresholds calculated to be effective. Sas pointed out that resilience patterns occur on two levels:

- Delayed response of in-lake total phosphorus concentrations to a reduction of input, due to the time required for flushing phosphorus out of the water body, and the time required for establishment of new sediment-water equilibria.
- Delayed response of phytoplankton biomass and species composition to reductions of in-lake total phosphorus concentrations, due to stability of prevailing biocoenosis structures and/or biotic enhancement of internal load.

Experience shows that several years (up to 10) may be necessary between the implementation of a restoration measure that substantially decreases inputs, and visible success in terms of reduction of phytoplankton biomass and cyanobacterial blooms. Monitoring of phosphorus inputs and phosphorus concentrations in the recipient water body during this time is recommended. Usually, a declining trend in total phosphorus concentration will be the first detectable response to a reduction in inputs and will indicate whether a particular measure can be expected to be successful, but it may take years for phosphorus concentrations to decline below the threshold effective for controlling phytoplankton biomass. Often, such time lags are due to the (sometimes substantial) phosphate storage capacity of anoxic sediments typical in hypertrophic waters. Flushing rates strongly influence the time necessary to reach a new equilibrium. In some cases, particularly in water bodies with low water exchange rates, supplementary "internal" measures may be advisable in order to accelerate a response.

With very few exceptions, internal measures are appropriate only after, or in combination with, an effective reduction of external inputs. In principle, reduction of inputs should be the actual restoration or resource protection measure, and internal measures should serve as a further boost to switch the ecosystem out of resilience and into a new balance. If possible, a few years of patience and observation of nutrient concentration trends within the lake will show whether internal measures are necessary.

Only rarely are internal measures without adequate reductions of inputs justified as an emergency approach; usually such measures require continuous operation or repeated application. Medium- to long-term success of restoration investments is at stake if this principle is not considered.

In-lake Phosphorus Precipitation

In lakes and reservoirs with high water retention times, decline of phosphorus concentrations may be very slow, even after external inputs have been reduced to levels which should ensure a mesotrophic or oligotrophic state. If the water body has a high phosphorus content that is flushed out only slowly, only some of the phosphorus within the biomass will settle to the sediments. Much of it is released from decaying organic material, entrained back into the water body by water circulation, taken up by cyanobacteria or algae and, in part, passed on to higher levels of the food web. Degradation of organic material at the sediment surface has often led to anoxic conditions which may accelerate phosphorus release rates dramatically. Thus, phosphorus within a lake can be recycled many times, and no decline of cyanobacterial biomass can be achieved without reducing this in-lake phosphorus pool. Sometimes, this situation also applies to lakes which are naturally eutrophic, such as lakes in western Canada situated on phosphorus-rich glacial till. Precipitation of phosphorus from the water body to the sediment can be a successful measure, if it is undertaken so that phosphorus remains permanently bound in the sediment. Prerequisites for lasting success are low external loading, sufficient depth to prevent sediment resuspension due to wind events, and adequate choice of flocculants. Experiments with precipitation of phosphorus have been undertaken with aluminium sulphate, ferric salts, ferric aluminium sulphate, clay particles and lime.

Ferric salts are effective in precipitating phosphorus, but difficult to handle because of their aggressive acidity. Furthermore, the iron-phosphorus complex is stable only under oxic conditions. Thus application of ferric salts usually requires subsequent continuous aeration to avoid re-dissolution of phosphorus under anoxic conditions. Due to the high mobility of iron ions, addition of iron frequently often has to be repeated at regular intervals. In addition, Prepas *et al.* point out that iron may be a limiting micronutrient in some systems and, in such situations, treatment with ferric salts may actually stimulate growth of cyanobacteria and algae.

Aluminium sulphate is poorly soluble under neutral and high pH conditions, but may decrease pH in waters with low buffering capacity, which leads to solubilisation and problems of alum toxicity.

Lime (both $Ca(OH)_2$ and $CaCO_3$) has been used as an algicide to coagulate and precipitate phytoplankton cells out of the water column. It is non-toxic, usually fairly inexpensive, and the pH-shock for the aquatic biota can be minimised by careful dosing over an extended time span. Unlike treatment with copper sulphate, the precipitation of cyanobacterial cells with $Ca(OH)_2$ does not appear to cause cell lysis and toxin release into the water. Lime also functions, to some extent, as a longer-term algal inhibitor, reducing eutrophication by precipitating phosphorus from the water. It appears that $Ca(OH)_2$ is more effective than $CaCO_3$ in precipitating phosphorus. Many of the studies of both the mechanism and effects of liming for algal control have been carried out in eutrophic, hard water lakes or farm dugouts (dams) in Alberta, Canada. It is possible that the technique may be more effective in these conditions than in soft water. The dose rates used are also quite high which would make the technique prohibitive for large lakes. Techniques for the application of lime, which involve pumping or spraying of a slurry.

Experience with in-lake precipitation of phosphorus is increasingly being compiled. A number of documented case studies show success either in terms of reducing phytoplankton biomass or in terms of shifting species dominance away from cyanobacteria. Nevertheless, numerous unsuccessful cases have also been documented, and further development of these techniques is ongoing. Furthermore, in some water bodies, the concentrations of iron or calcium compounds in the inflow are naturally high and regularly provide sufficient binding sites for phosphate to induce natural phosphorus precipitation. Measures in the catchment area or changes in inflow regime may have considerable impact in either increasing or decreasing this input, and thus may have a significant impact on the trophic state of the water body.

Sediment Dredging and Phosphorus Binding

Release from sediments may be a substantial source of phosphorus for many years after external inputs have been minimised. Water exchange rates, sediment chemistry, temperatures, mixing conditions, and bioturbation govern phosphorus release rates. Iron-bound phosphorus is highly sensitive to redox conditions; when sediment surfaces turn anoxic during summer stratification, phosphorus concentrations may increase dramatically, fertilising cyanobacteria in their optimum growing season. Under oxic conditions in shallow, unstratified systems, high pH (> 9.8) may strongly enhance oxic phosphorus release. Because high pH values are a result of intensive photosynthetic activity, this phosphorus release pathway is a positive feedback mechanism in favour of cyanobacterial blooms. Other aerobic phosphorus release mechanisms may also be significant, especially bioturbation by feeding fish and invertebrates. Options for measures to counteract sediment release are removal of sediment or treatment to bind phosphorus. Dredging is costly and will reduce release rates only if:

- It is carried out down to sediment layers with a lower or less mobile phosphorus content.
- Phosphorus-rich interstitial water is handled in such a fashion that it does not reach the water body and cause additional inputs.

— Dredged sludge can be deposited where it does not create a new external input with erosion and stormwater runoff into the lake.

In some urban and industrial regions, dredging is precluded or complicated by high concentrations of heavy metals and organic contaminants in the sediments which would then require disposal as hazardous waste. Dredging is particularly recommended for smaller water bodies where the trophic state can be further improved by gaining depth, or which also need to be cleared of dumped rubbish.

Sediment treatments aim at trapping phosphorus in the sediment, either by oxidation to insoluble iron compounds, or by adsorption onto calcium carbonate or clay particles. During the past two decades, broad experience collected with numerous failures and a few successful cases has shown that effective treatment requires careful design on the basis of profound understanding of the sediment chemistry and hydrology of the water body to be treated. Oxidisation may be achieved by aeration, artificial mixing, or the introduction of pure oxygen. It appears to be most effective if achieved with nitrate, which transports more oxygen and penetrates more readily into sediments. Well-treated sewage effluent (not contaminated with harmful substances, fully nitrified and after phosphorus removal) may be suitable for this purpose, if the process is controlled so that nitrate concentrations are not elevated in drinking water.

REMEDIAL MEASURES

The management and control of cyanobacteria in water supply storage facilities and of cyanotoxins in drinking water may be approached at a number of points and levels in the hierarchy of the total supply system.

Algaecides have been, and will continue to be, used as emergency measures for the control of cyanobacteria, and their role in management strategies needs to be assessed from practical and environmental viewpoints. The final option for management of cyanobacterial problems and cyanotoxins in water supplies is within the treatment system. Research on removal of algal and cyanobacterial cells has been widely published and recent work has generated quite detailed knowledge on cyanotoxin removal during drinking water treatment. Much of the work on cyanotoxin removal has focused on single treatment steps, and a few studies have investigated the common combinations of coagulation, clarification and filtration. As a research approach, this is useful because assessments of the performance of individual treatment steps may be generalised more readily than observations gained in complete, but individual, supply systems with their respective site-specific characteristics. However, management approaches aimed at providing safe drinking water from cyanobacteria-infested surface waters require considering the system as a whole, and using different combinations of resource management tailored to the specific locality and different treatment steps. In such an overall approach, steps that individually may be unsatisfactory can have their place in combination with others and thus contribute to a multi-barrier approach. Furthermore, the individual aspects of each drinking water supply necessitate local assessment of performance as well as local optimisation of resource management and treatment strategies.

Management of Abstraction

Management of raw water abstraction is effective in reducing the amount of cyanobacteria in the raw water, often by orders of magnitude. This can be achieved by choosing an optimum position for the off take, or by abstracting surface water through bank filtration.

Direct abstraction from surface water bodies

The horizontal and vertical distribution of cyanobacterial populations can vary enormously throughout a water body, whether in a lake, reservoir or river. This has obvious implications for both the siting of off takes and the choice of off take depth. Considerable contamination of raw water can be avoided by locating off takes away from sheltered bays where scums may accumulate. If this is not practical, it may be possible to employ temporary extensions to pipe intake points. Selection of off take depth can also be important in reducing contamination by avoiding surface or subsurface maxima of cell numbers. Many modern reservoir off take structures have the provision for multiple off take depths. If multiple off takes are not available it may be possible to install siphon off takes, at least as a temporary measure in small systems. In relation to cyanobacterial contamination, the choice of intake depth must take into account the time of daily maximum surface accumulation of cells and the amplitude of passive diurnal vertical sinking and rising of cells due to light-and photosynthetic-driven changes in cell buoyancy. In thermally stratified, mesotrophic reservoirs, attention must be given to the possibility of meta-limnetic maxima of *Planktothrix rubescens* (syn. *Oscillatoria rubescens*).

Operators need to become familiar with the amplitude of vertical movement of cyanobacterial populations, and also with the potential for the formation of metalimnetic peaks, in order to avoid high cell densities as much as possible. This requires multiple depth sampling to determine vertical profiles of cyanobacterial cell density. Collecting information and building up knowledge and understanding of local ecology and conditions can increase flexibility in the management of blooms. Another option to avoid contamination is to employ physical barriers or booms at the surface to prevent surface scums accumulating near the off take site. Surface booms or curtains, similar to oil-spill containment booms, have been used successfully in Australia, the UK and North America to keep surface scums away from off take structures. These physical barriers often only extend to a depth of 0.5-1.0 m, and do not affect bulk horizontal flow significantly. This technique is a worthwhile emergency measure for transient blooms and its use will depend upon the practical aspects of installation.

Bank Filtration and Groundwater Recharge

An abstraction method that has proved to be very effective in removing particles and many dissolved compounds in localities with suitable underground conditions is bank filtration or abstraction of groundwater artificially recharged with surface water. The process uses bore holes or infiltration galleries which are located near to the banks of a surface water supply. These wells fill with water which has infiltrated through the intervening porous soil materials. Depending upon the underground characteristics, water may travel for several hours, or even for weeks before it reaches the well. Longer retention times may enhance purification, but even retention times of between a few hours and days have substantially improved water quality. Many types of soils may be suitable, provided they allow water flow, are not too coarsely structured to achieve a filtering effect, or are not in contact with saline or otherwise unsuitable groundwater. Planning bank filtration requires local assessment of sites for their suitability.

Evaluation of bank filtration with respect to cyanotoxin removal is currently only beginning, and no published results are available. A study of elimination of algal and cyanobacterial taste and odour compounds has shown very effective removal by bank filtration at three study sites. Because of the generally positive experience with respect to removal of suspended materials, micro-organisms and a variety of chemical contaminants it may be expected that bank filtration will be a highly promising abstraction method to avoid contamination with cyanobacterial cells as well as dissolved toxins. This expectation is supported by the favourable results of a laboratory study from Finland which demonstrated good performance of experimental soil and sediment columns for both cell and toxin removal.

In this case, lake water was inoculated with both toxic and non-toxic cultures of cyanobacterial cells and pure microcystin-LR and filtered through soil and lake sediment columns. It was found that during the experimental period of one week, both cells and dissolved toxins were removed very efficiently, although there was some breakthrough in sediment columns at high loadings. The mean rates of removal for cells were 93.7-99.7 per cent and 97.5-99.5 per cent for extracellular toxins for both soil and sediment columns. It was suggested that the removal of microcystins in this filtration process was the result of both adsorption and biodegradation. However, the relative performance of the two processes would be very site specific and dependent upon local soil characteristics and microbial activity. The performance of bank filtration in relation to adsorption capacity, overloading and the potential for release over time of toxins from trapped cells would require monitoring.

Use of Algicides

Algaecides are used in reservoirs to control cyanobacterial growth and to prevent or reduce to some extent the problems of toxins in the associated drinking water supply.

Their role in the management scheme may be to provide effective short-term control of growth of cyanobacteria, at one point in time, particularly in circumstances where alternative drinking water sources are not available and preventive measures are not feasible or not yet effective. Algicide treatment has been proposed as being more cost-effective than toxin removal in drinking water treatment plants, as has been suggested for the control of off-flavour problems, because an extended period of persistent blooms greatly enhances the need for additional treatment for toxin removal. However, experience with abatement of off-flavours caused by cyanobacteria through algicide treatment has also demonstrated that this treatment may actually enhance the problem by supporting the development of species resistant to the treatment.

Environmental concerns have been raised because the most commonly used algicide, copper sulphate, has broad ecological impact. It should be used only in dedicated water supply reservoirs in special circumstances, but is nevertheless an unsatisfactory long-term solution. In many countries, national or local environmental regulations prohibit or limit the use of algicides due to their adverse environmental impact. This needs to be established prior to considering the use of algicides.

Algicides, like all management techniques, must be applied correctly to work effectively. If algicides are used they must be applied at the early stages of bloom development when cell densities are low, in order to reduce the potential for liberation of the high concentrations of intracellular toxin that may be associated with dense blooms. Early application will further enhance the effectiveness of treatment because cyanobacterial cells can form a major part of the "copper demand" along with other organic matter in natural water.

A major limitation of any agent which disrupts cyanobacterial cells is the release of toxins and of taste and odour compounds from the cells. A range of studies have indicated that cyanotoxins are predominantly intracellular in healthy cells, and are only released into the water at an advanced stage of bloom senescence, or following treatment with chemicals such as algicides. This release can be quite rapid and has been shown to occur within 3-24 hours in different studies. These dissolved toxins will then disperse and be diluted throughout the water body, but will not be removed by conventional flocculation and filtration procedures. Installation of additional treatment for removing cyanotoxins may be costly. The dangers of treating dense blooms with algicides was demonstrated in an incident which occurred on tropical Palm Island, Australia, where members of the community became ill with hepato-enteritis following treatment of the water supply reservoir with copper sulphate for a cyanobacterial bloom problem.

If algicides are used to control toxic cyanobacteria, the reservoir should be isolated for a period to allow the toxins and odours to degrade. Unfortunately, very little data exist on the withholding period in relation to toxin loss, but it could be in excess of 14 days.

In some cases algicide treatment may be unsuccessful or only partially successful. This can be due to inadequate dispersal and contact with the target organisms, variable sensitivity of cyanobacteria, and reduced toxicity due to complexation of the copper. The form of copper most toxic to aquatic organisms is the free cupric ion (Cu^{2+}) and this can be reduced by complexation with both inorganic ligands under alkaline conditions, and organic ligands present in natural waters.

Copper Sulphate

Chemical control of algae in water supply storage has been a widespread water quality management practice for over 100 years. Records of the use of copper sulphate date from 1890 in Europe, from 1904 in the USA, and at least since the mid 1940s in Australia. Copper sulphate has been regarded as the algicide of choice because it is economical, effective, relatively safe and easy to apply. It is also considered to be of limited significance to human health at the doses commonly used and has been considered not to cause extensive environmental damage. The latter point has been an issue of debate for some time because copper tends to accumulate in lake sediments. In some cases it appears not to be remobilised and is bound permanently to the bottom sediments. However, in a study of 10 drinking water dugouts in Canada, sediment copper was released back into the open water under low dissolved oxygen conditions in the hypolimnion in summer. It has also been suggested that sediment-bound copper could have an impact on the benthic macroinvertebrate community. It is important to remember that copper and other heavy metals differ from some other toxic contaminants in that they are not biodegradable, and once they have entered the environment their potential toxicity is controlled largely by their speciation or physicochemical form.

Copper sulphate treatment has been shown to cause short-term changes in phytoplankton abundance and species succession. Fish kills may also occur following copper sulphate treatment, although it is not clear whether this is as a result of copper toxicity or oxygen depletion. A recent extensive survey of water utilities in the USA and Canada indicated that copper sulphate is by far the most widely used algicide, although other alternatives are used under some circumstances.

McKnight *et al.* give an assessment of the use of copper sulphate for the control of nuisance algae and cyanobacteria. They also indicate that there are wide differences in copper sensitivity among species. The relative growth inhibiting concentrations for a range of phytoplankton are given in terms of cupric ion activity (i.e. [Cu^{2+}]), derived from laboratory toxicity studies. The toxic cupric ion activities range from greater than 10-6-10-11 M for species of diatoms, dinoflagellates, green algae and cyanobacteria - a difference of over four orders or magnitude. These toxic Cu^{2+} concentrations are very much less than the usual doses applied as total copper in copper sulphate treatments. The relative toxicity is given in terms of ionic copper because it is believed that phytoplankton react principally to the concentration of Cu^{2+} or loosely complexed copper rather than the total dissolved metal in the water.

McKnight *et al.* have used these findings to develop an experimental procedure to determine the required dose rates for target species in individual reservoirs, taking account of the particular water chemistry. This experimental procedure to determine dose rates is suggested as more effective than simple empirical formulae, based on pH and alkalinity, which were not very useful. The experimental approach requires access by the water supply operators to a good level of biological and chemical expertise and analytical capacity. The approach is based on first determining the cupric ion activity as a function of added copper, and thereby the complexing capacity of the reservoir water by a copper ion selective electrode. This is followed by a culture assay to determine the sensitivity of the particular nuisance algae to copper. The local nuisance species preferably need to have been isolated into laboratory culture. The required copper sulphate dose rates can be derived from a simple formula relating growth inhibition, in terms of cupric ion concentration, to the Cu^{2+} concentrations in the reservoir after complexation.

Copper Chelates

The problem of the reduced effectiveness of copper sulphate treatment in hard alkaline water has long been recognised. Chelated copper algicides were developed to overcome the problems of the complexation and precipitation loss of toxic copper under these circumstances. Examples of copper chelate algicides include copper ethanolamine complexes and copper citrate. Copper citrate has been used as an algicide in the USA. It is available either as a commercial preparation or by simultaneously dosing copper sulphate and citric acid. It is claimed that the use of citric acid as a chelating agent enhances the solubility of copper allowing it to remain in solution longer under alkaline conditions. Raman recommends applying copper sulphate: citric acid in the weight ratio 2:1 in high-alkalinity waters (> 40 mg l^{-1} $CaCO_3$). A study which examined equilibrium speciation of copper in water to determine the changes in distribution of Cu(II) in relation to pH, dissolved organic carbon (DOC) and citrate was carried out by Casitas Municipal Water District. This study demonstrated that citrate greatly enhances the solubility of copper even in the presence of appreciable alkalinity (100 mg 1^{-1}). McKnight *et al.* suggested that the advantage of using synthetic copper chelating agents in hard, alkaline waters probably results from decreasing the supersaturation of malachite ($Cu(OH)_2CO_3$) and tenorite (CuO) and thereby the rate at which equilibrium with these insoluble forms (precipitates) is approached. It is possible that a longer time taken to reach equilibrium would result in the maintenance of toxic ionic Cu^{2+} activities and the inhibition of algal growth for longer periods after dosing. It is acknowledged that, despite their relatively widespread use in the USA, the efficacy of chelated copper algicides in relation to water chemistry is poorly understood.

Use of oxidants

Potassium permanganate has been used as an algicide from as early as 1935. A survey of North American utilities indicated that a small number use potassium permanganate relative to those who use copper sulphate. Commercial formulations of potassium permanganate marketed specifically as algicides are available in the USA. Fitzgerald investigated the relative toxicity of potassium permanganate to eight species of algae and cyanobacteria and found the algicidal dose was in the range 1-5 mg l^{-1}, except for one green algae where up to 8 mg l^{-1} was required. Chlorine is used mainly for control of algae in water treatment works but is also known to have been employed in reservoir situations. The effective dose rates are dependent on the chlorine demand of the water, but most algae are reported to be controlled by residues of free chlorine between 0.25 and 2.0 mg 1^{-1}.

When to use algicides

Because cyanobacterial toxins are primarily intracellular, algicides must be used with particular caution to avoid release of intracellular toxins. Algaecides should be used when cell numbers are low to avoid excessive toxins or taints following rupture of the cells. This should be checked by post-dosing monitoring. Algaecides may be used at higher cell numbers only if the reservoir can be taken out of supply until the toxins and taints degrade, or if treatment for removal of the toxins and taints is available. In the latter cases the use of algicide should be assessed against the capability for whole cell removal offered by treatment processes, because cell removal may be safer. It is important to know how effective the chosen algicide is in the specific waters. For example, copper may be less effective in waters with high dissolved carbonate or at alkaline pH. Algaecides should only be used in waters where the environmental impacts are acceptable and this should be checked with the local environmental agency.

Efficiency of Drinking Water Treatment in Cyanotoxin Removal

Cyanobacterial toxins represent a challenge to drinking water treatment which involves removal of organic substances in both soluble and insoluble form. Water treatment processes may remove target substances by either separation or conversion. Separation processes are those which remove the target substance from the treated water, usually to a treatment residual which becomes a waste stream for disposal. Conversion processes involve transforming the target substance into a different chemical form, thereby reducing the water quality problem. Although conversion processes are sometime characterised as though they achieve destruction, there will always be reaction products and thus transformation is a more accurate description than destruction. Ideal conversion processes are those which yield innocuous reaction products.

A major factor in assessing water treatment for cyanobacterial toxin removal involves consideration of soluble and suspended substance removal. The primary toxins which have been studied, microcystins, nodularins and anatoxins, are all water soluble. However, laboratory observations for microcystins have shown that these toxins are produced within the cyanobacterial cells and are expected to be predominantly found within slow growing, healthy cells (Table 2).

Table 2. Distribution of microcystins during laboratory culture of Microcystis aeruginosa

Age of culture	*Distribution of toxins (%)*	
	Cells	*Water*
Young		
Slowly-growing cells	100	0
Rapidly-growing cells	75-90	10-25
Old		
Slowly-growing intact cells	70-80	20-30
Decaying cells	30-40	60-70

Recent work has shown that microcystin release from cultured *Microcystis aeruginosa* began to occur late in the exponential growth phase and increased significantly during the stationary phase. This release was linked to a decrease in the integrity of the cells as determined by staining with fluorescein diacetate. The amount of toxin release was influenced by the culture medium and reached as much as 50 per cent in most commonly used media late in the stationary phase of population growth.

Until a bloom collapses or is otherwise affected by some treatment practice, the majority of toxins will be retained within the cells, making removal of intact cells a high treatment priority. However, under bloom conditions, a substantial proportion of toxin would also be expected to be released to the water column, making removal of soluble toxin an unavoidable concern.

Physicochemical treatment has been shown to cause cell lysis and toxin release. Operational investigations in Africa demonstrated significant cell lysis during extended transport in pipelines. However, other experiments conducted with cultured *Microcystis* showed that the flow and mixing conditions associated with water treatment did not cause cell lysis or toxin release. In addition, changes in pH from 5 to 9, which can occur in the treatment of some waters, did not cause any release of the intracellular toxins. Effects of physical and chemical stress on toxin release from cyanobacterial cells should therefore be assessed in treatment and conveyance systems.

Screening and prefiltration

Water treatment facilities usually employ coarse screens to remove debris from the water intake. These screens have no effect on the removal of either cyanobacterial cells or soluble toxins. However, microstrainers or fine screens may be used to remove larger algae, cyanobacterial cells and aggregated cells. Mouchet and Bonnélye reported removal rates of 40-70 per cent for two cyanobacterial species but pointed out that smaller species are poorly retained. Concerns regarding possible cell rupture, lysis and toxin release resulting from pressure on the filter screen have not been sufficiently addressed.

Aeration and air stripping

There are a number of methods for contacting air with water in drinking water treatment that may be required for various purposes, such as to oxidise iron and manganese from soluble to insoluble forms, to prevent reducing conditions which may yield odorous compounds, and to remove dissolved gases such as carbon dioxide, hydrogen sulphide, other reduced sulphur compounds and other volatile organic compounds. Neither aeration nor air stripping will be effective for removing soluble toxins because they are non-volatile compounds. Nor would they be effective for removal of cyanobacterial cells.

Coagulation and clarification

Coagulation promotes the aggregation of small, dispersed particles into larger particles which can be separated by sedimentation, filtration or flotation. Coagulation differs from precipitation because the latter involves converting soluble substances into insoluble particles, whereas coagulation deals with pre-existing dispersed particles such as mineral turbidity (clay, silt), larger molecular weight natural organic matter, micro-organisms including cyanobacteria, and oxidised, insoluble forms of iron and manganese.

Common chemicals used for drinking water coagulation include various aluminium and ferric iron salts. More recently synthetic organic polymers have gained some acceptance. Coagulation with multivalent metal salts can also be aided by adding various organic polymers to promote floe growth. Leuschner reported substantially improved flocculation of *Planktothrix agardhii* after addition of a cationic polymer. Efficient removal of algae is dependent on optimisation of chemical doses and coagulation pH. Mouchet and Bonnélye have shown that the coagulant dose necessary for algal removal is proportional to the sum of alkalinity and the logarithm of cell number. They emphasise that minimising turbidity in ajar test is not a sufficient criterion for adjusting treatment to remove algae and cyanobacteria, and recommend measuring the electrophoretic mobility of the cells for optimising dosage. Bernhardt and Clasen have reported that coagulation of algal cells that are smooth and more or less spherical occurs largely by charge neutralisation. In contrast, filamentous algae, large algae or species with bristles on their cell surface can be dealt with effectively only by sweep coagulation, by encountering the algae with large amounts of metal hydroxide floe.

Coagulation, by its nature, offers some promise for removal of intact cyanobacterial cells. For neurotoxins, Falconer reported that alum dosed at 120 mg l^{-1} alone and in combination with a number of polyelectrolytes removed about 20 per cent of the toxicity from a neurotoxic bloom of *Anabaena circinalis*. For microcystins, a number of published studies have shown that coagulation has a negligible capability for removal of any soluble toxins present in water. This has been demonstrated with aluminium sulphate coagulation jar tests in which total toxin concentration was reduced as a result of the removal of algal cells rather than the extracellular toxin. Rositano and Nicholson also demonstrated this expectation by evaluating removal of purified, soluble microcystins by three coagulants: ferric sulphate, alum and polyaluminium chloride. In all cases they found essentially no toxin removal. Lambert *et al.* found inconsistent and low levels of microcystin removal across the coagulation-sedimentation stage of a small, full-scale water treatment plant using an alum dose of over 60 mg l^{-1}.

By contrast, it must be emphasised that a study on raw water treatment with high doses of alum found over 23 per cent of the cell-bound microcystin-LR was released, mostly within two days of treatment. However, at concentrations and conditions that would occur in water treatment plants, Velzeboer *et al.* found that aluminium sulphate did not appear to cause lysis of cells of cultured *Anabaena circinalis* or *Microcystis aeruginosa*. Flocculation under laboratory conditions, which simulated operating water treatment plants, resulted in removal of cells in a healthy state, with no additional release of geosmin or microcystin-LR. Further work by Chow *et al.* using ferric chloride as the coagulant showed similar results with some stimulation of growth of both algal species. There was no increase in concentration of microcystin in the water following treatment of *Microcystis aeruginosa*, although it appeared that *Anabaena circinalis* may be more susceptible to damage from chemicals. Later work using alum in a pilot plant with cultured *Microcystis aeruginosa* harvested at the late exponential phase of growth confirmed that the cells were not damaged through the treatment process and that no additional toxin was released during treatment. However, this study also confirmed that the low concentrations of extracellular microcystin present in the feed water (2-6 μg l^{-1}) are not removed during the treatment process. It was further found that the total cell number in sludge collected from the pilot plant decreased to half its initial value after two days, and that toxin release began virtually immediately, reaching almost 100 per cent after two days. After five days the toxin concentration began to decrease and was reduced by approximately 80 per cent after eight days and completely removed after 13 days. This corresponds to findings of Jones and Orr who observed that bacterial degradation of microcystin-LR occurred after nine days in a lake after chemical treatment of a *Microcystis aeruginosa* bloom. The importance of toxin release from sludge depends on the time that sludge is retained in sedimentation tanks and it could have implications for sludge management, particularly if supernatant is returned from sludge treatment processes to the head of the plant.

Selection of clarifier type will also affect cell removal rates. Mouchet and Bonnélye have summarised experience largely from warm climates and have shown that sludge blanket-type clarifiers are substantially more effective than static settlers, particularly if upflow pulsed systems are used. This achieved consistent reduction of total phytoplankton by 95-99 per cent at a plant treating Seine River water, 95-98 per cent elimination of cyanobacteria at a Philippine plant, and 96.7-99.5 per cent removal of *Anabaena* and *Microcystis* at an industrial-scale plant in Harare. In Cairo and Alexandria, Egypt, older settling tanks were successfully upgraded to upflow pulsed sludge blanket clarifiers, thus not only improving performance for algal and cyanobacterial removal, but also efficiency per unit area and a reduction in coagulant consumption by 15-45 per cent and chlorine consumption by 15-35 per cent.

Dissolved air flotation

Although coagulation is normally followed by a sedimentation step, in some waters where the content of the suspended matter is low it is often easier to float the floe rather than attempting to settle a light floe. Recycled water saturated with air under pressure is introduced following the flocculation stage. Following the release of pressure the air comes out of solution and forms tiny bubbles which attach to the floe and cause it to float to the surface. The floated sludge is then collected and removed. This process is called dissolved air flotation (DAF) and is more effective than sedimentation, particularly for water with low turbidity and high colour, because the resultant floe is lighter and floats easily.

Dissolved air flotation is also generally more effective than sedimentation processes for treating algal-rich waters; for example floe blanket clarification has been shown to remove 76.5 per cent of *Microcystis* cells whilst DAF removed 98 per cent in the presence of other algae. A Belgian DAF plant achieved 40-80 per cent removal *of Microcystis*, 90-100 per cent removal of Anabaena but only 30 per cent removal *of Planktothrix* have reported on the efficiency of algae removal at eight DAF plants. Like Bernhardt and Clasen, they observed that the characteristics of algae influence their removal by any clarification process. They found that most of the treatment plants produced more than 80 per cent removal and they expected this would be improved by optimisation. Vlaski *et al.* found that, in a pilot plant, DAF achieved high particle (algae) removal during a cyanobacteria bloom (mainly *Microcystis* aeruginosa).

Dissolved air flotation is unlikely to be more effective than conventional sedimentation processes for removing extracellular toxins. It may, however, remove more intact cells because the floating sludge tends to be removed more frequently than settled sludge in horizontal flow tanks, where the algae may die and then lyse. This assumption needs to be evaluated further.

Periods of high turbidity often cause problems for DAF, and any interruption in the process leads to an interruption in the treatment process. Thus a stock of spare parts and regular maintenance by qualified personnel are critical issues when using this approach.

Precipitation for hardness reduction

Conversion of soluble compounds into insoluble particulates for separation by sedimentation or filtration is commonly used for water softening and for iron and manganese removal. Some concurrent removal of soluble metals and dissolved natural organic matter may also be achieved. Lime is commonly used for adjusting hardness or for precipitation of soluble metals. This process typically uses rapid mixing followed by flocculation and sedimentation.

No studies evaluating lime precipitation as a separate process in a water treatment plant sequence are available. However, some insight into the expected removal of intracellular toxins has been provided by two studies looking at treatment of raw water blooms with lime. Kenefick *et al.* found that lime doses from 100 mg l^{-1} as $Ca(OH)_2$ precipitated the cells in cyanobacterial bloom material containing microcystin-LR without releasing toxin compared with control batches over 14 days, while Lam *et al.* found only 4 per cent release of microcystin-LR for the same lime dosage. These studies suggest that lime softening would be effective at removing intracellular toxin by removing the cyanobacterial cells without causing cell lysis, but that there is no evidence to suggest that lime softening can reduce extracellular toxins.

Direct rapid filtration

Filtration is a process for the removal of suspended particulate matter, typically including clay, silt, natural organic matter, coagulated floes, lime softening precipitates, iron and manganese precipitates, and microorganisms. Filters most commonly use granular media such as coarse sand, crushed anthracite coal, garnet and granular activated carbon (GAC). Direct filtration is applied for low turbidity waters by filtering directly after coagulation/destabilisation without an intervening clarification stage to remove the bulk of the floe. Conventional water treatment uses rapid filtration rates which require regular backwashing to maintain performance. Mouchet and Bonnélye reported poor removal rates of 10-75 per cent, depending upon phytoplankton species, by direct rapid filtration without prior chemical treatment.

Drikas *et al.* found that removal of *Microcystis aeruginosa* cells in the filtration stage of a pilot plant varied between 14 and 30 per cent following alum coagulation/sedimentation. Lepisto *et al.* evaluated full scale water treatment plants for their ability to remove cyanobacterial cells and found rapid sand filtration achieved only a 14 per cent reduction in cells. Rapid sand filtration, including GAC was somewhat better achieving 42 per cent reduction of cyanobacterial cells. These researchers expressed concern over the possible fate of intracellular toxins which may be released from degrading cells trapped in the filtration stage. Lambert *et al.* found inconsistent incremental removal of microcystins from 14-60 per cent across a dual media sand-anthracite filtration stage, following an alum coagulation-sedimentation stage, at a small, full-scale water treatment plant.

As an overall assessment of direct rapid filtration for elimination of algae and cyanobacteria, Mouchet and Bonnélye have indicated that direct filtration is generally not satisfactory, unless more sophisticated multimedia filters and adequate initial treatment are applied. They particularly emphasised the excellent results in algal removal after pre-ozonation. A potential issue of concern, which currently has been inadequately investigated, is the effect of long filter runs between backwashing. Death and lysis of cyanobacteria retained on filters could lead to substantial toxin release.

Combined coagulation, sedimentation and rapid filtration

Conventional water treatment commonly involves the combination of coagulation, clarification and filtration. Consequently, much of the limited research that has been published on water treatment performance for the removal of cyanotoxins has looked at overall removal across the common combinations of coagulation-filtration and coagulation-clarification-filtration, rather than looking at each stage individually.

Himberg *et al.* evaluated hepatotoxic fractions from *Microcystis wesenbergii*, *M. viridis* and *Planktothrix agardhii* in bench-scale treatment processes consisting of alum or ferric chloride coagulation combined with sand filtration and chlorination. Alum coagulation, at doses from 36 to 71 mg l^{-1}, with filtration achieved toxin removals from 11 to 32 per cent, while ferric chloride at 55 mg l^{-1} achieved from 9 to 16 per cent. The removal contribution of the low chlorination dosage in this case was apparently negligible. They also studied a similar conventional process at pilot scale using freeze dried *Microcystis* bloom material and found negligible toxin removal. Nonetheless, Lambert *et al.* found combined microcystin removal was 50-60 per cent across coagulation, sedimentation and dual media filtration in a small full scale plant.

Similar studies at bench scale with anatoxin-a have indicated no removal for either alum or ferric chloride process combinations at a toxin concentration of 20 μg l^{-1}, but at 10 times higher toxin concentrations, the alum process achieved a 14 per cent anatoxin-a removal and the ferric chloride process achieved a 49 per cent anatoxin-a removal.

Leuschner studied phytoplankton retention by flocculation, sedimentation and rapid filtration in a plant treating highly eutrophic river water. Whereas *Microcystis* spp. were rarely observed in the finished water, *Planktothrix agardhii* was poorly retained, showing an average breakthrough of 27 per cent of the filaments. As also reported by Mouchet and Bonnélye, addition of a cationic polymer during flocculation substantially improved retention. The removal of whole, intact cells presents the best opportunity to remove toxins in separation processes, whereas the literature indicates removal efficiencies are low with extracellular toxins. Some unsatisfactory results reported with lysis of entire cells may have been due to an excessive time delay between flocculation and analysis. In summary, currently available results indicate that conventional coagulation and rapid filtration processes assist in toxin removal, particularly if cyanobacterial cells are kept intact, but cannot be generally relied upon as the main removal process. Mouchet and Bonnélye have emphasised the need for:

> "... pilot scale investigation in order to estimate the technical and economical advantages of this choice in each case. Generally, a conventional treatment line, including coagulation, flocculation, settling or flotation, and filtration, is preferred to treat algae-rich waters. However, algae removal is somewhat more delicate than turbidity removal and, consequently, greater attention is required when selecting technology and adjusting the *chemical treatment*."

Slow Sand Filtration

In contrast to rapid filtration, slow sand filters operate at lower rates and develop a surface filter cake which performs most of the filtration together with biological treatment activity. These biofilms establish after some time of operation and contribute significantly to degradation of dissolved substances. Mouchet and Bonnélye reported a likely removal of 99 per cent of algal cells by slow sand filtration. Operation of these filters in the dark can prevent intensive algal growth on the filter. However, overloading of filters with algae or cyanobacteria from the raw water may lead to rapid blocking, requiring removal of the bioactive surface layer, thus temporarily reducing the efficiency for retention of dissolved substances. For removal of toxic cyanobacteria, this constitutes a dilemma because bloom-containing waters are likely to lead to rapid blocking and thus undermine the practicability of slow sand filtration. However, experiments have shown that before blocking, slow sand filters may be quite effective in the removal of toxic cyanobacteria and dissolved toxins. Keijola *et al.* evaluated laboratory-scale slow sand filters and reported over 80 per cent removal of toxins from *Microcystis*, 30-65 per cent removal of toxins from *Planktothrix* (syn. *Oscillatoria)* and about 70 per cent removal of anatoxin-a. Because filtration itself would not be expected to achieve any removal of extracellular toxin, these results suggest that the mechanisms were at least biosorption, and perhaps some biotransformation. Australian studies with roughing filters followed by slow sand filters showed that *M. aeruginosa* and some *Planktothrix* (syn. *Oscillatoria)* cells from toxic bloom material could be removed by physical means and biological processes. Superior microcystin removal, in one of two river water sources being treated with GAC filters, was attributed to biological activity. Freeze dried bloom material was used in this study.

Work on microcystin-LR degradation using an isolated bacterium for use in water treatment has been undertaken by Bourne *et al.* A pseudo-monad has been isolated which possesses an enzyme system capable of degrading microcystin, but the work is currently only at the laboratory scale. Pilot plant studies using a solid phase support for this bacterium to investigate this process are to be undertaken. Developments have occurred in the exploitation of slow sand filters at large treatment works, notably in the UK and Netherlands. These have included use of various pretreatments, such as conventional treatment by coagulation and filtration and preozonation to control the rate of blocking by algae and cyanobacteria. Whilst these processes will assist with removal of cells containing toxins, they have not been adequately assessed for their reliability in degradation of extracellular toxins.

A notable development has been the sandwiching of a layer of GAC within the bed of sand in slow sand filters in order to assist in removal of dissolved toxins. General experience with slow sand filters suggests that they are potentially very useful for removal of particles and dissolved substances, particularly if further developed or combined with other treatment steps to avoid blocking when loaded with waters rich in algae and cyanobacteria. New approaches to slow sand filtration are experimenting with horizontal rather than vertical water flow. This requires larger amounts of water but will remove most of the potentially filter-blocking particles and, in particular, would keep cyanobacteria suspended. Such systems may be developed locally, particularly to serve small communities. For large treatment facilities, bulk cell removal by coagulation and clarification before slow sand filtration may be an effective approach for obtaining the benefits while avoiding rapid blocking.

REFERENCES

Bowling, L.C. and Baker, P.D. 1996. "Major cyanobacterial bloom in the Barwon-Darling River", Australia in 1991, and underlying limnological conditions. *Aust. J. Mar. Freshwat. Res.*, 47, 643-657.

Chudek, E., Horn, A., Joo, M., McLaren, S. 1998 *Stratification Patterns in Queensland Dams.* Report by Water Monitoring Group, Queensland Department of Natural Resources, Australia, 76 pp.

Earthwatch 1992 *Chemical Pollution: a Global Overview.* United Nations Environment Programme, Nairobi.

Gerrard, S. 1995. "Environmental risk management". In: T. O'Riordan [Ed.] *Environmental Science for Environmental Management.* Longman Scientific and Technical, Harlow, 296-316.

Jones, G.J. and Korth, W. 1995. "*In situ* production of volatile odour compounds by river and reservoir phytoplankton populations in Australia". *Wat. Sci. Tech.*, 31, 145-151. NRA 1990 *Toxic Blue-Green Algae.* Water Quality Series No. 2. The National Rivers Authority, London, 125 pp.

Reynolds, C.S. 1984 *The Ecology of Freshwater Phytoplankton.* Cambridge University Press, Cambridge.

Skulberg, O.M., Codd, G.A. and Carmichael, W.W. 1984. "Toxic blue-green algal blooms in Europe: a growing problem". *Ambio*, 13, 244 - 247.

11

Ecological Management of Cyanobacteria

Marine microbes are capable of flourishing in all oceanic habitats, from several kilometers below the seafloor to the top millimeter of the ocean surface. They thrive in environmental conditions where other organisms cannot, ranging from supercooled brine channels of Arctic ice floes to near-boiling waters of hydrothermal vents. Consequently, marine microbes are the most numerous group of organisms on the planet. In addition to being abundant, the many different types of marine microbes carry out many different types of metabolism. As a consequence of this diversity, marine microbes are involved in virtually all geochemical reactions occurring in the oceans.

EUKARYOTIC PHYTOPLANKTON AND CYANOBACTERIA

A starting point for the carbon cycle is carbon fixation or the transformation of CO_2 to a "fixed," nongaseous form—organic carbon. Unlike terrestrial ecosystems, carbon fixation in marine systems is nearly exclusively by free-floating microbial "plants." The exceptions include a few near-shore environments such as salt marshes and mangrove stands where higher-plant production dominates, and in some shallow marine environments where much of the primary production can be by benthic algae. Aquatic ecologists use the term "phytoplankton" (or algae) rather than "plants," but in fact there are some important similarities between plants on land and phytoplankton in lakes and the oceans. Both land plants and phytoplankton are autotrophs and are the main primary producers in their respective ecosystems, using the same mechanism for fixing CO_2, the Calvin–Benson– Bassham cycle.

Both have chlorophyll a in reaction centers where light energy is converted to chemical energy. However, unlike land plants, in marine phytoplankton the main pigments absorbing light energy, "the light-harvesting pigments," are not chlorophylls. An example of these other pigments include the carotenoids, one being fucoxanthin, which is abundant in diatoms. For this reason, many phytoplankton are not green, because their dominant light-harvesting pigments absorb light with wavelengths that differ from that absorbed by land plants.

The most important difference, however, between land plants and phytoplankton is the most obvious one: land plants, such as California coastal redwoods and giant sequoias that tower 100 m above the ground, are among the largest creatures on the planet, while phytoplankton are among the smallest, some as tiny as a micrometer or less (10^{-6} m). This difference in size has many implications for how marine and indeed all aquatic ecosystems are structured. By "structured," I mean the size and number of organisms, biochemical composition, phylogenetic diversity, growth rates and net changes in population sizes, and trophic interactions.

Another huge difference between terrestrial plants and phytoplankton is that the latter includes cyanobacteria, in addition to eukaryotic algae. Especially in the nutrient-poor, oligotrophic oceans, cyanobacteria can account for a high fraction of primary production and of total phytoplankton biomass. Estimates for cyanobacteria may decrease as we learn more about small eukaryotic phytoplankton, but cyanobacteria will undoubtedly remain important in the oceans. Two groups of cyanobacteria are especially large contributors to primary production and phytoplankton biomass: Synechococcus and Prochlorococcus (Table 1). The cells in both cyanobacterial groups are small, smaller than eukaryotic phytoplankton. Phylogenetically, cyanobacteria are bacteria; they lack a nucleus, and their cell wall and membranes are like those of Gram-negative bacteria. Functionally, however, both Synechococcus and Prochlorococcus are members of the phytoplankton community because they are mainly photoautotrophic and use light energy to fix CO_2 by similar mechanisms as found in eukaryotic phytoplankton.

Table 1. Comparison of the Two Major Coccoid Cyanobacterial Genera Found in the Oceans

Property	*Synechococcus*	*Prochlorococcus*
Size (diameter)	0.9 μm	0.7 μm
Chlorophyll	Yes	Yes, but modified
Chlorophyll	No	Yes, but modified
Phycobilins	Yes	Variable
Distribution	Cosmopolitan	Oceanic gyres
N_2 fixation	Some species	No

Microbiologists knew about cyanobacteria for over a century, but the high abundances of oceanic Synechococcus and Prochlorococcus were discovered only around 1977 and 1986, respectively. Since then, we have learned much about these organisms. In contrast to most other marine bacteria, marine cyanobacteria have convenient markers, their pigments, for studying them. These unique pigments make it possible to examine Prochlorococcus by flow cytometry and Synechococcus by microscopy as well as by flow cytometry. Again unlike many other marine bacteria, ecologically relevant representatives from the two cyanobacterial groups can be isolated and grown alone in pure culture in the laboratory. Consequently, we have learned much about the physiology and biochemistry of these microbes, and the genomes of several of them have been completely sequenced.

The eukaryotic members of the phytoplankton community are also important in many oceanic waters. The large species (10–100 μm, which is large in the microbial world) are relatively easy to identify because they have distinctive shapes and sizes, in addition to their distinctive light-harvesting pigments. Large phytoplankton, such as diatoms and coccolithophorids, have been studied for years, and are well known for their importance in coastal waters, especially in spring when they form dense blooms. But often, much smaller eukaryotic phytoplankton species are much more abundant and dominate phytoplankton biomass, along with the cyanobacteria. These small eukaryotes are only slightly larger than bacteria and are members of the "picoplankton" community, which includes all microbes 2 μm or less in size. Unlike large phytoplankton, the picophytoplankton are hard to identify by traditional methods, because of their size and lack of distinguishing features.

PHOTOHETEROTROPHIC BACTERIA

All microbes, not just eukaryotic phytoplankton and cyanobacteria, are affected by light directly or indirectly in the surface layer of the oceans. An example of an indirect effect is the photochemical modification of dissolved organic matter (DOM) used by heterotrophic bacteria. The direct effects include the damage of microbial DNA by short-wavelength light, especially in the ultraviolet region (200–400 nm). These types of light effects are well known.

Less well known is the use of light by phototrophic bacteria to drive adenosine triphosphate (ATP) synthesis while also obtaining energy from other sources, most prominently the oxidation of organic material (Table 2). Except for cyanobacteria, these bacteria are not autotrophic and do not contribute to primary production. We think that they can synthesis ATP via both phototrophic and heterotrophic mechanisms, hence making them photoheterotrophic, but little else is known about them.

Table 2. Phototrophic Microbes in the Oceans

Microbial Group	*Pigment in Energy Production*	*CO_2 Fixation?*
Eukaryotic phytoplankton	Chlorophyll	Yes
Cyanobacteria	Chlorophyll	Yes
AAP bacteria	Bacteriochlorophyll	No
Several	Proteorhodopsin	No

We probably know more about the diversity of genes encoding proteorhodopsin and of the diagnostic gene (pufM) for AAP bacteria than we do about their ecophysiology and biogeochemical roles in the oceans. For instance, the number and diversity of proteorhodopsin genes in the oceans suggest that bacteria bearing them are abundant, but how proteorhodopsin benefits these bacteria is not entirely clear; light has no effect on the growth of Pelagibacter ubique, a cultured marine bacterium from the SAR11 clade with proteorhodopsin, whereas light does stimulate the growth of another marine heterotrophic bacterium in pure culture experiments. One hypothesis is that photoheterotrophic bacteria should be most abundant in oligotrophic oceans, where light energy can augment that gained from heterotrophy and the oxidation of organic material. However, AAP bacteria can be quite abundant in eutrophic estuaries and coastal waters. We know little about even basic parameters of photoheterotrophic bacteria.

One measure of the importance of photoheterotrophic microbes is the effect of light on bacterial production, as measured by leucine incorporation. This effect varies among environments. In some cases, visible light can inhibit leucine incorporation, whereas in other systems, it stimulates incorporation rates. One factor hypothesized to explain this variable effect is the abundance of Prochlorococcus. The role of Prochlorococcus in DOM fluxes has been demonstrated by flow cytometric studies of methionine assimilation. More recent flow cytometric studies confirmed that Prochlorococcus was responsible over 75 percent of light-stimulated leucine incorporation in the North Atlantic Ocean, but Synechococcus and probably other bacterial groups also accounted for substantial light-dependent leucine incorporation. While per cell rates can be quite high for Prochlorococcus and Synechococcus, their abundance is not as high as other microbial groups, such as heterotrophic bacteria and photoheterotrophic bacteria other than cyanobacteria. Consequently, heterotrophic bacteria and photoheterotrophic bacteria other than cyanobacteria account for most of leucine incorporation in the oceans regardless of light intensities.

DISSOLVED ORGANIC MATERIAL

This chapter is focused on organisms and viruses, but DOM is the "800-pound gorilla" in the ocean, making it hard to ignore: it is the largest pool of organic carbon in the ocean and one of the largest in the biosphere. Concentrations of DOM are not measured directly, but rather the key components are examined, given here in decreasing concentration and order of our understanding: dissolved organic carbon (DOC), dissolved organic nitrogen (DON), and dissolved organic phosphorus (DOP). The amount of C in the DOC pool is nearly equivalent to the carbon in atmospheric CO_2, but concentrations of individual DOM components are very low, often nanomolar (10^{-9} mol/L).

The large size of the DOC pool is perhaps the most obvious reason why it is included in oceanic carbon budgets and climate change models, but the flux through this pool is also quite large and in many ways is a more important parameter. By "flux," I mean the rate at which DOM components are produced or utilised. In spite of most oceanic DOC being old and refractory, the rest of the DOC pool is sufficiently labile that overall DOC fluxes are usually equivalent to about 50 percent of primary production, and in some marine systems, the fraction is even higher. Some DOM comes directly from phytoplankton, for example via release of small compounds or the sloughing off of large polymers associated with the outside of phytoplankton cells. However, most of the DOM appears to be produced by grazers, but every organism (and virus) in the ocean contributes to the production of DOM. Organic carbon from terrestrial sources can be a substantial fraction of total DOM in lakes and estuaries. The high fluxes of DOM help explain many features of biological communities in marine water columns. One important feature is the retention of material in the surface layer of the oceans. Because most of the DOC used by bacteria is respired and because of the small size of microbes, as much as 90 percent of primary production is mineralised by grazers and heterotrophic bacteria in the upper surface layer of the open oceans and only 10 percent or less sinks out to deeper waters. These percentages vary greatly, especially in coastal waters where less primary production is mineralised in the surface layer and more is exported to deep waters or horizontally to less productive waters. Figure 1 illustrates the fates of primary production and the possible relationships between DOM fluxes and export. At the risk of oversimplifying complex relationships, the figure suggests that any carbon going into the microbial food webs is less likely to end up in larger organisms and higher trophic levels or to sink out of the surface layer, leading to the prediction that there should be an inverse relationship between DOM fluxes and export out of the surface layer. Any organic carbon going into the dissolved pool has to pass through too many transfers among trophic levels before it reaches organisms large enough to produce large particles that sink out of the surface layer.

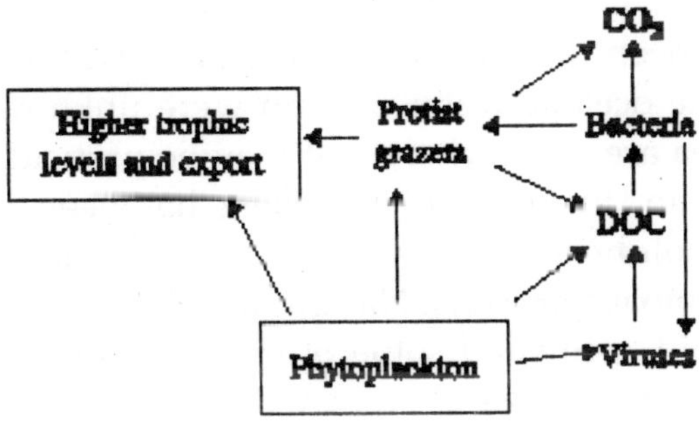

Figure 1. Main trophic interactions and the fate of primary production in the oceans. DOC" is dissolved organic carbon. "Higher trophic levels" refers to organisms much larger than microbes, such as zooplankton and fish. "Export" is the sinking of organic material out of the surface layer.

There are some data supporting the inverse relationship between DOM fluxes and export examined bacterial biomass production in the Yellow Sea, off the coast of South Korea, and calculated the ratio of bacterial production to primary production, which is a useful index of the relative size of the DOM flux. Heterotrophic bacteria are the main users of DOM in the oceans, as discussed below, although respiration by bacteria is usually much larger than biomass production by bacteria. Cho and colleagues found an inverse relationship between an index of organic carbon export and the ratio of bacterial production to primary productio, consistent with the negative relationship between export and the phytoplankton–DOC–bacteria pathway implied by Figure 1. While photochemical and other abiotic reactions can modify and even mineralise DOM (especially from terrestrial sources) back to CO_2 and other inorganic compounds (e.g., NH_4^+), the dominant reactions affecting marine DOM are catalysed by microbes. Phytoplankton, grazers, and viral lysis all contribute to DOM production, and heterotrophic bacteria are the main users of DOM. Microbes thus set what goes into the DOM pool and what remains.

MARINE ARCHAEA

Archaea share some rather superficial similarities with bacteria, the other major domain of prokaryotes. Bacteria and archaea cannot be distinguished by simple microscopy, and cells in both microbial groups do not have a nucleus and lack the organelles found in eukaryotes. The old term for these microbes, archaebacteria, hints at this commonality. However, in many respects, archaea are as different from bacteria as both prokaryotic groups are from eukaryotes. In fact, archaea may be phylogenetically more closely related to eukaryotes than to bacteria. Of the many differences between bacteria and archaea, perhaps the most important one is the composition of the cell wall and membrane. Archaea do not have the peptidoglycan found in bacterial cell walls, and the archaeal membrane has unique lipids not found in either bacteria or eukaryotes.

Oceanic archaea had been discovered in the early 1990s, but their abundance and biogeochemical role in the oceanic water column were unknown until recently. These studies found that archaea are everywhere, but may be particular abundant, relative to bacteria, in the deep ocean. Even more surprising has been the evidence that these oceanic archaea are chemolithoautotrophs that oxidise ammonium (chemolithotrophy) to drive CO_2 fixation (chemoautotrophy). Natural-abundance ^{14}C data and incorporation of $^{13}CO_2$ into archaeal lipids provided some of the first clues that marine archaea are autotrophs and derive their biomass C from CO_2 rather than from organic material. Subsequent microautoradiographic studies demonstrated fixation of $^{14}CO_2$ by archaea from the deep North Atlantic Ocean, albeit in decompressed samples. Ingalls et al., suggested that over 80 percent of archaeal carbon is from CO_2 assimilated via chemoautotrophy in 670 m water of the North Pacific gyre. We still know very little about marine archaea, about chemolithoautotrophy, and about the many other microbes and processes in the deep ocean. Like most biological oceanographers, microbial ecologists have focused on the surface layer and the euphotic zone where sunlight is available for primary producers. The water masses below the euphotic zone—the "dark ocean"—are greatly undersampled, especially considering their vastness. Over 75 percent of the ocean is below 1000 m, making the dark ocean the largest habitat in the biosphere.

HETEROTROPHIC PROTISTS

Marine protists, which are single celled eukaryotes, are very diverse, in both form and function. Some are members of the phytoplankton community and contribute substantially to primary production and phytoplankton biomass, especially in oligotrophic oceanic regimes. Others are heterotrophic and eat other microbes by phagotrophy; they engulf other microbes into food vacuoles, where the prey is digested. Still other eukaryotic microbes probably are mixotrophic, carrying out both light-driven photosynthesis and grazing. Some protists used to be called protozoa, but this term, which refers to the lowest division of the animal kingdom, is misleading because many of these microbes are mixotrophic, if not strict autotrophs. We are beginning to understand the phylogenetic diversity of eukaryotic picoplankton and other protists, but we know little about their ecological roles in the oceans. However, we do know some things about the role of several protist groups that are distinguishable by size and other characteristics observable by simple microscopy.

Nanoflagellates (2–20 μm)

Microscopy studies revealed that the oceans harbor a large number of small eukaryotic cells without any chlorophyll, bearing one or more flagella. These microbes have been called "colorless microflagellates," "heterotrophic nanoplankton" (HNAN), or simply "flagellates." As with all such groups, nanoflagellates are taxonomically diverse, with probably equally diverse ecological and biogeochemical roles. Given the cell size of nanoflagellates, they probably graze on heterotrophic bacteria, cyanobacteria, and picoeukaryotes because these potential prey have roughly the same size.

Microzooplanktonic Protists (20–200 μm)

Microbial ecologists first focused on one group of microzooplankton protists, tintinnids, a type of ciliate with elaborate houses (loricae) surrounding them. However, improved methods for studying microzooplankton protists led to the discovery of many "aloricate" or "naked" ciliates. These microbes can be abundant in the microzooplankton community and are quite important as grazers of phytoplankton and predators of flagellates that in turn feed on bacteria, cyanobacteria, and picoeukaryotes in the oceans. Ciliates are critical links in the microbial food chain and are likely important prey for carnivorous zooplankton. Some ciliates, such as Laboea spiralis, are mixotrophic and carry out photosynthesis with the help of chloroplasts from phytoplankton ingested by the ciliate. Still others, such as Mesodinium rubrum, are strictly autotrophic and can form large blooms in coastal waters. The term "ciliate" gives a general idea about the appearance and mode of locomotion for these protists, but their ecological roles in the oceans vary greatly.

Dinoflagellates

These fascinating microbes vary greatly in size and shape and in their roles in estuaries, coastal waters, and the oceans. The size of dinoflagellates ranges from unarmored dinoflagellates, such as Gymnodinium, which are 8–15 mm, to Noctiluca which are 200–2000 mm. A "typical" dinoflagellate is armored by cellulose plates (theca) with two grooves to accommodate the two flagellates that are used for locomotion. Many dinoflagellates, such as Gymnodinium, are pigmented photoautotrophs and are members of the phytoplankton community. However, about 50 percent of dinoflagellates, including Noctiluca, are colorless and probably are heterotrophic. Along with ciliates, heterotrophic dinoflagellates are important members of the microzooplankton community, and can account for a large fraction of grazing on phytoplankton.

Some heterotrophic dinoflagellates feed on microbial prey by phagocytosis, but others use feeding tubes to suck out the cytoplasm of their prey, and still other dinoflagellates capture their prey with a feeding veil. Dinoflagellates include several pathogens of higher marine life and humans. Red tides are well known to be caused by dinoflagellates such as Gonyaulax polyhedra and Gymnodinium splendens, which secrete potent neurotoxins called saxotoxins. Less well understood are the dinoflagellates, such as Pfiesteria piscicida, that are potential fish pathogens. Some are even parasites. Amoebophrya, for example, is a genus of dinoflagellates that parasitise other dinoflagellates. These topics and the original literature are discussed by Sherr and Sherr.

Marine Fungi

Filamentous fungi and yeasts are found in the oceans, but these heterotrophic eukaryotes do not appear to be abundant and are not ecologically important in aquatic ecosystems, except as decomposers of dry detritus from vascular plants in coastal regions. Fungi and yeasts live on DOM or particulate organic detritus, and thus would have to compete with heterotrophic bacteria and archaea in the oceans. The low abundance of fungi and yeasts indicate that they do not fare well in this competition, probably because they are larger than prokaryotes and thus have much higher surface area-to-volume ratios, putting them at a disadvantage in transporting dissolved compounds. Filamentous fungi do well in the standing dead plants in salt marshes, because their hyphae are able to penetrate and propagate up the dry, dead plant, something that prokaryotes cannot do. A similar reason explains why fungi are so important in the carbon cycle of terrestrial systems.

MARINE VIRUSES

In addition to grazing by protists, lysis by viruses is the other main type of mortality inflicted upon marine microbes. Viruses are the most abundant biological entity in the biosphere, and even though they are small, much smaller than bacteria, their biomass in the oceans is substantial. There are probably viruses for every micro- and macroorganism on the planet. Since viral abundance follows host abundance, viruses for heterotrophic bacteria (bacteriophages or just phages) are probably the most common type in the oceans. However, viruses also attack eukaryotic phytoplankton, and there is some evidence of viruses stopping phytoplankton blooms from growing even further. Viruses may be especially critical in controlling blooms of harmful algae in coastal waters.

We expect viral lysis to be the dominant form of bacterial mortality in anaerobic environments such as sediments where few protists can grow. In contrast, viral lysis may be lower in oligotrophic environments where bacterial abundance is low. In support of the latter hypothesis, Strom pointed out that estimates of grazing rates matched bacterial production in oligotrophic systems, but not so in eutrophic environments, where bacterial growth exceeded grazing, necessitating another form of bacterial mortality, such as viral lysis. Very few studies have examined both grazing and viral lysis, in part because the methods for both are difficult and time-consuming.

Viruses and protists have similar but also different impacts on bacteria and other microbial prey. The release of DOM from cells lysed by viruses probably leads to higher growth rates for the surviving bacteria. Like viral lysis, grazing also can release potentially limiting organic carbon and inorganic nutrients, but any released DOM is probably more altered, due to enzymatic attack in food vacuoles, than the DOM released by viruses. However, unlike viruses, grazers may have a direct negative effect on bacterial growth rates. Since grazing rates vary with prey size and since bacterial cells must get bigger before dividing, grazers should tend to graze more heavily on rapidly growing cells. This mechanism could explain why growth rates are often higher in treatments that minimise grazing (e.g., dilution and filtration), although artificial breakage of cells and other DOM contamination during sample preparation would have the same effect.

Laboratory experiments provide more insights into how grazers and protists affect their microbial prey. In the absence of larger organisms and other predators, a protist grazer will reduce prey numbers to vanishingly low levels, although some bacteria can form filaments and flocs that are resistant to grazing. In contrast, viruses can apparently keep bacterial abundance constant over time, but they cannot reduce their host numbers to low levels, unlike a grazer feeding on bacteria. The difference is because microbes develop resistance against viruses. A field study of Synechococcus provides a good example of this coexistence. In the end, mortality via viral lysis or grazing matches bacterial growth most of the time, since bacterial abundance varies relatively little over time and space in the oceans.

N_2 FIXERS

Just as the carbon cycle starts with the fixation of CO_2 to organic C, so too can the nitrogen cycle be said to start with the fixation of N_2 to ammonium. There the similarity between carbon and nitrogen fixation ends. Unlike carbon fixation, N_2, fixation is carried out only by prokaryotes and in the oceans mainly by some cyanobacteria. Fixation by heterotrophic bacteria is thought to be rare in the oceans because the supply of labile DOM is too low to support the high expense of N_2 reduction to ammonium. Once N_2 is fixed, there are several similarities between the exchange of C and N among trophic levels and between fixed inorganic and organic compounds. Similar to C, N is transferred to the rest of the marine biota by a variety of mechanisms, ranging from the direct release of ammonium and DON to grazing on diazotrophs by protist grazers and zooplankton.

Two elements other than N are thought to potentially control N_2 fixation and microbial production in some oceans. Chemical oceanographers have always argued that phosphorus is the ultimate limiting element in the oceans, and now biological oceanographers have recognized the importance of P limitation in regimes such as the Mediterranean Sea and the North Pacific Gyre, arguably the largest biome on the planet. The second element, iron, has also been examined extensively by both oceanographers and microbial ecologists. The role of Fe in limiting primary production in high nutrient–low chlorophyll (HNLC) oceans is now well established. The impact of low Fe and P concentrations in the oceans on N_2 fixation is being intensively examined.

NITRIFIERS AND OTHER CHEMOLITHOTROPHS

We have assumed that oceanic bacteria gain energy either from light, which fuels the metabolism of cyanobacteria and drives the synthesis of organic material from CO_2, or from the oxidation of organic material back to CO_2, which defines what heterotrophic bacteria do. But already we have seen that some of these bacteria may really be photoheterotrophic and fuel ATP synthesis with light energy as well as by organic carbon oxidation. Another possible mechanism for ATP synthesis is by chemolithotrophy, which is the oxidation of reduced inorganic compounds coupled with the reduction of an electron acceptor, which is often oxygen. We have encountered one form of chemolithotrophy, the oxidation of ammonium. Other inorganic compounds used by chemolithotrophic bacteria include various reduced sulfur compounds, carbon monoxide, and methane. Ammonium oxidisers were discovered decades ago, but microbial ecologists thought bacteria carrying out this reaction were not abundant in the surface ocean because of slow growth rates. Other forms of chemolithotrophy had been found previously only close to sources of reduced inorganic compounds, such as anoxic sediments. These bacteria were not thought to be common in the surface ocean, because concentrations of reduced inorganic compounds are very low in the oceans.

Genomic and experimental data are changing this picture. In addition to the work on ammonium oxidation by archaea in the deep ocean, new data now suggest that marine bacteria may be mixotrophic, harvesting energy from whatever reduced compound, organic or inorganic, they can obtain. While concentrations of reduced inorganic compounds are very low in the oceans, so too are concentrations of labile DOM components.

DENITRIFIERS

While factors such as DOM quantity and quality, grazing, and viral lysis all affect the activity and composition of microbial communities, oxygen has probably an even more dramatic impact. The lack of oxygen excludes nearly all eukaryotes from anoxic systems, the exceptions being some protists, and there are large shifts at the division (phylum) level in bacteria as well. The shift is due in part to the rise in organisms that use inorganic sulfur compounds in ATP synthesis, the most important being the sulfate-reducing bacteria. Many of these bacteria are Deltaproteobacteria, which are common in anoxic systems but not in oxic waters. Although sulfate as an electron acceptor is not energetically favorable, the high concentrations of sulfate ensure the dominance of sulfate reduction in the mineralisation of organic carbon in anoxic marine systems. After oxygen is depleted, sulfate reduction often accounts for a large fraction (20 percent > 50 percent) of organic carbon oxidation in salt marshes, estuaries, and shallow coastal sediments. In contrast to sulfate, nitrate respiration does not oxidise much organic carbon in most ecosystems, because nitrate concentrations are too low, even though it is an energetically-favorable electron acceptor.

CYANOBACTERIA AND ECOSYSTEM MANAGEMENT

In freshwaters, scum formation by cyanobacterial phytoplankton is of concern to human health. Freshwater algae proliferate quite intensively in eutrophic waters and may contain irritative or toxic substances. Nevertheless, incidents of impairments of human or animal health caused by algae are rarely reported. One example was the closure of a number of bathing sites in Sweden because of mass occurrences of the flagellate *Gonyostomum semen* which causes skin irritations and allergies. Incidents attributed to cyanobacteria are far more numerous and, in most cases, have been caused by species of cyanobacteria that may accumulate to surface scums of extremely high cell density. As a result, the toxins they may contain reach concentrations likely to cause health effects.

Surface aggregations of planktonic cyanobacteria occur because of their capability to regulate their buoyancy, enabling them to seek water depths with conditions optimal for their growth. Regulation of buoyancy is a slow process, and cells adapted to ambient turbulence may take several days to adapt their buoyancy when conditions change. Thus, cells or colonies may show excessive buoyancy and accumulate at the water surface. Light winds drive such accumulations to leeward shores and bays, where the resulting scums become thick. In extreme cases, such agglomerations may become very dense, with cells frequently concentrated by a factor of 1,000 or more, eventually reaching in some cases, one million-fold concentrations with a gelatinous consistency. More frequently, surface accumulations are seen as streaks or slimy scums that may look like blue-green paint or jelly. Such situations can change rapidly within hours with changes in the wind direction. Monitoring strategies must take into account this highly dynamic variability of cyanotoxin occurrence in time and space.

Scum formation is influenced by the morphological conditions of the water body, such as the water depth from which cyanobacteria can rise to the surface (i.e. the thickness of the stratum in which they are dispersed) and the length of wind fetch over which surface aggregations can be swept together to form shoreline scums. Accumulated scum material may take a long time to disperse especially in shallow bays. Dying and lysing cells within the scum release their contents, including the toxins, into the water. However, toxin dissolved in the water is rapidly diluted and probably also degraded. Cellbound cyanotoxin concentrations usually are the greater cause for concern in recreational waters.

Most of the problems reported with nuisance and toxin-containing aquatic cyanobacteria in freshwaters have involved planktonic species, i.e. those distributed in the water body or forming surface scums. However, benthic (i.e. bottom-dwelling), species have occasionally surfaced, and been washed ashore where they have caused the death of dogs scavenging upon the material. Benthic cyanobacteria can grow as mats on sediments in shallow water. Some mats become detached and are driven onshore where they result in acutely toxic accumulations.

Severe illness due to direct dermal contact with such mats has been reported from tropical marine bathing sites. In coastal marine environments many toxic species of dinoflagellates, diatoms, nanoflagellates and cyanobacteria occur, and have led to several forms of human health impacts mainly after consumption of shellfish and fish, i.e. syndromes such as Paralytic Shellfish Poisoning, Diarrhetic Shellfish Poisoning, Amnesic Shellfish Poisoning, Neurotoxic Shellfish Poisoning and Ciguatera. Nevertheless, there exists very little scientific evidence that marine toxic algal blooms cause health problems for recreational users of water. This evidence is reviewed in the WHO *Guidelines for Safe Recreational-Water Environments*.

FIELDWORK SITE INSPECTION AND SAMPLING

Fieldwork, including site inspection, sample collection and, in most programmes, some on-site analysis, determines to a large extent the quality of information obtained and represents a significant proportion of the total cost of a cyanobacteria or cyanotoxin monitoring programme. A well-designed and implemented fieldwork programme enhances the quality of the data obtained and may assist greatly in containing overall costs. Inclusion of members of the public as active participants in monitoring programmes is rapidly gaining acceptance and can contribute significantly to the quantity and quality of information obtained from a monitoring programme. Special interest groups and also concerned local persons in sensitive or affected areas can provide useful information. Proper orientation and training, for example in site inspection, in flexible responses to the results of inspections, and in conducting basic tests such as transparency, can assist in providing valuable additional data.

Planning for Fieldwork

The principal components of effective planning for fieldwork comprise:

- Timing and preparation (in the context of the monitoring programme plan and information needs for management).
- Prior liaison with the laboratory that will receive and process samples.
- Logistic preparation.
- Local co-ordination as required.
- Prior liaison with other information recipients (local or central).

Preparation of the monitoring programme principally concerns the timing of visits and careful cataloguing of sampling sites.

Before routine field visits are performed, a period of pilot testing should be implemented. This will help ensure that time requirements for inspection and sampling are reconciled, and that activities are planned to make the best use of staff time and other resources. Realistic estimation of travelling time is important to avoid exceeding allowable sample storage times prior to analysis. Pilot testing should lead to the development of a detailed inventory and description of sampling sites. If changes in water quality with time are to be interpreted with confidence, samples must be taken consistently from the same locations and/or from precisely identified locations. Pilot testing also provides an opportunity for training personnel and allows their familiarisation with particular aspects of the monitoring programme itself. Co-ordination with the laboratory is an important aspect in determining the final value of the sampling expedition. In some cases, the laboratory will be responsible for preparation of sample containers and chemical additives for sample preservation, and may also be responsible for the provision and maintenance of equipment for on-site testing.

Laboratory capacity is an important area of concern which should be addressed in programme design and in pilot testing. It is essential that the workload generated by a sampling expedition is properly managed within the laboratory; it is therefore vital that analysts know how many samples will be arriving, the approximate time of arrival and the analyses that are to be carried out. Excessive delays before sample processing and analysis may render the sample results invalid for the management purposes for which they had been collected. Therefore, the timing of sample delivery to the laboratory and the workload management within the laboratory should be co-ordinated prior to fieldwork.

Good logistical preparation prior to fieldwork requires that equipment is checked to ensure that it is functioning properly. The correct number, size and type of sample containers must be prepared and transport must be arranged, ensuring permission and local access to any restricted sites. It is essential to prepare a sampling checklist which includes maps of sampling site locations, a list of equipment required and a detailed explanation of the methods for sample collection. The checklist should also include lists of the types and numbers of samples to be taken at each site, as well as of the required volumes. Good preparation for sampling involves previous labelling of sampling containers with at least the site, date and depth from which the sample is to be taken.

Frequency of field visits

The frequency of site inspection and sampling must be adapted to the local situation. Key criteria are:

— Potable water supply reservoirs may need to be monitored regularly throughout the year if perennial persistence of cyanobacteria cannot be ruled out. This applies particularly to warm climates and in temperate zones to water bodies populated by certain taxa such as *Planktothrix*.

— Monitoring of recreational lakes in tropical countries may cover the whole year, while in temperate zones it can be focused on the warm season from early summer to autumn.

— The frequency of site inspection and sampling should be increased during development of cyanobacterial populations or when persistent blooms occur.

Safety

Caution and attention are appropriate while working with cyanobacteria, particularly when they are highly concentrated in scums. It is wise to treat all blooms as highly toxic. Contact with water should be minimised during sampling and gloves and rubber boots should be worn because cyanobacteria might contain toxins and could also have a high allergenic potential.

In some areas of the world other water-based hazards such as schistosomes may also be present. In such circumstances water contact should be minimised and following contact the skin should be vigorously dried. Caution during fieldwork should also apply during the use of boats or other vessels and whilst wading, especially in waters with low transparency where underwater hazards may not be readily visible.

Site Inspection

Sites used for drinking water abstraction or recreation should be subject to programmed inspection by trained professional staff, and preferably in conjunction with sampling expeditions. Careful inspection can assist interpretation of results from sample analysis. Moreover the development of personal expertise in relation to specific water bodies can provide the best form of early warning system.

A protocol for site inspection should be established that includes the ambient data of temperature, wind and other weather conditions as well as an estimate of the situation during the previous 24 hours.

High nutrient input from, for example sewage outlets or run-off from excessively fertilised areas favours the development of cyanobacteria. Site inspection protocols should also address the identification of sources of nutrient input and significant land uses, as well as land use changes that may assist in the interpretation of findings. Such assessment is not necessary during every sampling tour but is particularly recommended during the pilot phase, and at moderate intervals, such as annually, when changes in the catchment area are suspected.

When scums appear on the water surface, cyanobacteria may be present in densities hazardous to human health, and thus appropriate responses should be initiated quickly and samples for further analysis should be taken. Sampling of scums outside designated or habitual bathing sites is also of great value for determining and predicting further risk, e.g. for when wind directions change during the following days.

Sampling

Sampling may address both cyanobacterial population development and hazardous accumulations. Informed on-site decision-making is necessary to refine sampling programmes. Samples addressing population development as a basis for assessing the potential for hazardous concentrations are usually taken at one or several points in the water body. These points should be representative for the whole water body and are often where it is deepest. For assessing hazards, samples must also be taken in areas where accumulations of cyanobacteria can affect both humans and livestock, or they should be taken at the raw water intake of drinking water reservoirs. The potential for spatial heterogeneity demonstrated by some species, i.e. horizontal and vertical variations in both cell numbers and toxin content, must also be considered when selecting the number and location of sampling sites. Sample collection should always be accompanied by a site inspection because the data obtained during the inspection will be important in the interpretation of the results of sampling.

Sample collection and storage procedures differ depending on the type of analysis which will be carried out. The three principal categories of analysis usually performed

— Nutrient analyses (phosphorus and nitrogen).
— Cyanobacterial identification and quantification.
— Cyanotoxin analysis, e.g. toxicity testing and analysis of cell-bound and dissolved toxin.

Sample containers

Containers and bottles for the transport of samples should ideally be provided by the laboratory that will conduct the analyses. This helps to ensure that they are of a suitable volume, are properly prepared and that due consideration has been given to the need for pre-treatment and chemical addition. Field work is easier if bottles are pre-labelled and well-arranged in suitable containers. For routine sampling of the same sites, it is advisable to always use the same bottle for each site and each parameter. This avoids cross-contamination, which is a particular concern for phosphorus analyses. For most samples, glass bottles are appropriate but often plastic containers can be used that are considerably lighter and unbreakable. It should be decided in advance whether it is more practical to subdivide a water sample into aliquots for each analysis prior to transportation, or whether a single sample can be divided on receipt in the laboratory. The following containers are recommended for the transport of cyanobacteria and related samples:

— *Phosphorus analysis.* Use 100 ml glass bottles pre-washed with and stored containing sulphuric acid (4.5 mol l^{-1}) or hydrochloric acid until usage. Phosphate is indicative of the potential for cyanobacterial growth even when it is at very low concentrations (µg l^{-1}) and, therefore, special care must be taken to avoid contamination of samples. Contamination may arise from phosphate-containing detergents or from previous storage of samples with very high phosphate concentrations. Phosphates are easily adsorbed to glass surfaces and may be released later when the bottle is filled with a new sample with low concentration.

— *Nitrate, ammonia and total nitrogen.* Use clean 100 ml glass or polyethylene bottles.

— *Microscopic identification of cyanobacteria.* Wide-mouth polyethylene bottles are appropriate for studying living material in a fresh grab or net sample.

— *Microscopic identification and quantification of cyanobacteria.* Brown glass bottles (100 ml) are preferable. These should already have about 1 ml of Lugol's iodine or formaldehyde solution added, or the preservative may be added immediately after filling the bottles with the sample. Clear glass bottles may be used but must be stored in the dark because sunlight destroys iodine, thereby allowing the sample to degrade.

- *Cell material for analysis of cell-bound toxin concentrations.* Use 1 litre (minimum volume) containers, preferably made of glass although good results have also been obtained with plastic. Containers should be pre-cleaned to prevent contamination.
- *Cell material in large amounts (usually enriched with a plankton net) for structural identification or some toxicity assays.* Use plastic containers with wide necks to facilitate filling. If samples are to be freeze-dried, the sample must be frozen in a layer not thicker than 1-2 cm. Specimen containers (100 ml) for urine samples are particularly suitable. Well-sealed, heavy-duty household plastic bags can also be used, but care must be taken to avoid puncture because highly concentrated cyanobacterial material can present a safety hazard. Bags should be placed in watertight ice boxes for cool storage and as a precaution against leakage.
- *Dissolved cyanotoxins.* Use 1 litre (minimum volume) containers, preferably made of glass although good results have also been obtained with plastic. Containers should be pre-cleaned to prevent contamination.
- *Chlorophyll* a *analysis.* Brown glass bottles of 1 litre capacity are recommended to avoid degradation of chlorophyll by sunlight. Clear bottles may be used if the samples can be stored in the dark.

Preparation of Lugol's iodine solution for preserving phytoplankton samples Dissolve 20 g of potassium iodide in 200 ml of distilled water, mix thoroughly and add 10 g of sublimated iodine. The solution must not be supersaturated with iodine because this can result in crystal formation which interferes with cell counting. Supersaturation can be tested by diluting 1 ml of stock solution to 100 ml with distilled water to give concentrations similar to those used for preserving samples. If iodine crystals appear after standing, more potassium iodide should be added and the test repeated. If no crystals appear, 20 ml of glacial acetic acid is added. Use about 1 ml of Lugol's iodine to preserve 100 ml of phytoplankton sample. The resultant sample should be the colour of whisky.

Dense scum samples may rapidly consume the iodine from the Lugol's solution. If samples lose the whisky colour, and particularly if they no longer smell of iodine, they are likely to decay. Samples stored for more than 10 weeks should be periodically checked for sufficient preservation; further drops of Lugol's solution may be added if necessary. If substantially more than 1 ml per 100 ml are required, the volume of Lugol's solution added to the sample should be determined because the sample volume will have changed appreciably and a corrective calculation is required for the cell counts.

Types of sample

Two different types of sample can be taken from lakes or similar surface waters. The simplest, a "grab" sample, is a discrete volume of water taken at a selected location, depth and time. Grab samples are also known as "spot" or "snap" samples. In contrast, composite or integrated samples are made up of several subsamples from different parts of the water body. These are aimed at representative sampling of a water body. Whereas grab samples are suitable for analysing situations at specific sites, composite samples are preferable for assessing the water body's total content of a substance or the population of an organism. Composite samples are particularly important if the variables to be assessed are unevenly distributed. If knowledge of the precise distribution is required, each sample can be evaluated individually. However, the integration of samples prior to analysis is often far more cost-effective.

Depth-integrated samples may be obtained either by continuously sampling the total column of water from the surface to just above the sediment, or by discontinuously taking grab samples from representative depths and then mixing them together. The latter is particularly appropriate for deep lakes. However, in order to choose representative depths and to achieve meaningful integration, knowledge of thermal stratification of the water body is necessary.

The thermal stratification of a water body is influenced by the morphology of lakes and reservoirs, the latitude, weather conditions and the physical nature of the water. It can be determined by measuring vertical profiles of temperature within the water body. Where thermal stratification occurs, it results in a water body functioning as two separate masses of water with different physicochemical characteristics and cyanobacterial populations, and with a transitional layer sandwiched between. In temperate climates, thermal stratification generally occurs seasonally in water bodies of appropriate depth, whereas in tropical climates it often follows diurnal time patterns. Thermal stratification has important implications for the depth at which cyanobacteria are likely to be found, as well as for concentrations of nutrients and interpretation of phosphorus and nitrogen data. Usually, shallow (2-3 m), wind-exposed lakes are unstratified, whereas in temperate climates deeper lakes usually exhibit a stable stratification from spring to autumn. Lakes of intermediate depth (e.g. 5-7 m) may develop transient thermal stratification for a few calm and sunny days, and which is then disrupted by the next event of rain or wind.

Thermal stratification usually results in inhomogeneous distributions of oxygen, nutrients, and populations of algae, cyanobacteria and other organisms. However, even when temperature is uniform throughout depth, stratification of organisms may develop on calm days. Depth gradients of oxygen concentration and pH are good indicators of this. Depth-integrated samples are more adequate than surface grab samples for the assessment of population size and nutrient concentration in such homothermous situations.

In deeper lakes or reservoirs with thermal stratification, depth-integrated sampling of representatively chosen depths is necessary - taking samples evenly spaced over depth may not yield fully representative results. When background information on the typical stratification characteristics of a given lake is available, sample numbers can be reduced by selecting adequate depths to represent specific strata. If depth intervals are unequal and samples are to be integrated, the volume of each sub-sample must be chosen to represent the actual fraction of the vertical gradient it represents. Continuous depth-integrated samples are often quite adequate for shallow and moderately deep water bodies. They can be obtained using a water pump attached to a garden hose which is operated at a steady pumping rate while the water inlet is drawn upwards between the desired depths at a uniform speed. The sample obtained is therefore representative of the water column.

A very simple hose-pipe sampler may be used to depths of 30-35 m. This sampler is made of a piece of flexible plastic piping of about 2 cm diameter and appropriate length; the pipe is weighted at one end and open at both. The weighted end, to which a cord is attached, is lowered slowly into the water so that when the tube is fully extended it encloses a columnar section of the water. Before hauling in the lower end with the attached line, the upper end should be closed in order to avoid loss of water once the lower end emerges from the surface. Hoses need to be stored clean and dry between sampling trips to avoid microbial growth inside the tube.

Sampling scums

Scums of cyanobacteria often occur near shorelines at low water depths, and therefore working with a grab-sampler or a plankton-net may be difficult. Sampling scums is carried out more easily with a wide-necked plastic or glass container. When sampling scums, a suitable approach must be developed which allows for their observed heterogeneous density. Two samples can be useful for different purposes. One sample can be aimed at assessment of the maximum density of cyanobacteria or highest toxin levels by taking a sample from where the scum is thickest. The other sample can aim to simulate conditions where shallow waters are mixed by bathers and playing children. Both types of sample can be used for comparison

Plankton net sampling

Sampling of cyanobacteria with a plankton net is mainly carried out when large quantities of cell material are required or when only a qualitative analysis of the phytoplankton is necessary. In contrast to the use of depth samplers which quantitatively trap all of the particles in a defined volume, the water volume filtered through a plankton net cannot be determined precisely. Calculations based on the area of the net opening and on the length and the distance it has been hauled are not recommended because they strongly overestimate the amount of water actually filtered. The depth at which the plankton net is deployed is dependent on the taxa of algae and/or cyanobacteria present. Floating cells (e.g. *Microcystis, Anabaena)* are harvested within the upper metres of the water column, while the sampling of well mixed or stratified water bodies with distinct depth distributions of cyanobacteria may include deeper water layers.

Nutrients, Cyanobacteria and Toxins

Nutrient analysis (phosphorus and nitrogen)

Mass developments of cyanobacteria are associated with high nutrient concentrations. Phosphorus is usually the key nutrient controlling proliferation, although the availability of nitrogen may be an important variable to assess because it can influence whether or not nitrogen-fixing species dominate. Total phosphorus determines the capacity of a water body to form cyanobacterial blooms. If dissolved phosphate is detected at concentrations of only a few micrograms per litre, cyanobacterial growth and biomass are not limited by phosphate. Because such low concentrations are critical for interpreting the situation, sample contamination may very easily lead to misinterpretation. In contrast, samples for total phosphate analysis are less sensitive. If SRP is important in a sampling programme, rapid filtration in the field or even on the boat may be desirable, particularly in warm climates. Transformations between nitrate and ammonia may occur, if samples are not stored properly and analysed rapidly.

Identification and quantification of cyanobacteria

Cyanobacteria tend to accumulate in layers of sampling containers. When samples are integrated or when sub-samples are taken, the sample must be mixed immediately before sub-sampling.

In addition to the samples preserved with Lugol's solution, fresh unpreserved samples are useful to aid microscopic identification, because Lugol's solution masks the characteristic colours of the cyanobacteria, making it more difficult to recognise some species. Unpreserved samples can be qualitative and are easily obtained by hauling a plankton net through the water column, or by taking a grab sample if the density of cyanobacteria is high. Unpreserved samples for identification may be stored for up to 24 hours, provided ambient temperature and light conditions are kept similar to those in the field. If longer storage is necessary, preservation with formaldehyde may be useful.

Samples for chlorophyll a *analysis*

When collecting samples for chlorophyll *a* determination it should be born in mind that this pigment will start to degrade rapidly after collection; therefore it is essential to keep storage times to a minimum. Filtration at the sampling site and storage of the filters in an ice box is recommended for warm climates. If samples are to be filtered and analysed in the laboratory, they should be transported as soon as practically possible (e.g. within 4 hours) and processed immediately on arrival.

Apparatus

- Measuring cylinder
- On-site filtration equipment, including vacuum hand pump and filter cup
- Glass fibre filters, GF/C, 47 mm
- Plastic Petri dishes and/or aluminium foil

Procedure

1. Thoroughly mix the sample and pour immediately into a measuring cylinder. Pass the measured sample through the glass fibre filter with the aid of a hand vacuum pump. If samples are dense, begin by filtering a small volume. If this is easily filtered, add more sample, mixing every time a sub-sample is poured from the bottle. Use as much sample as will easily pass through the filter leaving a distinctly visible greenish layer on the filter. Record the volume of water filtered.
2. Once filtration is complete, carefully remove the filter from the filtration apparatus and either place it in a Petri dish or wrap it in aluminium foil.
3. Fold the filter with the cells innermost because this both protects the pigment from light and also prevents the sample from being dislodged from the filter surface.
4. Keep filters at near freezing temperatures and in the dark prior to their delivery to the analytical laboratory. Deep-freeze filters if storage of several hours or more is necessary.

Bulk samples of cyanobacteria for cyanotoxin detection

For some purposes, it is important to collect a large amount of cell material using a plankton net; toxicity tests for example require large quantities because bioassays are less sensitive than chemical analysis. Large amounts of cell material may also be needed for some detailed chemical analyses. If the results are to be related to the dry weight of cellular material, a quantitative sample is not necessary.

For later expert identification of the cyanobacterial taxa, and particularly for relating toxin concentrations to cell numbers or biomass of specific taxa, a subsample for cell counts may be taken directly from a well-mixed net sample before it is frozen. It is important to realise that plankton net hauls accumulate particles selectively and are therefore likely to contain a different quantitative distribution of taxa than the sample used for assessing the composition of cyanobacteria. Such subsamples for microscopy should be preserved in 10 per cent formaldehyde or Lugol's solution in a separate container; small volumes are sufficient.

High biomass densities in such concentrated samples may rapidly consume the preservation capacity of Lugol's solution. If not analysed within several days, samples should be periodically checked for their preservation capacity; if they no longer show a brown colouration, more Lugol's solution should be added. Dilution of the sample may be useful to avoid rapid consumption of preservative.

Quantitative analysis of cell-bound and dissolved toxins

For recreational sites and for assessment of the total toxin concentration, cell-bound toxins are of primary importance. The concentration of dissolved toxin in the water is of special interest for drinking water reservoirs and where large amounts of cyanotoxin such as anatoxin-a are released into the water. Large volumes may be required for dissolved toxin analysis and this should be checked with the analytical laboratory. If the concentration of cell-bound and dissolved cyanotoxin is to be related to the water volume from which the cells were collected, a defined volume must be filtered. Essentially samples are handled in a similar manner as those prepared for chlorophyll *a* analysis. The filtration step can be readily performed at the sampling site but with the following modifications:

- Glass fibre or membrane filters may be pre-weighed so that the amount of material retained can be determined. Thus, toxin concentration can be related both to dry weight and to water volume.
- Filters and filtrate should be kept cool.

Laboratory studies have shown that the microcystin content of material collected and dried on filters is unaltered for several months, when the filters are stored in the dark at room temperature and low humidity. So far this method has been shown to be useful for microcystin as well as for anatoxin-a.

On-site Analysis

Analyses for many variables can be carried out in the field. A significant advantage of on-site testing is that tests are carried out on fresh samples that have not been contaminated or otherwise changed as a result of storage. Variables generally analysed on-site are:

- *Transparency*. Cyanobacteria may occur as floating streaks or mats on the water surface making it difficult to obtain representative transparency data. It may be useful to determine the transparency in areas without floating cells as well as within scums. The Secchi disc has to be lowered very carefully so as not to destroy the formation of accumulated cyanobacterial cells, and before taking the measurement the surface scums should be given time to return to their original position. Bathers can easily perform an improvised transparency determination while standing knee-deep in the water and being careful to avoid suspending the sediment. If greenish turbidity obscures their view of their feet, a significant cyanobacterial bloom could be present corresponding to a level at which some authorities advise against bathing.

- *Temperature.*
- *Dissolved oxygen.*

Field Records

As with every sampling programme, careful field records are of critical importance for interpretation of results. Field records involve preprepared lists and questionnaires, but which also leave plenty of space for additional comment and for the recording of observations. Such observation should include:

- Presence of scums.
- Weather conditions on the day of sampling and, if available, an indication of conditions over the previous 2-3 days, because this affects the accumulation of cyanobacteria.
- Any additional or unusual observations, such as scum distribution, estimation of numbers of people bathing despite the presence of scums, defect warning or prohibition signs regarding bathing in scum areas, suggestions for improvement of the sampling programme, previously undetected inlets, changes in agricultural practices or other use, etc.

Sample Preservation and Transport

Samples must be clearly labelled with sampling site, date and time of sampling and depth of sampling. In general, samples should be stored cool and dark in a storage box, taking into consideration the following criteria:

- Cool storage is most important for all samples to be filtered because biological activity may enhance shifts between particulate and dissolved fractions of the variable to be analysed. This is particularly important for nutrient samples, chlorophyll *a* samples and samples for cellbound cyanotoxin analysis. If storage time exceeds 2-3 hours, particularly in warm climates, cooling during sampling and transportation is recommended.
- Unpreserved samples for identification of cyanobacteria, fixed samples for determining cyanobacterial cell numbers and biomass, and samples fixed with Winkler's reagents for oxygen determination should be stored at ambient temperatures but protected from direct sunlight.
- Storage times should be kept as short as possible, generally in the range of a few hours. Storage times over 24 hours should be avoided.

DESIGN OF MONITORING PROGRAMMES

Many cyanobacterial species frequently forming mass developments may contain hepatotoxins or neurotoxins, and all of them contain lipopolysaccharides (LPS) in their cell wall. Lipopolysaccharides may be the cause of irritations of the skin, digestive tract, respiratory membranes, eyes and ears that are frequently associated with cyanobacteria. Research in pharmacology and ecotoxicology indicates that cyanobacteria contain a variety of substances not yet identified, but that may have a potential impact on people. The implications of the present state of knowledge for surveillance and management are that any mass development of cyanobacteria may be a potential health hazard. If the cyanobacterial cells contain hepatotoxic microcystins, cause for concern may be higher because of the chronic effects of this potent toxin. Therefore, monitoring should address primarily the occurrence of cyanobacterial mass developments, whereas microcystin analysis may be adequate in specific situations.

Visual inspection of a bathing site is of crucial importance because it shows immediately whether cyanobacteria occur in potentially hazardous densities. However, as scum formation and dispersion may occur within hours and thus too frequently for monitoring, assessment of the risk of cyanobacterial exposure during recreational activities is greatly enhanced by an understanding of the population development of these organisms in a given water body. Good knowledge of the local growth conditions for cyanobacteria can greatly enhance predictability of bloom formation. As knowledge and understanding of a given site accumulate, regular patterns of cyanobacterial growth may be noticed, and surveillance may as a result be focused upon critical periods. Involving limnological expertise may be very useful, particularly during the development of monitoring programmes and for their periodic assessment for efficacy.

Monitoring strategy for freshwater cyanobacteria

A structured, quantitative investigation approach aims at focusing surveillance efforts upon those sites that are likely to present a risk. It further provides a scheme for immediate assessment and action by discerning three steps of action.

- Determination of the carrying capacity of the ecosystem for cyanobacteria.
- Site inspection to detect mass developments.
- Quantitative assessment of biomass as a basis for risk assessment when mass developments occur.

Preliminary identification of water bodies potentially harbouring high densities of cyanobacteria is possible on the basis of simple transparency measurements with a Secchi disc. If transparency is high and no significant discolouration of the water can be seen, cyanobacterial densities are unlikely to be high. It should be noted, however, that in large and deep lakes with a large volume from which cells can accumulate to scums, high densities in some parts of the lake cannot be excluded. In water bodies with Secchi disc readings of less than 2 m, the following three steps may be undertaken to assess the risk of toxic cyanobacteria.

Step 1. Determination of the carrying capacity of the ecosystem for cyanobacteria

Algal and cyanobacterial population growth requires phosphorus and nitrogen. The concentrations of these nutrients determine the maximum amount of algae and cyanobacteria that can develop in a given water body, or the "carrying capacity" of an ecosystem for these organisms. The carrying capacity is more often limited by the availability of phosphate but sometimes, particularly in marine ecosystems, it may also be limited by nitrogen. If total phosphorus is not limiting, it may be worthwhile to analyse nitrogen to check whether the carrying capacity may be lower than assumed from the total phosphorus concentrations. Total phosphorus should be measured rather than dissolved phosphate because algae and cyanobacteria can store sufficient amounts of phosphate to increase their population 10-fold, even if no dissolved phosphate can be detected. Thus, cellbound phosphate is more meaningful for the assessment of carrying capacity.

Total phosphorus should be assessed several times during the cyanobacterial growth season in order to check temporal variability. If variability of the concentrations is low (less than 50 per cent), assessment twice a year may prove sufficient. If total phosphate concentrations are below 0.01-0.02 mg l^{-1} P, mass developments of cyanobacteria are unlikely and high turbidities, may have other causes. If total phosphate concentrations are higher, monitoring should move to Step 2 in order to check for the presence of phytoplankton mass developments.

Step 2. Monitoring to detect possible mass developments of phytoplankton (algae plus cyanobacteria) by visual inspection of bathing sites and immediate actions to prevent health hazards

Monitoring should generally be performed at fortnightly intervals. The areas most likely to be affected should be assessed first, such as the downwind shorelines. Information on changes in wind direction or strength in the preceding 24 hours may be valuable for understanding the movement of surface scums in the water body. Visual inspection should consider the three following conditions. Each condition may be considered a prerequisite for the following condition.

1. Determine if Secchi transparency is less than 1 m or, in absence of a Secchi disc, if the bottom of the lake cannot be seen at 50 cm depth along the shore line; if so
2. Determine if cyanobacteria are visible as a greenish discolouration of the water or at the water's edge or as green or blue-green streaks on the water surface; if so
3. Determine if a green or blue-green scum is visible on the water surface in any area.

High nutrient inputs favour the development of cyanobacteria and algae. Therefore, if cyanobacteria are present, inspect the catchment area for signs of sewage outlets, excessive fertilisation close to the shoreline, erosion, or other potential sources of phosphate input. The identification of such sources provides the basis for measures addressing the cause of the problem.

Step 3. Quantitative assessment of cyanobacterial biomass and further actions to prevent health hazards

Upon detection of cyanobacteria at bathing sites, their quantification may be desirable for risk assessment. Two quantitative measures are equally valuable: microscopic cell counts or determination of chlorophyll *a* concentration as a simple measure for algal density. The choice of methods depends on equipment, expertise and personnel available. If chlorophyll *a* is used and its concentrations remain below 10 $\mu g\ l^{-1}$, hazardous densities of cyanobacteria are unlikely. At higher concentrations, chlorophyll *a* analysis must be supported by qualitative microscopic investigations for dominance of cyanobacteria.

If the results of the measurements exceed the guideline values, immediate management action are suggested. Furthermore, the detection of cyanobacteria at potentially hazardous concentrations should initiate planning of measures for restoration of bathing water quality. Because of the complexity of factors leading to cyanobacterial proliferation, flexible approaches are important, involving further development of monitoring and of protection measures as information on a given water body is accumulated.

Regular measurements of transparency in Step 3 can greatly enhance understanding of the system. If transparency is high and no significant discolouration of the water can be seen, cyanobacterial densities are unlikely to be high. However, samples should be taken to determine the phytoplankton community and the water chemistry, in order to estimate the capacity of the water body for cyanobacterial bloom formation. In deep and stratifying lakes, samples from different levels within the vertical stratification are required because some cyanobacteria accumulate near the thermocline. If transparency is low and accompanied by a greenish to bluish discolouration, high cyanobacterial densities are likely. In addition, greenish streaks formed by buoyant cyanobacteria during warm and calm weather may be visible on the water surface. Inspection of downwind areas of the water body is essential when these characteristics are observed, because wind action can readily lead to accumulations of these buoyant organisms. Samples for further analysis (taxonomical and toxicological) should be taken. Assessment of toxicity or chemical analysis for specific cyanotoxins is not generally recommended for two reasons: (i) the results of toxin analysis provide only a partial basis for risk assessment because only some of the substances causing health outcomes are known and can be analysed, and (ii) the results of an epidemiological study indicate that some health outcomes are not due to known cyanotoxins.

Assessment of toxicity using a bioassay may circumvent this problem, but the results are not easily interpreted in terms of human exposure during swimming, particularly with respect to skin reactions. Many cyanotoxin survey studies in different parts of the world have shown that more than half of the field populations investigated did contain toxins, particularly microcystins. Therefore, cyanobacteria are likely to be toxic.

Toxin analysis or toxicity assays may be useful under specific circumstances, because if concentrations of the known cyanotoxins prove to be low, some critical health outcomes may be excluded. Warning notices regarding potential irritative effects of cyanobacteria should nonetheless be posted while the cyanobacterial population density is above the guideline values. Toxin analysis or toxicity assays may be required in advance of a sports event to improve understanding of the potential risk due to cyanobacteria. If cyanotoxin concentration is low, a local authority may decide to proceed with the event in spite of high cyanobacterial density. Recreational facilities may also choose to invest in monitoring of the known cyanotoxins in order to avoid temporary closure of the facility whenever cyanobacteria proliferate.

Sample site selection

Additional factors to be considered in site selection are: the history, if available, of cyanobacterial and algal population development and occurrence of toxins at the water body or coastal strip; local characteristics of the catchment and water body which influence the development and fate of cyanobacterial populations, or even their current location in parts of the water body; and the wider knowledge of the characteristics of cyanobacterial and algal population development and fate and the production and fate of cyanobacterial and algal toxins.

The heterogeneous and dynamic nature of many planktonic population developments may require sampling several sites, such as locations that are prone to accumulation or scum formation, particularly if these are in areas used for recreation; locations at beaches or in the open water body which are used for immersion sports or those involving accidental immersion; a central reference site in open, mixed water; and decaying accumulations for dissolved toxins. A useful approach for large water bodies is to sample individual zones that differ in some physical, chemical, geomorphological or biological features. Preliminary sampling may help to define these zones, based on their spatial variability and gradients in environmental and plankton properties.

Sampling to assess the total population size may require information on the horizontal distribution of cyanobacteria within the water body, as well as on their distribution with depth. Particularly in thermally stratified water bodies, some cyanobacteria form pronounced population maxima at the depth with optimal light intensity and/or nutrient concentrations. Stratification leads to a water body functioning as two separate masses of water with different physico-chemical characteristics, with a transitional layer sandwiched between. Thermal stratification can be determined by measuring vertical profiles of temperature within the water body. In temperate climates, thermal stratification generally occurs seasonally in water bodies of appropriate depth, whereas in tropical climates it often follows diurnal patterns. Thermal stratification has important implications for the distribution of the concentrations of nutrients and the interpretation of phosphorus and nitrogen data.

Usually, shallow (2-3 m), wind exposed lakes do not stratify, whereas in temperate climates deeper lakes usually exhibit a stable stratification from spring to autumn. Lakes of intermediate depth (e.g. 5-7 m) may develop transient thermal stratification for a few calm and sunny days; the stratification is then disrupted by the next event of rain or wind. However, even if temperature is uniform throughout the water column, stratification of organisms can occur on calm days. Depth gradients of oxygen concentration and pH are good indicators of such stratification. Depth-integrated samples are more adequate than surface samples for the assessment of population size and nutrient concentrations in such situations.

Monitoring Frequency

Monitoring at time intervals should aim (i) to give warnings of developing cyanobacterial and algal populations and associated toxin levels; (ii) to provide information on the duration of cyanobacterial and algal populations and toxin levels that exceed guideline values; (iii) to provide information on the decline of cyanobacterial and algal populations and toxins due to natural processes; or (iv) to enable assessment of the persistence or reduction in cyanobacterial populations and toxin levels due to intervention, such as eutrophication control.

In recreational waters where bloom formation is suspected, the frequency of monitoring should be sufficient to provide data to enable an appropriate Alert Levels Framework system to operate. For example, monitoring may begin on a fortnightly basis and be increased to twice-weekly whilst alert levels are exceeded, before being reduced again after alert levels and guideline values for cyanobacterial cells and toxins are no longer exceeded.

If a water body prone to cyanobacterial mass development is to be used for watercontact sports on a seasonal basis, or for a single event, monitoring should begin not less than two weeks before the beginning of the season or the event. Monitoring should then continue, with the frequency adjusted to enable decisions to be made about access to the facility throughout the season, or whether to proceed with the event.

Laboratory and Staff Requirements

Monitoring for cyanobacterial and algal health hazards makes a range of demands upon analytical resources, some of which are different from those made by other aspects of water quality monitoring. Although a higher level of sophistication will provide more information, cyanobacterial and algal monitoring can be highly effective at a very low level of demand on facilities.

Background monitoring of physical and chemical variables reflecting bloom-forming potential makes limited demands upon analytical resources and capacities and may be readily decentralised. While any capable chemical laboratory can carry out laboratory analysis, some limnological or oceanographic expertise is necessary for planning field work, quality control of data, and interpretation of results.

Basic monitoring for indicators of toxic cyanobacteria focuses mainly on critical site inspection and requires almost no facilities. If performed by local staff with observation skills and increasing experience, regular monitoring and recording of simple variables such as transparency, discolouration and scum formation, provides much information for management.

Monitoring of populations of cyanobacteria and algae requires a microscope and some skill in its use. Health authority staff with experience in microscopy can easily learn to recognise the most important toxin-producing cyanobacteria and algae in the waters under their responsibility, provided occasional training by experts is provided.

Sampling

Samples taken directly by immersion of a sample bottle or sampling device are termed "grab" samples; they are also known as "spot" or "snap" samples. For sampling cyanobacteria, grab samples are often taken from the surface. Composite or integrated samples consist of several sub-samples collected separately and then mixed together. They are taken for quantitative, representative samples when the variables to be assessed are unevenly distributed, for example, when assessing the total content of a substance in a water body or the total population of an organism. If knowledge of the precise distribution is required, each sub-sample can be evaluated individually. Composite samples may be:

— *Depth-integrated.* These are most commonly made up of two or more equal subsamples collected with a sampler at predetermined depth intervals from the surface to just above the bottom. Selection of depths for subsampling must be adequate to account for stratification of temperature, substances and organisms in the water body. Continuous depth integration can be obtained with a tube sampler or a pumping system.

— *Area-integrated.* These are made by combining a series of samples taken at various sampling points spatially distributed in the water body.

— *Time-integrated.* These are made by mixing equal volumes of water collected at a sampling station at regular time intervals.

— *Discharge-integrated*. These are integrated over time intervals adapted to the discharge at regular intervals over the period of interest. A common arrangement is to sample every 3 hours over a 24-hour period. The composite sample is then made by mixing portions of the individual samples that are proportional to the rate of discharge at the time the sample was taken.

A depth sampler, sometimes called a grab sampler or bottle, is designed in such a way that it can retrieve a sample from any predetermined depth. It consists of a tube that can be closed at its ends by spring-loaded flaps that are triggered by dropping a weight down the lowering rope. A sample obtained in this way can be used for all chemical analyses except dissolved oxygen. This common sampler is relatively inexpensive, robust and can be deployed from almost any vessel. It gives samples for quantitative analysis. The hosepipe sampler is a piece of flexible plastic piping of several metres in length, weighted at the bottom, and provides a simple mechanism for collecting and integrating a water sample from the surface to the required depth in a lake. The hosepipe is lowered with its upper end open, trapping a water column as it is lowered.

The upper end is closed before hauling up the lower (open) end by means of an attached rope. The total length of this tube can be up to 30-35 m and it is suitable for relatively calm waters. Another device suitable for taking depthintegrated samples for shallow water columns or surface waters of deeper water bodies is a simple pipe sampler. The segmented tube sampler is a similar alternative for relatively shallow and calm waters. It consists of lengths of PVC (polyvinylchloride) pipe linked with valves, with a total length of up to 20 m. Integrated samples can also be obtained using a battery-operated water pump and flexible plastic piping which is operated at a steady pumping rate while the water inlet is drawn upwards between the desired depths at a uniform speed. This apparatus may also be operated to sample many litres of water from the same depth or to filter, for example through a plankton net, large quantities of water from a fixed depth for qualitative and for quantitative analyses. General discussions are given in Beers and Powlik *et al.* for peristaltic pumps, and Voltolina and Taggart and Legget describe diaphragm pumps.

Sampling scums

Scums usually occur near shorelines at low water depths and therefore working with a grab sampler or a plankton net may be difficult. Sampling scums is carried out more easily with a wide-necked plastic or glass container. When sampling scums their heterogeneous density must be taken into account. Two different approaches may be developed. The first aims to assess the maximum density of cyanobacteria and/or highest toxin level by taking a sample where the scum is thickest. The second approach aims to simulate conditions where shallow waters are mixed by bathers and playing children. Both types of approach may be used for comparison.

Plankton net sampling

Sampling with a plankton net is mainly performed when large quantities of cell material are required or when only qualitative analysis of phytoplankton is necessary. Sampling with a net causes a bias according to the size and shape of the organisms, i.e. the finest mesh size of 10 μm will miss small cells, such as unicells of *Microcystis* spp. and picoplankton. Furthermore, filamentous cyanobacteria may be under-represented because some filaments may slip through the net. The sampling depth is dependent on the taxa of algae and cyanobacteria present. Floating cells *(Microcystis, Anabaena)* are harvested within the upper few metres, whereas sampling of well mixed or stratified water bodies showing a depth distribution of cyanobacteria (e.g. *Planktothrix*) may include deeper water layers. For sampling the water column (or parts of it) lower the plankton net (25-50 μm mesh) to the desired depth, wait until the rope is taught, and then draw it back slowly to the surface. The net should be drawn very slowly out of the water to allow the water to run through it because large nets are sometimes heavily loaded with water when the pores are clogged with plankton. Rinsing plankton off the netting can be assisted by shaking the net slightly while raising it out of the water slowly.

For sampling surface blooms horizontal net hauls are more appropriate in order to filter floating cells. The plankton net should be moved parallel to the water surface. It can also be towed behind a boat moving slowly.

A disadvantage of collecting material with a plankton net is that the water volume filtered through the net cannot be determined precisely. Calculations based on the area of the net opening and length and distance hauled are not recommended because they overestimate strongly the amount of water actually filtered.

Determination of phosphorus and nitrogen

Water samples are collected with a clean sampler, a water pump or directly with the sample bottle. In shallow, unstratified lakes the sampling depth is less important than in deep, stratified lakes, where at least one sample from the epilimnion, one from the metalimnion and one from the hypolimnion should be taken. If this is not possible, a single surface sample will provide useful information, but only an incomplete picture of the growth conditions for cyanobacteria and algae.

A 100 ml sample container should be filled, immediately closed and stored cool. If analysis aims at differentiating between the different forms of phosphorus and nitrogen, the storage time should be as short as possible, and no more than 24 hours if the samples can be stored cool. Risks during extended storage involve transformations between dissolved and particulate fractions as well as between nitrate and ammonia. Preferably, the samples should be filtered in the field using membrane filters (0.45 μm pore diameter) pre-washed with a few millilitres of sample, and the filtered fractions should be stored separately. Alternatively samples should be filtered within 4 hours after sampling.

During filtration of samples for analysis of the dissolved nutrient fractions it is particularly important to avoid contamination. Filtering devices may be contaminated with higher concentrations from previous samples, especially if nutrient-enriched deep water layers had been filtered previously. This is especially important for phosphate, because it tends to adsorb to materials. Rinsing the equipment with double-distilled water between localities or with the sample to be filtered is recommended. It is helpful to take samples with low concentrations first and to move on to samples in which higher concentrations are expected.

Quantitative and qualitative determination of cyanobacteria and algae

For microscopic determination of cyanobacteria and algae, and for their microscopic quantification grab samples of 50-200 ml are put into a brown glass bottle and fixed immediately with Lugol's iodine solution or formaldehyde solution. Lugol's solution renders cells heavier, thus facilitating enumeration. Addition of 1-2 ml of Lugol's iodine per 100 ml of sample results in a 1 per cent final concentration. Formaldehyde should be avoided because it presents health risks to the user, or it should be used only under conditions of excellent ventilation in the laboratory and at the microscope. It has the advantage of not causing discolouration of the sample, but the disadvantage of not enhancing settling in counting chambers as effectively as Lugol's solution. Additional fresh, unpreserved samples are useful for microscopic identification because the iodine in the Lugol's solution covers the characteristic colour of the cyanobacteria. Such samples need not be quantitative and may be collected with a plankton net or as a grab sample at a site with high cyanobacteria and algal density. Unpreserved samples for identification may be stored for several hours without appreciable deterioration if kept cool during transportation to the laboratory. If biomass is to be quantified by chemical analysis of chlorophyll *a* concentration, samples of 1 litre (or less if cell densities are high) of water are taken and filtered as soon as possible. Filtration in the field involves problems of filter transport, which is possible either on ice in an icebox, or submerged in ethanol used for extraction. Direct sunlight must be avoided during filtration and transport.

Lugol's solution

Dissolve 20 g potassium iodide into 200 ml of distilled water, mix and add 10 g of sublimated iodine. Supersaturation can be tested by diluting 1 ml of stock solution to 100 ml with distilled water to give concentrations similar to those in preserved samples. If iodine crystals appear after standing, more potassium iodide (approximately 5 g) should be added and the test repeated. If no crystals appear, 20 ml of glacial acetic acid must be added. Store stock solution in a dark bottle and use within one year.

Cyanotoxin analysis

For toxicity testing, a large amount of cell material may be collected (without determining the water volume from which it originates) with a plankton net. If the concentration of cell-bound toxin is to be related to the water volume from which the cells were collected, the best approach is to filter a defined volume as described by Lawton *et al.* through a 0.45 μm mesh membrane filter. The volume chosen should be sufficient to provide a pronounced greenish layer of material on the filter without clogging the filter. Care must be taken to stir the sample to disperse the cells evenly every time immediately before a sample is poured onto the filter. The filtrate may be used for the analysis of toxins dissolved in the water.

Sample containers

The laboratory that conducts the analyses should ideally provide containers and bottles for the transport of samples. These should be pre-labelled and well-arranged in suitable containers . For routine sampling of the same sites, it is advisable always to use the same bottle for each site and each variable. This avoids crosscontamination, which is a particular concern for phosphorus analyses. For most samples glass containers are appropriate, but often plastic containers can be used instead. The following containers are recommended for the transport of cyanobacteria and related samples:

- *Phosphorus analysis.* 100 ml glass bottles pre-washed and stored with sulphuric acid (4.5 mol l^{-1}) or hydrochloric acid (2 mol l^{-1}) until use, rinse well before use.
- *Nitrate, ammonia and total nitrogen.* Glass or polyethylene bottles (100 ml).
- *Microscopic identification of cyanobacteria.* Wide-mouth polyethylene bottles (100 ml) are appropriate for studying living material in a fresh grab or net sample.
- *Microscopic identification and quantification of cyanobacteria.* Brown glass bottles (100 ml) already containing preservative. Clear (plastic or glass) bottles may be used if the samples can be stored in the dark.
- *Chlorophyll a analysis.* 1 litre brown (plastic or glass) bottles are recommended to avoid degradation of chlorophyll by sunlight. Clear (plastic or glass) bottles may be used if the samples can be stored in the dark. If filtration for chlorophyll *a* analysis is performed in the field, moist filters are best transported either on ice, by folding them with the cell layer inside and wrapping them in aluminium foil, or by immersing them in ethanol immediately after filtration. For the latter, 50 ml wide-mouth brown glass bottles or 10 ml tightly sealing test tubes may be used.

For toxin analysis, cell material is either collected with a plankton net or by filtration. Material enriched with a plankton net is best transported in wide-mouth plastic containers, in which it may be frozen for subsequent freeze-drying. Material on filters is best transported either dry, if filters can be dried rapidly without direct sunlight, or on ice by folding them with the cell layer inside and wrapping them in aluminium foil. Samples for analysis of dissolved cyanotoxins are acquired by filtration either in the laboratory or in the field. They are transported (filtered or unfiltered) in 1 litre plastic containers.

Sample transport and preservation

Samples must be labelled clearly with sampling site (station), date and time of sampling and depth of sampling. Preservation of samples, filtered volumes and any irregularity should be noted in the field record. In general, samples should be stored cool and dark in a storage box (if necessary, with solid coolant) during sampling and transportation. Storage of samples between 2°C and 5°C may preserve many types of samples, but checks should be made to confirm this with each sample type. Preferably, the samples should be analysed immediately after sampling. If a storage time longer than 12 hours is necessary, quick-freezing of samples to -20°C is recommended. Samples to be filtered must be filtered before freezing.

For the analysis of ammonia, storage is particularly critical. Samples can be cooled in a refrigerator but should be analysed within three hours of collection. Preservation for longer periods is not recommended. Filtration of samples should be avoided. It is nearly impossible to obtain filters free of ammonia and filtration may also evaporate the ammonia contained in the sample.

If samples for chlorophyll *a* are filtered in the field, filters must be either transported frozen, or submerged in ethanol. The ethanol should be boiling (at 75°C) when put onto the filters. If this is not possible in the field, transporting filters submerged in cold ethanol and heating the ethanol later in the laboratory may be preferable to the risk of degradation occurring on filters transported dry but poorly cooled. The suitability of this approach should be checked for any given situation.

Samples for microscopic enumeration preserved with Lugol's iodine at the time of collection are relatively stable and no special storage is required, although they should be protected from extreme temperatures and strong light. Nevertheless, it is recommended that samples are examined and counted within a few weeks because some species of phytoplankton are sensitive to prolonged storage and Lugol's iodine solution and may degrade if stored for many months or even years. Unpreserved samples for microscopic quantification require immediate attention, either by addition of preservative or by following alternative counting methods that do not use preserved cells. Where unpreserved samples cannot be analysed immediately they should be stored in the dark with the temperature kept stable, at about +4 °C.

On-site Analysis

A significant advantage of on-site testing is that tests are carried out on fresh samples which have not been contaminated or the characteristics of which have not otherwise changed as a result of storage. Some analyses such as temperature and transparency can only be carried out in the field.

Transparency

The transparency of water is strongly influenced by turbidity due to particles such as phytoplankton (cyanobacteria and algae) or suspended silt. It can be measured easily in the field. If transparency is low (less than 1-2 m) and accompanied by greenish to bluish discolouration, streaks, or even scums, high cyanobacterial densities are likely. Such results may initiate immediate inspection of further downwind sites and the collection of samples for cyanobacterial analysis.

Transparency can be obtained approximately with a Secchi disc. The disc is made of rigid plastic or metal, but the details of its design are variable. It may be 20 or 30 cm or even larger in diameter and is usually painted white. Alternatively, it may be painted with black and white quadrants. The disc is suspended on a light rope or chain so that it remains horizontal when it is lowered into the water. The suspension rope is graduated at intervals of 0.1 and 1 m from the level of the disc itself and usually the rope does not need to be more than 5 m in length. A weight fastened below the disc helps to keep the suspension rope vertical while the measurement is being made. Transparency estimated in this way (submersible photometers are also available for these measurements) is taken to be the mean of the depths at which the disc disappears when viewed from the shaded side of the boat and at which it reappears upon raising after it has been lowered beyond visibility. Observation of the disc through a tube (painted black inside) with a transparent pane at the lower end and held just below the surface improves precision, particularly if the water surface is perturbed.

If cyanobacteria occur as floating streaks or mats on the water it is difficult to obtain representative transparency data. Depending on the measuring site, values can vary from 0 to greater than 2 m. It may be useful to determine transparency in areas without floating cells as well as within scums. The Secchi disc has to be lowered very carefully to prevent destroying the formation of accumulated cyanobacterial cells, and before taking the Secchi disc measurement the surface scums should be given time to return to their original water coverage again.

An improvised transparency determination may be recommended by local authorities for users of recreational sites known to be affected frequently by cyanobacterial mass developments. If bathers cannot see their feet while standing knee-deep in the water because of greenish turbidity, bathing should be avoided.

Temperature

Temperature is best measured *in situ* with a probe because water samples gradually reach the same temperature as the surrounding air. If this is not possible, it may be measured with a thermometer or probe in a water sample of at least 1 litre, immediately after taking the sample. Graduations of 0.1 °C are appropriate.

Procedure

1. If measuring in a sample, immerse the thermometer in the water until the temperature reading is constant. Record the reading to the nearest 0.1 °C.
2. If using a probe, lower the probe to the required depth. Hold it at that depth until the reading on the meter is constant. If a complete profile of temperatures is to be taken, measurements should be made at 1 m intervals from the surface to the bottom. Near the surface, or in areas of large thermal discontinuities, measurements should be made at intervals of less than 1 m.

In situ ***fluorometric analysis of chlorophyll*** **a** ***and remote sensing***

Submersible fluorometers exist which can provide fine scale profiles vertically and horizontally. This is valuable, particularly for monitoring large water bodies and coastal areas for highly variable patterns of chlorophyll concentration (algal and/or cyanobacterial biomass). This approach has been used successfully to monitor variability in phytoplankton biomass and species composition, as well as surface temperature, salinity and nutrient concentrations, in the Baltic Sea. Fully automated analyser systems are installed on three passenger ferries. The system allows high frequency sampling with a spatial resolution of about 100 m and a temporal resolution of 1-3 days. The project uses, especially during the cyanobacterial bloom period, satellite images to detect the extent of the algal surface accumulations.

Real-time data for chlorophyll *a* distribution and concentrations, and potentially for cyanobacterial phycobiliprotein pigments in freshwaters, can be generated by the remote sensing of water optical properties by high resolution airborne scanners. However, flight times may be infrequent and data collection depends on factors such as cloud cover. Nevertheless, the remote sensing of cyanobacterial populations as a contribution to water body management has excellent potential.

Visible satellite imagery provides a synoptic perspective but, with few exceptions, does not yet have the ability to discriminate between different phytoplankton taxa nor is it effective during inclement weather. New techniques like the use of Advanced Very High Resolution Radiometers (AVHRR) on board the polar orbiting National Oceanic Atmospheric Administration (NOAA) satellites have been applied to monitor large-scale algal blooms, for example in North America and the Baltic Sea.

REFERENCES

Chorus, I. and Bartram, J. [Eds] 1999 *Toxic Cyanobacteria in Water: a Guide to Public Health Significance, Monitoring and Management.* E & FN Spon, London.

Cracknell, A.P., Wilson, C.C., Omar, D.N., Mort, A. and Codd, G.A. 1990. "Toxic algal blooms in lochs and reservoirs in 1988 and 1989". In: *Proceedings of the NERC Symposium on Airborne Remote Sensing.* Natural Environment Research Council, Swindon, UK, 203-210.

Hedges, J. I., and J. M. Oades. 1997. "Comparative organic geochemistries of soils and marine sediments". *Org. Geochem.* 27: 319–361.

Ingalls, A. E., S. R. Shah, R. L. Hansman, L. I. Aluwihare, G. M. Santos, E. R. M. Druffel, and A. Pearson. 2006. "Quantifying archaeal community autotrophy in the mesopelagic ocean using natural radiocarbon". *Proc. Natl. Acad. Sci.* USA 103: 6442–6447.

Karl, D. M. 1999. "A sea of change: Biogeochemical variability in the North Pacific subtropical gyre". *Ecosystems* 2: 181–214.

Karner, M. B., E. F. DeLong, and D. M. Karl. 2001. "Archaeal dominance in the mesopelagic zone of the Pacific Ocean". *Nature* 409: 507–510.

Kirchman, D. L., and P. J. L. B. Williams. 2000. Introduction. In D. L. Kirchman (ed.), *Microbial Ecology of the Oceans,* 1st edn. Wiley-Liss, pp. 1–11.

Bibliography

Anagnostidis, K. and Komárek, J. 1985 Modem approach to the classification system of cyanophytes. 1 Introduction. *Arch. Hydrobiol. Suppl. 71, Algological Studies,* 38/39, 291- 302.

Astrachan, N.B., Archer, B.G. and Hilbelink, D.R. 1980. "Evaluation of the sub-acute toxicity and teratogenicity of anatoxin-a". *Toxicon,* 18, 684-688.

Bagu, J.R., Sykes, B., Craig, M., Holmes, C.F.B. 1997. "A molecular basis for different interactions of marine toxins with protein phosphatase 1". *J. Biol. Chem.,* 272, 5087-5097.

Borowitzka MA, Borowitzka LJ. "Vitmins and fine chemicals from micro algae". In: *Microalgal Biotechnology* 1992. Cambridge University Press. Great Britain p 179

Bowling, L.C. and Baker, P.D. 1996. "Major cyanobacterial bloom in the Barwon-Darling River", Australia in 1991, and underlying limnological conditions. *Aust. J. Mar. Freshwat. Res.,* 47, 643-657.

Bryant, D.A. [Ed.] 1994 *The Molecular Biology of Cyanobacteria.* Kluwer Academic Publishers, Dordrecht, 879 pp.

Carmichael, W. 1997. "The cyanotoxins". *Adv. Botan. Res.,* 27, 211-256.

Chen, D.Z.X., Boland, M.P., Smillie, M.A., Klix, H., Ptak, C., Andersen, R.J. and Holmes, C.F.B. 1993 "Identification of protein phosphatase inhibitors of the microcystin class in the marine environment". *Toxicon,* 31, 1407-1414.

Chorus, I. and Bartram, J. [Eds] 1999 *Toxic Cyanobacteria in Water: a Guide to Public Health Significance, Monitoring and Management.* E & FN Spon, London.

Chudek, E., Horn, A., Joo, M., McLaren, S. 1998 *Stratification Patterns in Queensland Dams.* Report by Water Monitoring Group, Queensland Department of Natural Resources, Australia, 76 pp.

Cohen, S.G. and Reif, C.B. 1953. "Cutaneous sensitization to blue-green algae". *J. Allergy.* 24, 452-457.

Cracknell, A.P., Wilson, C.C., Omar, D.N., Mort, A. and Codd, G.A. 1990. "Toxic algal blooms in lochs and reservoirs in 1988 and 1989". In: *Proceedings of the NERC Symposium on Airborne Remote Sensing.* Natural Environment Research Council, Swindon, UK, 203-210.

Davio, S.R. and Fontelo, P.A. 1984 "A competitive displacement assay to detect saxitoxin and tetrodotoxin". *Analyt. Biochem.,* 141, 199-204.

Dillenberg, H.O. and Dehnel, M.K. 1960. "Toxic water bloom in Saskatchewan 1959". *Can. Med. Assoc. J.,* 83, 1151-1154.

Earthwatch 1992 *Chemical Pollution: a Global Overview.* United Nations Environment Programme, Nairobi.

Edwards, C., Lawton, L.A., Beattie, K.A., Codd, G.A., Pleasance, S. and Dear, G.J. 1993 "Analysis of microcystins from cyanobacteria by liquid chromatography with mass spectrometry using atmospheric-pressure ionization". *Rapid Commun. Mass Spectrom.,* 7, 714-721.

Ely R, Supriya T, Naik CG. "Antimicrobial activity of marine organisms collected off the coast of South East India". *J. Exp. Biol. And Ecol.* 309: 121-127, 2004.

Eriksson. J.E, Meriluoto, J. and Lindholm, T. 1986 "Can cyanobacterial toxins accumulate in aquatic food chains?" In: *Proceedings of the 4th International Symposium of Microbiol Ecology,* Ljubljana (Yugoslavia), 658-658.

Falconer, I.R. 1991. "Tumor promotion and liver injury caused by oral consumption of cyanobacteria". *Environ. Toxicol. Water Qual.*, 6(2), 177-184.

Febles CI, Arias A, Gil-Rodriguez MC et al. "In vitro study of antimicrobial activity in algae (Chlorophyta, Phaeophyta and Rhodophyta) collected from the coast of Tenerife (in Spanish)". *Anuario del Estudios Canarios* 34: 181-192, 1995.

Fogg, G.E., et.al.. 1973. *The Blue-green Algae.* Academic Press, London and New York.

Gerrard, S. 1995. "Environmental risk management". In: T. O'Riordan [Ed.] *Environmental Science for Environmental Management.* Longman Scientific and Technical, Harlow, 296-316.

Harada, K.-I. 1996 Chemistry and detection of microcystins. In: M.F. Watanabe, K.-I.

Hedges, J. I., and J. M. Oades. 1997. "Comparative organic geochemistries of soils and marine sediments". *Org. Geochem.* 27: 319-361.

Ingalls, A. E., S. R. Shah, R. L. Hansman, L. I. Aluwihare, G. M. Santos, E. R. M. Druffel, and A. Pearson. 2006. "Quantifying archaeal community autotrophy in the mesopelagic ocean using natural radiocarbon". *Proc. Natl. Acad. Sci.* USA 103: 6442-6447.

Isobe, M., Sugiyama, Y., Ito, T., Ohtani, I.I., Toya, Y., Nishigohri, Y. and Takai, A. 1995 "New analysis method for protein phosphatase type 2A inhibitors using the firefly bioluminescence system". *Biosci., Biotechnol. Biochem.*, 59, 2235-2238.

Jones, G.J. and Korth, W. 1995. "*In situ* production of volatile odour compounds by river and reservoir phytoplankton populations in Australia". *Wat. Sci. Tech.*, 31, 145-151. NRA 1990 *Toxic Blue-Green Algae.* Water Quality Series No. 2. The National Rivers Authority, London, 125 pp.

Karl, D. M. 1999. "A sea of change: Biogeochemical variability in the North Pacific subtropical gyre". *Ecosystems* 2: 181-214.

Karner, M. B., E. F. DeLong, and D. M. Karl. 2001. "Archaeal dominance in the mesopelagic zone of the Pacific Ocean". *Nature* 409: 507-510.

Kirchman, D. L., and P. J. L. B. Williams. 2000. Introduction. In D. L. Kirchman (ed.), *Microbial Ecology of the Oceans,* 1st edn. Wiley-Liss, pp. 1-11.

Kiviranta, J., Sivonen, K. and Niemelä, S.I. 1991 "Detection and toxicity of cyanobacteria" by *Artemia salina* bioassay. *Environ. Toxicol. Water Qual.*, 6, 423-426.

Kukkonen, J. 1999: *Comparison of cyanobacterial detection methods. - M. Sc. thesis, Department of Applied Chemistry and Microbiology,* University of Helsinki. 64 pp.

Lampert, W. & Taylor, B. E. 1985: "Zooplankton grazing in a eutrophic lake: implications of diel vertical migration". *Ecology,* 66: 68-82.

Lawton, L.A., Campbell, D.L., Beattie, K.A. and Codd, G.A. 1990. "Use of a rapid bioluminescence assay for detecting cyanobacterial microcystin toxicity". *Lett. Appl. Microbiol.*, 11, 205-207.

Lima-Filho JVM, Carvalho AFFU, Freitas SM et al. "Antibacterial activity of extracts of six macroalgae from the Northeastern Brazilian Coast". *Brazilian Journal of Microbiology* 33: 311-313, 2002.

Maneiro, I., Frangópulos, M., Guisande, C., Fernández, M., Reguera, B. & Riveiro, I. 2000: "Zooplankton as a potential vector of diarrethic shellfish poisoning toxins through the food web". - *Mar. Ecol. Prog. Ser.* 201: 155-163.

Matveev, V., Matveeva, L., and Jones, G.J. 1994a Study of the ability of *Daphnia carinata* King to control phytoplankton and resist cyanobacterial toxicity: implications for biomanipulation in Australia. *Aust. J. Mar. Freshwat. Res.*, 45, 889-904.

Moore, R.E., Ohtani, I., Moore, B.S., de Koning, C.B., Yoshida, W.Y., Runnegar, M.T. and Carmichael, W.W. 1993 Cyanobacterial toxins. *Gazzetta Chimica Italiana*, 123, 329- 336.

Moreau J, Pesando D, Bernad P et al. "Seasonal variations in the production of antifungal substances by some Dictyotales (brown algae) from French Mediterranean coast". *Hydrobiology* 162: 157- 162, 1988.

Mundt S. and Teuscher E. "Blue-green algae as a source ofpharmacologically-active compound". *Pharmazie.*, 43: 809-815 (1988).

Namikoshi, M., and Rinehart, K.L. 1996 Bioactive compounds produced by cyanobacteria. *J. Ind. Microbiol.*, 17, 373-384.

Niemistö, L., Rinne, I., Melvasalo, T. & Niemi, Å. 1989: "Blue-green algae and their nitrogen fixation in the Baltic Sea in 1980, 1982 and 1984". - *Meri* 17: 1-59.

Ohtani, I., Moore, R.E. and Runnegar, M.T.C. 1992 Cylindrospermopsin, a potent hepatotoxin from the blue-green alga *Cylindrospermopsis raciborskii. J. Amer. Chem. Soc.*, 114, 7941-7942.

Osanai, T., Ikeuchi, M. and Tanaka, K. (2008) "Group 2 sigma factors in cyanobacteria". *Physiol. Plant.* 133: 490-506.

Ostensvik O., Skulberg O.M., Underal B., and Hormazabal V. "Antibacterial properties of extracts from selected planktonic freshwater cyanobacteria- a comparative study of bacterial bioassays". *J. Appl. Microbiol.*, 84: 1117-24 (1998).

Paerl, H. W. 1990: "Physiological ecology and regulation of N2 fixation in natural waters". - *Adv. Microb. Ecol.* 11: 305-344.

Papke U., Gross E.M., and Francke W. "Isolation, identification and detrmination of the absoluteconfiguration of Fischerellin B. A new algicide from the freshwater cyanobacterium *Fischerella muscicola"* (Thuret). *Tetrahedron Lett.*, 38: 379-382 (1997).

Patterson G.M.L. and Carmeli S. "Biological effects of tolytoxin (6-hydroxy-7-o-methylscytophycin b), a potent bioactive metabolite from cyanobacteria". *Arch. Microbiol.*, 157: 406-410 (1992).

Rai, A.N. 1990 *CRC Handbook of Symbiotic Cyanobacteria.* CRC Press, Boca Raton, 253 pp.

Raivio, T.L. and Silhavy, T.J. (2001) "Periplasmic stress and ECF sigma factors". *Annu. Rev. Microbiol.* 55: 591-624.

Raziuddin, S., Siegelman, H.W. and Tornabene, T.G. 1983 Lipopolysaccharides of the cyanobacterium *Microcystis aeruginosa. Eur. J. Biochem.*, 137, 333-336.

Reynolds, C.S. 1984 *The Ecology of Freshwater Phytoplankton.* Cambridge University Press, Cambridge.

Rodger, H.D., Turnbull, T., Edwards, C. and Codd, G.A. 1994 Cyanobacterial bloom associated pathology in brown trout *Salmo trutta L.* in Loch Leven, Scotland. *J. Fish Dis.*, 17, 177-181.

Rudolph-Böhner, S., Mierke, D.F. and Moroder, L. 1994 Molecular structure of cyanobacterial tumour-promoting microcystins. *FEBS Letters,* 349, 319-323.

Schagerl, M. and Müller, B. (2006) "Acclimation of chlorophyll a and carotenoid levels to different irradiances in four freshwater cyanobacteria". *J. Plant Physiol.* 163: 709-716.

Servant, P. and Mazodier, P. (2001) "Negative regulation of the heat shock response in Streptomyces. *Arch. Microbiol.* 176: 237-242.

Skulberg, O.M., Codd, G.A. and Carmichael, W.W. 1984. "Toxic blue-green algal blooms in Europe: a growing problem". *Ambio,* 13, 244 - 247.

Sudhir, P. and Murthy, S.D.S. (2004) "Effects of salt stress on basic processes of photosynthesis". *Photosynth.* 42: 482-486.

Whitton, B.A. Phylum Cyanophyta (Cyanobacteria). in *The Freshwater Algal Flora of the British Isles.* Cambridge University Press, Cambridge.

Wille, N. 1904 *Die Schizophyceen der Plankton Expedition. Ergebnisse der PlanktonExpedition der Humboldt-Stiftung.* Bd. II. M.f. Verlag von Lipsius & Tischer, Kiel, 88 pp.

Zevenboom, W. and Mur, L. 1980 N2 fixing cyanobacteria, why they do not become dominant in dutch hypertrophic lakes. *Dev. Hydrobiol.,* 2,123-131.

Index